GAGING WITH VISION SYSTEMS

Nello Zuech
Editor

Robert E. King
Manager

Published by

Society of Manufacturing Engineers
Machine Vision Association of SME
Publications Development Department
Marketing Division
One SME Drive
P.O. Box 930
Dearborn, Michigan 48121

GAGING WITH VISION SYSTEMS

First Edition

First Printing

Library of Congress Catalog Card Number: 87-60607
International Standard Book Number: 0-87263-275-X
Manufactured in the United States of America

SME wishes to acknowledge and express its appreciation to the following contributors for supplying the various articles reprinted within the contents of this book. Appreciation is also extended to the authors of papers presented at SME conferences or programs as well as to the authors who generously allowed publication of their private work.

American Machinist
McGraw-Hill Publishing Company
1221 Avenue of the Americas
New York, New York 10020

Marc Bothel
Opcon
720 80th Street Southwest
Everett, Washington 98203

Cognex Corporation
72 River Park Street
Needham, Massachusetts 02194

Electro-Optical Systems Design
Pennwell Publishing
119 Russell Street
Littleton, Massachusetts 01460

Lasers & Applications
High Tech Publications, Inc.
23868 Hawthorne Blvd.
Torrance, California 90505

Machine & Tool Blue Book
Hitchock Publishing Company
25W550 Geneva Road
Wheaton, Illinois 60188

Manufacturing Engineering
Society of Manufacturing Engineers
One SME Drive
P.O. Box 930
Dearborn, Michigan 48121

Robotics World
Communication Channels, Inc.
6255 Barfield Rd.
Atlanta, Georgia 30328

Society of Photo-Optical Instrumentation Engineers
P.O. Box 10
Bellingham, Washington 98227

Veeder South III
Matrix Videometrix, Inc.
5321 Sterling Center Drive
Westlake Village, California 91361

Vision
Society of Manufacturing Engineering
One SME Drive
P.O. Box 930
Dearborn, Michigan 48121

Nello Zuech
Vision Systems International
3 Milton Drive
Yardley, Pennsylvania 19067

Cover photo reprinted from the MVA/SME Technical Paper
The EndSpector™ System for Automated Inspection of Beverage Can Ends
By Robert L. Jackson, Russell S. DeMuth and Boyd Eldridge

PREFACE

Much has been made about applications of machine vision in metrology. However, attention must be paid to many issues associated with components, techniques, and system design. Absolute gaging implies traceability to the National Bureau of Standards as well as calibration procedures to assure traceability. Frequently "system" issues such as traceability and coefficients of expansion are overlooked and, consequently, accuracy and repeatability of measurements are compromised or rendered useless.

Chapter One of this book, titled Overview, includes definitions and has been designed to review the above issues and alert prospective machine vision users to them. After reviewing the implications of machine vision as a gaging data acquisition system to provide intelligence to a CIM system in Chapter Two (Quality Assurance), Chapter Three (Measuring Issues) speaks to these issues. Chapter Four, Measuring Techniques, examines alternative machine vision techniques including active systems using laser scanners for ranging based on points and sheets (structured light technique) and passive linear and area array-based systems.

Chapter Five (Off-line Metrology) includes several articles that describe off-line dimensional measurement systems including three-dimensional gaging. As a rule, more complex parts can be examined off-line where time is not as critical. Off-line measurements are well-suited to inspecting the "first piece" to verify setup or die cavity integrity as well as to perform sample inspection which, in conjunction with SPC techniques, can, for example, spot trends from nominal conditions resulting from tool wear.

In Chapter Six, On-line Metrology, a number of on-line applications are described for 100% inspection of critical dimensions. These include gaging fish fillets as well as steel billets for automatic shearing, sheet metal assemblies and discrete parts, such as forgings. Significantly, with continued computer price/performance improvement, the application of machine vision techniques to gaging will become more viable and the possibility of a single machine vision system having adequate configurability to handle a family of gaging tasks will improve. For now, however, as the articles in this chapter suggest, applying machine vision to on-line gaging applications involves a significant custom content.

I wish to thank all of the companies, organizations, publishers, and authors who gave permission to have their articles reprinted in this volume. Thanks also to the Publications Development Department Staff at SME for their assistance in the research and development required in making this book possible.

Nello Zuech
Vision Systems International

ABOUT THE EDITOR

Nello Zuech, founder of Vision Systems International, has been engaged in the field of machine vision, electro-optics and factory automation for more than 18 years. He has been involved in the application of advanced manufacturing technology in virtually all manufacturing industries and in all phases of the production process including: incoming inspection, forming and assembling operations, testing, packaging, warehousing, and work-in-process monitoring. He is experienced in installations of both dedicated and flexible machine vision.

His professional experience encompases a wide variety of assignments in engineering, production, quality control, program management, and sales and marketing. Mr. Zuech has held executive positions with Object Recognition Systems, the EMR Group of Schlumberger, CBS Laboratories, National Science Foundation and Ratheon.

Mr. Zuech holds masters degrees in management and electrical engineering from the Baruch School of the City University of New York and New York University. He has lectured and written widely on machine vision and related automated electro-optical inspection techniques. He is a member of the Board of Directors of the Machine Vision Association/Society of Manufacturing Engineers, and a member of the Institute of Electrical and Electronic Engineers and Tau Beta Pi.

MVA/SME

Founded in 1984, the Machine Vision Association of the Society of the Manufacturing Engineers (MVA/SME) is a technical society that provides leadership in education and information exchange for the manufacturing, research and academic communities. MVA/SME promotes individual professional development to effectively utilize machine vision technology for quality and productivity improvement.

MVA/SME members have access to the latest information and developments in machine vision technology through a wide variety of opportunities including seminars, clinics, conferences, expositions, films, videotapes, and publications, including the association's quarterly entitled *Vision*.

This book is one example of its wide-ranging activities.

Members of MVA/SME are employed in one of four areas:

- planning, selecting and applying machine vision systems for manufacturing or processing;
- designing vision systems and/or machines and equipment used in machine vision systems;
- research and development leading to the creation of new or improved machine vision equipment or processes, and
- administrative, educational, or governmental activities relating to machine vision.

The world headquarters for MVA/SME is in Dearborn, Michigan.

MANUFACTURING UPDATE SERIES

Published by the Society of Manufacturing Engineers and its affiliated societies, the Manufacturing Update Series provides significant up-to-date information on a variety of topics relating to manufacturing. This series is intended for engineers working in the field, technical and research libraries, and as reference material for educational institutions.

The information contained in this volume doesn't stop at merely providing the basic data to solve practical shop problems. It also can provide the fundamental concepts for engineers who are reviewing a subject for the first time to discover the state of the art before undertaking new research or applications. Each volume of this series is a gathering of journal articles, technical papers, and reports that have been reprinted with the expressed permission from the various authors, publishers, or companies identified within the book. Educators, engineers, and managers working within industry are responsible for the selection of material in this series.

We sincerely hope that the information collected in this publication will be of value to you and your company. If you feel there is a shortage of technical information on a specific manufacturing area, please let us know. Send your thoughts to the Manager, Publications Development, Marketing Division at SME. Your request will be considered for possible publication by SME or its affiliated societies.

TABLE OF CONTENTS

CHAPTERS

6 ON-LINE METROLOGY

CHAPTER 1

INTRODUCTION

Reprinted courtesy of Nello Zuech
Yardley, Pennsylvania

… and Gaging Application Considerations

by Nello Zuech
Vision Systems International

Gaging or dimensional metrology applications represent about 30% of all automatic optical inspection applications. Gaging being defined as the inspection task that specifically verifies that dimensions agree with design data. Requirements for deriving dimensional data using remote noncontact techniques also can be found in many of the balance of applications; location analysis for robot guidance or wire bonder guidance are examples.

Conversely, requirements for performing other types of complementary inspection simultaneously while gaging a part is often a case--flaw detection, for example. In fact, a flaw is often characterized in terms of dimensions, consequently gaging is involved. Gaging does not always require absolute measurements--the traces and spaces printed circuit board application is an example of this.

Gaging is most often associated with the forming applications which cut across all industries. Forming often involves a single setup based which produce many products. To avoid scrap, first piece inspection is critical. This is generally performed off-line and involves a detailed analysis of all dimensions of the part. Coordinate Measuring Machines (CMMs) and TV-based dimensional metrology systems have been developed in response to this market need. The former are generally used for la… …nalysis and the latter for small parts (less than 12"…

Once the forming process has b… …tion is generally adequate to assure that mold c… …s do not occur. The rationale for gaging stems fro… …ality control. Quality control is a sorting ap… …eeks to separate out the bad products. This is typi… …ecking 100% of the product, and hence is an on-lin… ….

Process control, on the other hand, can be either an in-process, or post-process analysis and can be either on-line or off-line and reflects inspection of a sample of the products either while they are being manufactured or immediately after. The objective is to spot trends so one can take corrective action before producing reject parts. Process control, strictly speaking, involves a system in which the difference between the process variable and a reference is measured almost simultaneously with the forming operation. This information changes the control elements in the system to restore operation to original prescribed parameters.

The term "integrated gaging" is frequently used to connote in-process or post-process instrumentation designed into the line or work cell. Post-process gaging is performed after the part has been formed, usually in an immediately following station. Generally, this is mandated by the manufacturing environment; i.e. inaccessibility or interferences.

The rationale for automating gaging are virtually the same as for most quality automation:

1. Avoid consumerism issues (warrantees, liabilities, lost customers);
2. Improve quality;
3. Improve productivity;
4. Reduce scrap;
5. Ability to perform 100% inspection at high rates;
6. Process control;
7. Eliminate subjectivity;
8. May eliminate need to capture the part;
9. Where noncontact, no gaging surfaces on the gaging tool will wear for use;
10. Permits inspection of parts that can not be touched;
11. By automating data collection, traceability is possible for consumerism issue;
12. Automated gaging may be dictated by part complexity;
13. Automated gaging and in process control may permit reduction of manufacturing tolerances, and
14. Detection of broken tools and other machine mechanical tolerance breakdown.

Comments About Metrology Terms

Accuracy is a term that refers to conformance of a measurement to the actual size of a part. Accuracy should reflect back to a standard traceable to the National Bureau of Standards. Accuracy can be regarded as the difference between the value of the measurement representing the center of the precision distribution and some reference standard. If the reference standard is an NBS standard, then absolute accuracy is implied.

Precision is the ability of a measuring system to repeat a measurement on a particular part and is often referred to as repeatability. A gaging system of high precision is not necessarily accurate; however, precision is needed to achieve accuracy.

Reliability is the probability of performing a specific measurement successfully.

Reproduceability is the variation of a measurement from operator to operator employing the same edge.

Sensitivity is the minimum input into a transducer that will result in a discernable change in the output of the measuring system.

Discrimination is the finest scale division of a measuring instrument. It is related to, but not necessarily a measure of, precision or accuracy.

Resolution is synonymous with discrimination, insofar as machine vision related approaches are concerned. In the case of machine vision, resolution reflects the property of the sensor. In the strictest sense, resolution is the ability of the sensor to detect two closely spaced objects. For gaging application, detectivity (which is the ability to detect an edge) is more important.

Stability is the ability to repeat measurements over time and temperature. Thermal factors such as radiant energy, conductive heating, drafts, and room temperature differentials should be compensated for by a gaging system.

In any gaging application, variations in geometry and surface finish of the measured part directly affect the repeatability of a measuring system. Such part variations can manifest themselves as an apparent error in the measuring systems. Insofar as machine vision is concerned, this may require compensation by taking the average of many readings to assess the location of an edge.

Rules of Thumb

The accuracy of a system should be 10% of the tolerance value. So, if the tolerance associated with a specific measurement is +/-0.010", the accuracy should be +/-0.001".

Recognizing that the accuracy of a measurement reflects a system's ability to make a measurement between a minimum of two points (edges), the accuracy with which a system detects an edge should be better.

The resolution (or detectivity--ability to detect an edge) of a system should be equal to accuracy or 10% of the measurement tolerance. In a brute force machine vision system (no sub-pixel processing), the accuracy becomes synonomous with pixel size.

In the automotive industry, a general rule of thumb is that repeatability plus accuracy should be less than 30% of the measurement tolerance.

Another point, as tolerances get smaller, (1 mil or less) maintaining a 10:1 ratio is generally compromised to 5:1 or 3:1.

Performance Envelope of Gaging Systems

The performance envelope of a gaging system is a function of the object being gaged and the application for which the object is intended. In general, measurements are made between points on a part defined by edges. Consequently, machine vision systems should have an edge enhancement capability to permit the detection of an edge in space reliably and repeatably. It is this ability along with the number of spatial pixels of the machine vision sensor that dictate the accuracy of a machine vision system.

When comparing measurements made by people with conventional instruments (calipers, micrometers) to those based on TV techniques, the differences associated with the techniques must be understood. Because of the nature of the instrument used, the measurement will reflect a condition averaged over an area on the surface corresponding to measuring across peaks. A TV-based system generally makes the measurement based on an edge reflected in a single pixel or across the area associated with the pixel. Because of surface roughness properties, measurements made across the same apparent points may not agree.

When solid state sensors are used, since the value of the intensity corresponds to the average intensity across a pixel site, a system with robust grey scale processing would have the ability to develop an intensity profile across the pixel site. Using a variety of mathematical techniques then, such a system would have the ability to detect an edge to subpixel repeatability.

Thus, detectivity is the more important property of a machine vision system being used for gaging, rather than resolution. Resolution corresponds to the ability to detect two closely spaced objects. This is generally not the case unless you are dealing with a very small object or making a measurement between edges 5-6 pixels apart. In that case resolution then becomes the limiting factor in the accuracy of a gaging system.

Machine Vision System Considerations

Optics. Every component in a machine vision system contributes to performance in a gaging application. The optics, for example, must be as distortion-free as possible or otherwise the system must be able to compensate for varying pixel sizes due to geometric distortions in the lenses. Alternatively, one might only use the center of the lens and avoid making measurements beyond a certain distance from the center of the lens. Where edges may be defined by color boundaries, chromatic aberrations in the lenses may contribute to errors.

Distortions in the optics will contribute to the error budget and should be understood especially insofar as full field of view measurements are concerned. Pixels can be a different size and shape at the edges than in the center.

An optical system has limiting resolution dictated by the wave nature of light and the aperture through which the light passes. The larger the aperture, the closer two small objects can be observed and still be detected individually. It is unrealistic to expect to measure below 0.5 microns or 20 micro inches with visible light. Two associated optical contributions to an error budget are: (1) depth-of-focus or range over which the image does not appreciably vary from the sharpest image observed and (2) magnification. Both can contribute to the error budget. In the latter case, variations in object to optics distance as well as optics to sensor distance will result in dimensional variations when conventional optics are used.

Telecentric optical designs can minimize these contributions to error. These are appropriate where the silhouette of the part includes the features to be measured. Such a system is more tolerant of working distance and depth of field variations.

Another optical consideration involves field of view. This will have to be larger than part sizes to make allowances for positional variations of the part, especially if motion is involved. Hence, the smallest detail on an object that can be seen in a full field of view measurement will be proportionately larger.

Extender tubes are frequently used as a means of filling the sensor area with the image. Significantly, this introduces distortions as most lenses are not designed for use with such extender tubes.

Some systems have the capacity to handle measurements of large parts and make efficient use of the pixels in the frame store by handling the input from multiple cameras via a frame splitter, (commonly used in broadcast television). With such a device the inputs from several cameras are merged onto a single frame store. Again because the absolute position of cameras relative to each other is known; gaging between features on large parts can possibly be more readily handled this way. An alternative is to use image quality fiber optics to bring attributes from different locations on a part back to a common camera.

Light fall off, a condition which is experienced when lenses are less efficient at collecting light at the edges than in the center of the lens, introduces optical error. Extension tubes are significant contributors to this problem.

Lighting. Lighting is also a very important component in a machine vision system. Many of the systems that are presently used for gaging applications that are on the market today capitalize on the use of structured light techniques and triangulation techniques to provide an enhancement of the edges of three dimensional objects.

Structured light involves creating a pattern of imaged light (usually a line or sheet of light) and superimposing the pattern on the object. The effect of the object on the pattern of light is vie[illegible]d by a sensor. Gradient changes caused by geometric attributes [illegible]ggerated by the use of structured light techniques so ed[illegible]e pronounced upon which to base measurements.

The use of monochromatic [illegible]ditional advantage that one can filter out all other [illegible] properties such as those that vary with the geomet[illegible]ct being gaged.

If front lighting arrangeme[illegible] use of polarizers on either or both the lighting and op[illegible] advantage to avoid specular reflections from the object [illegible]ing to errors associated with edge definition. Some systems on [illegible] market have the ability to vary the light or control the lighting on an object in an attempt to bring out edges that might otherwise be in shadows.

The spectral output of the illumination source can be a factor in characterizing spatial features on an object. Keep in mind that solid state sensors are sensitive to infrared phenomena. Consequently, without attention paid to filtering, they may see part attributes (such as shading and edges stemming from the shading due to temperature gradients) that the eye does not see.

In a similar way, unstable illumination sources--those that change in output over time or otherwise drift - can cause a machine vision system to characterize features differently and, therefore, introduce a measurement error. Strobes are notorious for wide output intensity variations.

Throughout this, one must recognize that the illumination must be adequate to obtain good signal-to-noise properties out of the sensor and not excessive so as to cause blooming, burn-in or saturation. The lower the wavelength of the light, the more accurate can be the system due to the Rayleigh criteria stemming from diffraction limits of optics and wavelength relationship.

Sensors. A machine vision system's ability to make useful measurements based on examining an object with a full field of view is very limiting because of the finite number of resolvable pixels in a sensor. Linear array sensors have pixel arrangements as large as 4096. With robust subpixel processing ability, the use of such sensors might permit measurements to accuracies of one part in 40,000 by moving the array past the object or the object under the scanner or using a mirror stepper arrangement to "dissect" the object one line at a time across the array.

The most significant property relating to applications in gaging is detectivity or the ability to detect an edge repeatably in space. The ability to resolve two closely spaced objects, or resolution, is not a factor. Detectivity does relate to the number of pixels available in a sensor as well as the ability to dissect a pixel. For example, image dissector tubes are available that can resolve an image into 2048 x 2048 pixels. Significantly, with the proper camera electronics (coils, amplifiers, A/D's, etc) they can repeatably go to a point in space to one part in 40,000!

The same thing is true of vidicon cameras. There are some with superior camera electronics that can go to a point in space to one part in 1000. Solid state cameras can only go repeatably to a point in space consistent with the array elements; i.e. for a 256 x 256 array, to one part in 256.

Significantly, the ability to detect an edge is a function of the ability to perform subpixel processing as well as return to the same point in space.

Aliasing in solid state cameras is a phenomenon experienced due to the fact the image is formed by an array of picture elements rather than a continuous surface. Consequently, there are discontinuities between picture elements where light is not detected. This becomes more noticeable when viewing scenes with lots of edges.

In some cases, they are fabricated with readout circuits interleaved between photosites, their "optical fill factor" is less than 100%. In other words, there are significant discrete locations on the imager in which there is no photosensitivity. If an edge falls in this region, it would be assigned to a contiguous pixel site resulting in a spatial error. Some solid state sensors experience "dead" photosites and again edges falling on them would be assigned to a contiguous pixel. Some cameras get around this by calculating a sensitivity for the pixel based on the neighbors' sensitivity. This "look up table" for the matrix is stored permanently in the camera.

Nonuniformities in sensitivity can introduce encoded value distortions which could lead to gaging errors when using systems with only pixel threshold capabilities. Vidicons experience nonuniformity on the order of +/-20-40%, while solid state sensors are +/-10% or less but may suffer from the above mentioned dead pixel site phenomena.

Another camera/sensor consideration involves gamma correction and automatic light gain circuits. Significantly, cameras have been in general designed to be compatible with displays and to perform signal correction to compensate for display signal distortions associated with phosphor phenomena. The result in a given application is that the camera output going into the image processor is not proportional to the image input photon energy. The result could be grey scale distortions which introduce errors.

When vidicons are used one also has to contend with geometric distortions and nonlinearities stemming from deflection circuit properties. One to two percent distortions are possible with high-quality cameras, otherwise distortions of 5% are possible. Another factor with cameras is the aspect ratio. Generally 3(V) x 4(H) aspect ratio is standard. This means the pixel is a different size in its horizontal dimension than in its vertical dimension. This has to be accounted for in a gaging application. Significantly, the aspect ratio in vidicon cameras can be adjusted by changing the sweep circuit bandwidth.

Certain time constants can be a problem in certain applications. Lag in vidicons is evidenced by "tails" on bright moving objects, and an impression of "multiple exposures" in moving scenes. It is caused by incomplete target charging in one scan, and is a result of local electrical field phenomena in the vicinity of the target. The more an area is recharged toward equilibrium, the less affinity it has to accept further charge. Solid state sensors may suffer from this same phenomena but to a lesser degree. Vidicons may also suffer from image "burn-in" a phenomena in which a continuous image may permanently degrade the local photosurface sensitivity. Blooming is another phenomena experienced when a saturated pixel influences contiguous pixels, utlimately causing them to saturate resulting in defocusing of regions in the picture.

Mounting of the sensor within the camera itself should be rigid. Vibration should not result in variations in the distance between the lens and the plane of the sensor, otherwise magnification errors will be introduced.

A/D. An error can be introduced from pixel jitter due to the difference between camera frequency and the sampling rate of the A/D. This stems from the fact that "off the shelf" cameras are designed to be RS 170 compatible for closed circuit TV purposes. These signals lack clock information to indicate where in the analog output signal the center of each photosite is represented. As a result, the precise location to perform a sample and A/D conversion to maintain an exact 1:1 relationship between the digitized samples and the corresponding photosite coordinate in the image sensing array is unknown. This can be overcome by driving the sensor's scan control ICS and the A/D converter from the same clock source.

Nyquist theory suggests that one sample at least twice for each data point. Considering a sensor with a 512 x 512 array and a camera frequency of 15.75 KHZ (horizontal rep rate for EIA RS 170 camera) one has per pixel integration rate of approximately 100 nanoseconds. At 50 nanoseconds per sample, one should have an A/D that operates at 20 MHZ.

System Considerations

Application system design also must have the capacity to compensate for thermal coefficients of expansion associated with the object being measured in those cases where absolute measurements are critical. In otherwords, compensation for temperature and its impact on the system and the object. Systemic considerations surrounding the mounting of cameras also may become a factor in some applications. Again, for example, floor to ceiling differences in temperature may impact the position of the cameras with respect to the object as a result of coefficients of expansion associated with the brackets or fixturing.

Temperature excursions experienced in the camera and system electronic housings can also lead to drift in the video amplifier and A/D circuitry. This can be avoided by proper heat sinking or air conditioning of the electronics. Another systems issue is the demand for repeatably positioning a part. Even when it appears that part presentation is repeatably organized, even a slight mispositioning can introduce an error in measurement. The result is that the measurement is actually not being made between the original programmed points on a part print and the dimension could be different, especially if curvature is involved. Any reliable gaging system must correct for part mispositioning before attempting a comparative measurement.

Conclusion

Using machine vision for gaging is not easy. With proper attention to staging details, however, one will at least be assured of reliable and repeatable data entering into the vision engine for processing. Since, in general, dimensions are characterized as the distance between two edges, it makes sense that the vision engine have the capacity to enhance the definition of edges to optimize the measurement. Because of the limited number of pixels in any sensor, a vision engine with subpixel processing capabilities offers advantages.

References

Lapidus, S. "Gauging with Machine Vision", Vision, May, 1986.

Stafford, R.G., "Induced Metrology Distortions Using Machine Vision System", MVA/SME Vision 85 Conference Proceedings.

Chu, Y.C., "Resolution and Accuracy in Gray Level Machine Vision", MVA/SME Vision 86 Conference Proceedings.

MacDonald, J.A., "Solid State Imagers Challenge TV Camera Tubes", Information Display, May, 1985.

Hofmeister, R., "Solid State Imagers", Sensors, June, 1986.

Harold, P., "Solid State Area Scan Image Sensors Vie for Machine Vision Applications", EDN, May 15, 1986.

Bloom, L. "Interfacing High Resolution Solid State Cameras To Digital Imaging Systems", Digital Design, March 25, 1986.

CHAPTER 2

QUALITY ASSURANCE OVERVIEW

Reprinted from *Manufacturing Engineering*, April 1984

An integrated metrology system is the key to minimizing the cost of quality—a new hidden opportunity to improve your company's profits

Boosting Product Quality for Profit Improvement

BY L. SCOT DUNCAN AND GARY L. BOWEN

Often overlooked in the cost-of-goods-sold equation—after labor, materials, energy, and capital plant—is the cost of quality. It is not a negligible consideration. Quality costs amount to 5 to 20% of sales in defense manufacturing, while in other industries, like electronics and medical products, the cost is even higher.

The "cost" of quality can be segregated into two elements with distinctively different characteristics—the cost of failure and the cost of control. Many companies err in the direction of too little control, with too many resulting failures in their attempt to minimize cost in the short term. Advances in control technologies coupled with rising failure costs will make this trend even more pronounced in the future.

These conclusions are based on an in-depth study of quality control among heavy equipment manufacturers conducted by Booz, Allen's Productivity Technology Center (Cleveland, OH) as part of an assignment for the U.S. Army Tank and Automotive Command (TACOM). The Army sponsored the development of an Integrated Metrology System (IMS) because metrology (dimensional inspection) is considered an essential part of the factory of the future, and incorporating it in the manufacturing process is considered essential to reducing the failure rate of products. The results of the program will provide the guidelines and direction for the future implementation of integrated metrology systems in TACOM vendor facilities and in other plants producing components and systems requiring similar manufacturing technologies. Twelve manufacturer/users and three suppliers of metrology equipment participated in the Booz, Allen study as shown in *Figure 1*.

The study definitively established that a factory-wide integrated approach to metrology is a key to minimizing the cost of quality, and it resulted in a conceptual design for an integrated metrology system that will be compatible with future factory operations.

The cost of quality

Simply defined, the cost of quality is the sum of all costs associated with maintaining quality control in the plant and absorbing the external costs of loss of business, warranty, and liability claims associated with a defective product. The cost of quality assumes that there is an optimum level of defect.

Industry Coalition Participants

- GENERAL DYNAMICS
- CINCINNATI MILACRON
- CUMMINS
- DETROIT DIESEL
- CATERPILLAR
- FMC
- GARRETT
- FEDERAL GAGE
- TELEDYNE
- AVCO
- TRW
- BENDIX
- MARPOSS
- KEUFFEL & ESSER
- DIFFRACTO

FIGURE 1

Except in rare applications, the goal of zero defects is economically infeasible. To put the economics in perspective, it should be noted that quality costs and profitability are the same order of magnitude; therefore, a given percentage drop in the cost of quality relates directly to a similar percentage increase in corporate profitability.

The cost of quality can be visualized as a curve formed by adding two component costs, one pertaining to control and the other to failure, as shown in *Figure 2*. This composite quality cost curve is based on the most cost-effective inspection and quality control systems.

The cost of control, relative to other quality costs, is generally easy to quantify and includes prevention and appraisal measures employed in the factory to keep defects from occurring and to find defects before products are shipped to a customer—inspection and quality control labor costs and inspection equipment costs.

The cost of failure is much more difficult to quantify and includes internal failures, resulting in materials scrap and rework, and external failures, resulting in warranty claims, liability, and recall orders, as well as hidden costs, such as the loss of customers.

While the manufacturers studied were able to estimate many elements of their control costs quite accurately, they tended to underestimate the difficult-to-quantify failure costs. In addition, management tends to cut highly visible control costs in times of financial difficulty, trading short-term control costs for the inevitable failure costs. As a result, the current cost of quality for heavy equipment manufacturers tends to be skewed to the left—too high on the failure curve. The optimum quality point is significantly to the right, favoring increased control.

This deviation from the optimum will become even more pronounced in the future. While costs for both control and failure are increasing in absolute terms, the cost of failure is rising dramatically due to increased labor and materials costs for scrap and rework and due to skyrocketing litigation, warranty and recall costs, and the increasing penalty for defective products in a highly competitive marketplace. At the same time, the improved cost performance of new metrology technologies and practices can reduce the relative cost of control.

The systems approach

Because control and failure are related, better control reduces the frequency of failure and makes lower quality cost a directly achievable objective. As shown in *Figure 3*, control has

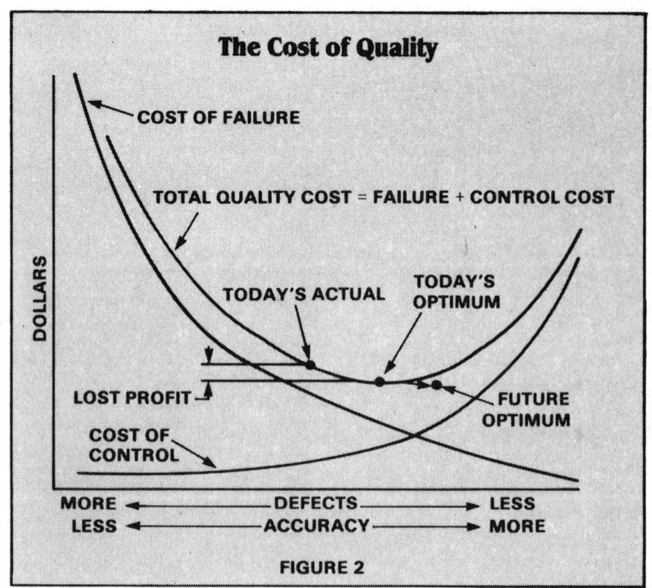

FIGURE 2

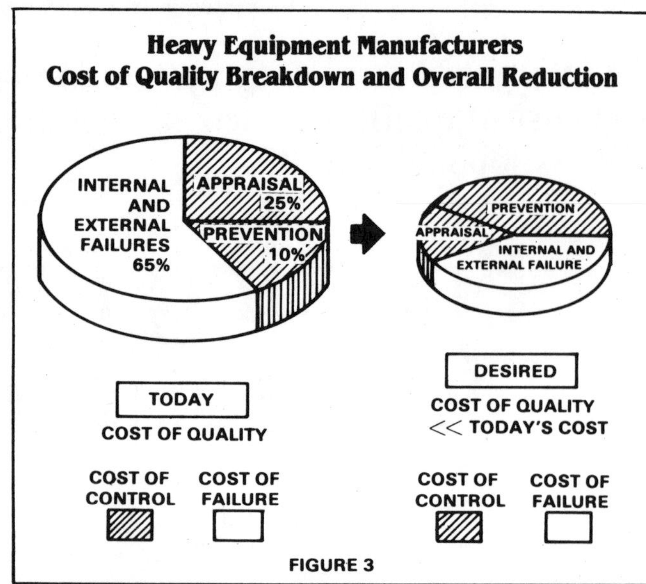

FIGURE 3

two components—appraisal and prevention—appraisal focusing on detecting errors in the product and prevention focusing on improving the accuracy of the manufacturing process. Input from the coalition members and from statistical literature indicates that an increased emphasis on prevention of defects will most directly reduce the overall cost of quality.

Thus, the focus of future process/manufacturing efforts should be to "make it right the first time"—taking the proper steps to make a good part in the process in order to minimize post-process part inspection (appraisal). An integrated approach to metrology offers an alternative to conventional parts inspection, one which concentrates on qualifying and maintaining the parameters of the processes (machinery maintenance, tool calibration, and tolerances) in real time so the resulting part will be right.

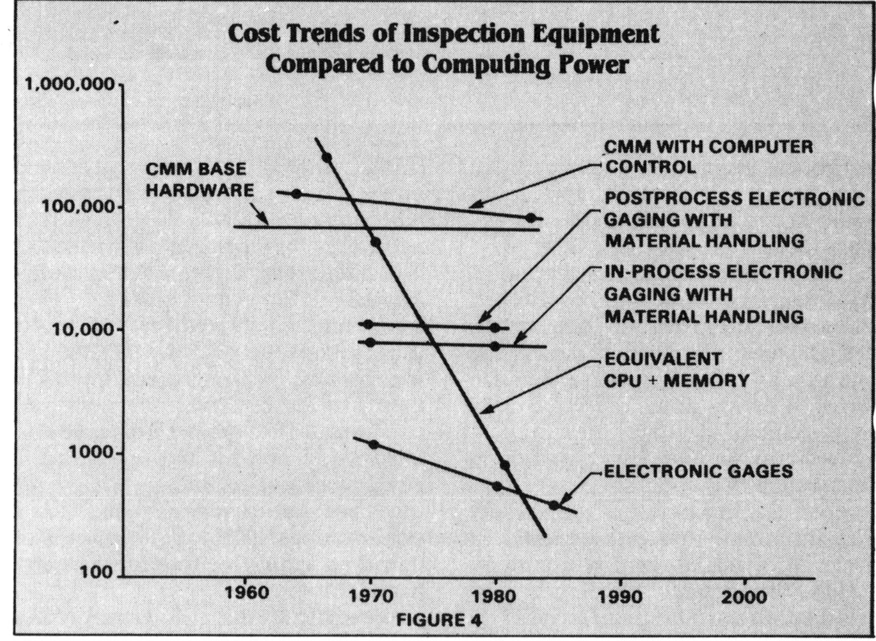

FIGURE 4

Conceptual design

The Integrated Metrology System design, funded by the Army and supported by participation of an industry coalition, is a tool to assist manufacturers in developing an integrated inspection system that minimizes the cost of quality. It integrates and organizes all activities related to dimensional inspection: production parameters, inspection parameters, and product specifications. It provides a means for continuous control and monitoring of these activities and a means for collecting and organizing all dimensional data, plus providing the information to both management and shop floor personnel.

The key to the IMS concept is the integration of state-of-the-art metrology hardware, statistical techniques, and computerized data handling to significantly improve the control side of the quality cost equation. By and large, the metrology and statistical technologies for inspecting and maintaining process parameters have been in existence before. Through the widespread availability of computers, these technologies have become cost effective and have provided quality control information throughout the factory as shown in *Figure 4*. In addition to real-time monitoring of process parameters, an integrated metrology system closes the loop on process information, providing the machine operator with product/process specifications and the responsibility to maintain those specifications.

Input from industry

Booz, Allen's work on the integrated metrology system consists of three major products: a survey of current metrology systems, technologies, and practices; a series of factory models relating to metrology—function model, information model, and dynamic model; and a simulation package that can project the impact of new metrology technologies on quality costs. The coalition of users and suppliers was used to get an accurate picture of existing systems and to ensure that the concepts developed for the integrated metrology system can be readily implemented in the factory floor environment. The coalition members provided input, criticism, and access to their plants for analysis.

Conducting the metrology survey of existing manufacturing plants revealed that the majority of metrology work-hours were spent on postprocess inspection of parts. In the future factory, other metrology applications, such as

Summary of the Integrated Metrology System IMS Costs and Benefits

TODAY'S MAJOR COST DRIVERS	PROPOSED IMPROVEMENTS	BENEFITS	IMPLEMENTATION COST AREAS
INTERNAL • SCRAP • REWORK • LABOR • CAPITAL EQUIPMENT • INVENTORY (WIP) • ASSEMBLY DIFFICULTY **EXTERNAL** • LIABILITY/WARRANTY	**PROCESS** • IMPROVED PREPROCESS • CONTROL OF TOOLS, FIXTURES, MACHINERY • IN-PROCESS DIAGNOSTICS • AUTOMATED/INSPECTION WHERE FEASIBLE —IN-PROCESS/IN CYCLE —POSTPROCESS —PREPROCESS **SYSTEM INTEGRATION** • IMPROVED DATA GATHERING AND INFORMATION PROCESSING • IMPROVED CLOSED-LOOP CONTROL	• REDUCED SCRAP • REDUCED REWORK • REDUCED LABOR • REDUCED CAPITAL EQUIPMENT —INVENTORY OF GAGES —OBSOLESCENCE • REDUCED WIP • REDUCED ASSEMBLY DIFFICULTY • IMPROVED PRODUCTION EQUIPMENT UTILIZATION • REDUCED MATERIAL HANDLING • REDUCED EXTERNAL FAILURES • IMPROVED TOOL UTILIZATION • POSSIBLE REDUCED PRECISION REQUIREMENT FOR INSPECTION EQUIPMENT	• INTEGRATION OF INSPECTION TECHNOLOGY TO PROVIDE CLOSED-LOOP CONTROL TO PROCESS —IN-PROCESS/IN CYCLE —POSTPROCESS • SENSOR DEVELOPMENT FOR IN-PROCESS INSPECTION • CONTROL/COMMUNICATION SYSTEM DEVELOPMENT • SOFTWARE DEVELOPMENT • INCREASED TOOLING, FIXTURE, AND MACHINE ACCURACY CERTIFICATION COST

FIGURE 5

in-process inspection, tooling inspection, fixtures inspection, and equipment inspection, will become the primary focus.

The series of models developed for the Integrated Metrology System provides the blueprint for the system design. The function model details how IMS relates to and supports manufac- ...eaks down the three major ... inspection of ...eters, ...s par- ...oduct. ...les an ...mmuni- ...trology ...in the ...ter simulatio... to assi... ...new ... cost-... ...ative chang... ...ty over time.

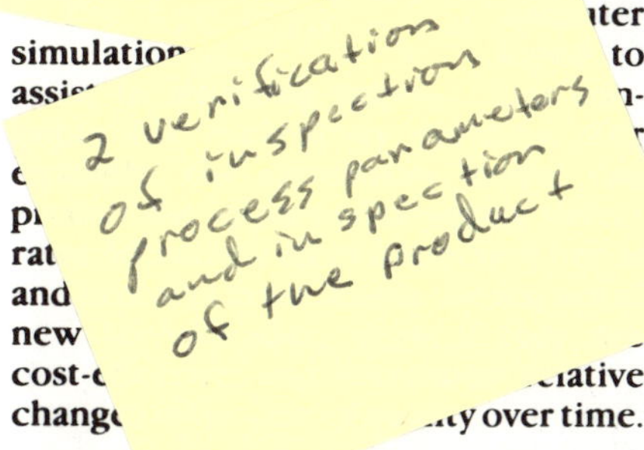

Summa...

A detailed summary of the costs and benefits of the Integrated Metrology System is shown in *Figure 5*. Coalition members were anxious to apply the metrology system to their existing operations—not wait for a factory of the future—and, in fact, many elements of the IMS can be implemented today. Perhaps the most significant outcome of the study is unanimous agreement on the need for a long-term perspective with regard to quality costs in general and metrology systems in particular. The IMS study furthered the understanding of the cost of quality.

Technologies for automated inspection systems are readily available, and the results of postproject questionnaires to the coalition members bear this out. The major obstacles to metrology system implementation cited were not technology shortcomings, but rather management issues, organizational problems, marketplace demands, employee training, and cost justification.

Virtually all the manufacturer/user members of the coalition indicated that they expect to try the IMS concept on a test-case basis in specific areas of their operations. Machining of large components, shaft grinding, and new product lines were mentioned by several manufacturers as target areas for applying elements of the metrology system.

On the basis of this project, a growing emphasis on an integrated approach to metrology as a key to lowering quality costs in discrete parts manufacturing can be predicted, as well as a needed shift in quality control from product-oriented appraisal and failure detection to process-oriented prevention. Because control costs, such as implementing an IMS, involve "up front" capital expenditures and organizational changes, this shift in costs must be accompanied by a long-term perspective on the part of management.

Additionally, nondimensional metrology, involving the monitoring of process parameters such as horsepower, temperature, and vibration, can be expected to play a significant role in future integrated inspection systems. While nondimensional inspection was not within the scope of the IMS program discussed above, the models and analytical techniques used to evaluate metrology approaches can be extended to these nondimensional technologies as well. **ME**

ABOUT THE AUTHORS

L. Scot Duncan is a vice president of Booz, Allen & Hamilton. He leads Booz, Allen's producibility activities, specializing in the area of new products and manufacturing process development. He has also led or participated in numerous programs involving technology assessments, venture risk analysis, and the business implications of technological alternatives. Duncan had overall responsibility for the TACOM integrated metrology program.

Gary L. Bowen is a senior associate with Booz, Allen. His area of specialization is technology assessment and the development of new products and identification and development of manufacturing technologies to fabricate them. Bowen served as the program manager of the TACOM assignment. **ME**

AMERICAN MACHINIST Special Report 775, April 1985

Integrated QA: closing the CIM loop

Integrating quality assurance into the process is essential to successful computer-integrated manufacturing. The technology exists to do the job

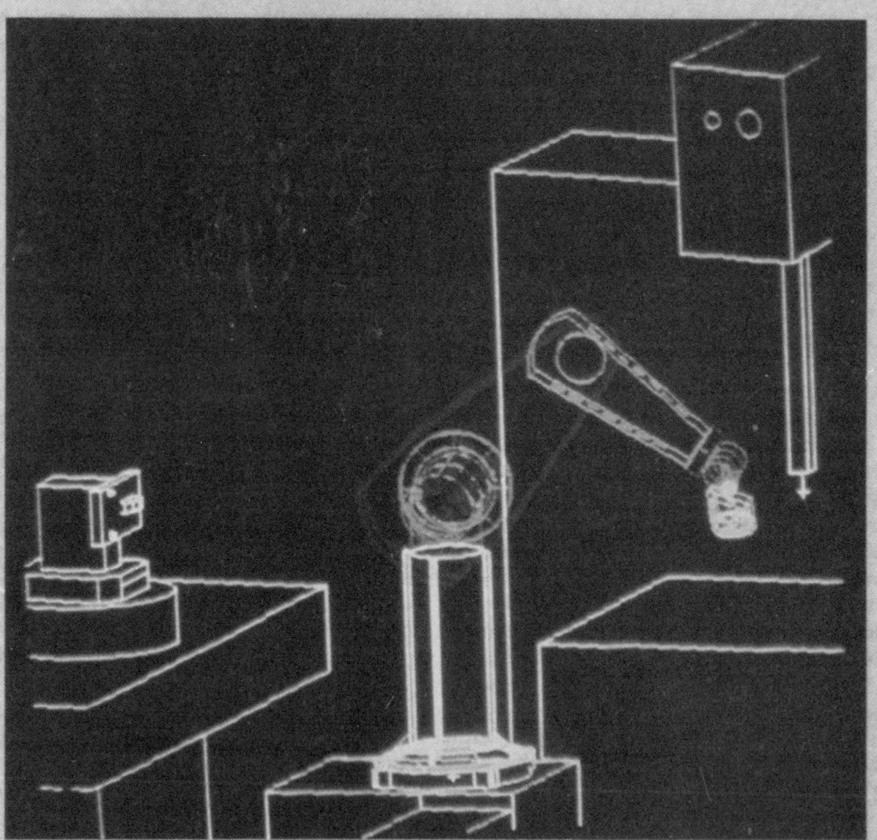

UNLESS QUALITY ASSURANCE becomes an integral part of the life cycle of a product, instead of serving strictly as an audit function, computer-aided manufacturing (CIM) may never flourish.

By George H. Schaffer, senior editor

Sooner or later, manufacturing managers realize that, although CIM holds out the greatest potential for improving manufacturing capability, that promise may very well remain unfulfilled unless QA is an intimate element of every step in the design/manufacturing/distribution cycle of a product. Simply increasing productivity through the application of CIM is futile unless the output is defect-free and meets the original design intent.

According to Edward Suor, president of Quality Information Systems (Rochester), a successor company to Hansford Data Systems, the goals of integration are higher-quality products and better utilization of fixed assets. However, he points out, for many manufacturers, QA represents the last incremental problem in the integration process. "In the past, QA was viewed as just one part of a factory's operation," says Suor. " 'Make it, inspect it, sort it' is the way it has been. The more modern understanding is that overall quality assurance is the end result of a carefully thought out and controlled production process."

In fact, he views QA as essential for achieving the benefits that CIM promises: increased productivity, greater flexibility, faster response time, reduced work-in-process inventory (WIP), economy of scale through the application of group technology in short-run production environments.

By the same token, CIM imposes new challenges on the QA function, and it is up to manufacturing engineers and QA specialists to turn these challenges into opportunities for improved quality and increased productivity. Some of these major challenges can be summarized as an increasing demand for the following requirements:

- Automated inspection and testing, including faster inspection rates.
- Noncontact inspection, particularly machine vision.
- In-process and on-the-machine inspection.
- Application of and reliance on statistical-analysis techniques, including statistical process control (SPC).
- Effective, integrated quality-data management and reporting.
- QA/CAD/CAM links.
- Standardization of QA-programming and -control languages.
- Development of flexible inspection systems (FISs).
- Earlier QA involvement in the product cycle.
- Designing for inspectability.
- Inspection planning at the design stage.
- Computerized analysis and modeling techniques.
- New vendor relationships based on performance rating and tracking.

According to most experts, there are no real technological barriers to achieving any of these requirements. The real challenge to QA professionals is to ensure that the quality function assumes top priority within the corporation. "Quality professionals must understand the changes that are affecting the factory and must become active to see that quality is a prime consideration," says Keith E McKee, director of the Manufacturing Productivity Center at IIT Research Institute (Chicago). "If quality professionals do not assume this leadership, this challenge will be accepted by others."

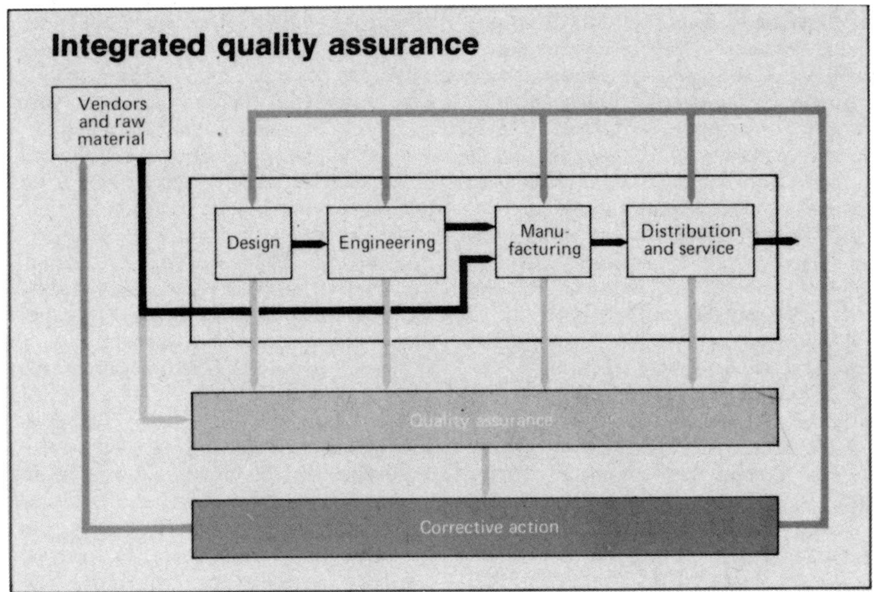

In fact, as QA becomes integrated into the manufacturing process, manufacturing engineers, instead of QA specialists, are increasingly involved with specifying and implementing QA systems. "One would think that quality professionals would be leading the push toward automated inspection," reports McKee, "but this is not the case. When automated inspection is being introduced, manufacturing engineers are leading the way; quality professionals are frequently the antagonists."

Automation has been a vital ingredient of CIM because it strives for a smooth and continuous manufacturing process with minimum human tending and intervention, particularly in the growing batch-oriented facility. Concurrently, the constant striving for increased productivity coupled with a growing concern with quality throughout industry has led to the rapid proliferation of automated inspection.

Historically, inspection has frequently been a bottleneck in the manufacturing cycle. In many instances, it simply takes longer to completely inspect a part, using traditional techniques, than it does to manufacture it on modern automated production equipment. Also, the increasing complexity of modern products makes nearly impossible demands on conventional quality-control techniques. The evolution of electronic gaging and digital techniques, the application of programmable controllers and computers, the development of noncontact inspection, and, more recently, the development of machine vision (See Special Report 767, "Machine vision: a sense for CIM," AM—Jun'84,p101) have all contributed to removing the inspection bottleneck through automation of the QA function.

Machine vision is particularly well suited to CIM automation because it can usually keep up with the production process and can provide information to institute corrective actions before the process gets out of control. The technology is suitable for dimensional inspection and gaging, as well as checking for such attributes as flaws, presence or absence of features, location, recognition, and identification.

Many vision systems are operating in the automobile industry, with General Motors probably leading the way, "As familiarity and confidence in the technology are improved, both the numbers and diversity of applications will increase," says Gerry L Elson, executive director of Machine Intelligence Technology Implementation, Advanced Product & Manufacturing Engineering Staff, GM Tech Center (Warren, Mich). "Other sensors, such as tactile, force, and torque, will be used; often in combination, with vision expected to continue as the most frequently used sensor device." Elson also predicts that, in the future, the essentially computer-based nature of machine-intelligence equipment (such as machine vision) will be integrated into plant data-reporting systems and "factory of the future" applications.

In addition to speed, machine vision is

capable of collecting and processing a vast amount of disparate data from one instantaneous viewing. For example, a machine-vision gaging system under construction for the Ford St Louis assembly plant by Automatix (Billerica, Mass) has the capability of 100% inspection of minivan underbodies. Several checks are made at production-line speeds:

- Width measurements at seven locations on each side of the body.
- Location, in two axes, of the two front cross-member support holes.
- Location, in three axes, of the seven rear-suspension-support holes.
- Detection of significant burrs in six of the rear-suspension-support holes.
- Detection of significant cracks in the one flanged rear-suspension-support hole.

Some 39 distinct measurements are taken to arrive at a possible total of 54 significant features or dimensions; the user can choose 15 arbitrary calculated dimensions based on the measurements. The microprocessor-based system stores all data on hard disk for anaylsis by Automatix statistical process control (SPC) software. Continuous trend analysis alerts the line operator to developing problems. All fixture operations, including the gaging cycle, are controlled by a programmable controller that also provides all interfacing to the rest of the automated assembly line and advises both operator and the main line controller of part status.

Automation is not enough

But, unless automated techniques are an integral part of CIM and provide timely and meaningful information for keeping the manufacturing process under control, the effort is in vain. McKee notes, "The concept of building parts or products and then inspecting them to see whether they are acceptable is ludicrous when you consider available technology. The automated factory must achieve quality by building it into the system."

After all, product inspection at the end of the line can only catch defects after they have occurred and prevent shipment of bad products. No amount of inspection though, regardless how automated, can, in itself, prevent the manufacture of scrap. Defect prevention, on the other hand, can be achieved only through a systems approach, through integrated quality assurance (IQA).

Another compelling reason for integrating quality assurance into the system is the need for accurate scheduling information. Typically, schedules are based on an assumption of 100% quality, but the reality is quite different. William R. Thurston, president of GenRad (Waltham, Mass), a manufacturer of automated test equipment (ATE) for the electronics industry, points out that no company can maintain good overall business control if its management-information system is fueled by inaccurate or obsolete information. "Without taking into account the real-time, actual state of quality, cost information and scheduling information based on an assumption of 100% quality are misleading instead of useful," says Thurston.

"If you have a device or subassembly and its quality is unknown because it hasn't yet been inspected or because the inspection results are not known to the computer or to the people doing the scheduling, then whatever schedule they come up with is probably not valid," he continues. "If that device or subassembly is correct, then the schedule is valid. But, if it does not meet specifications, then something has to be done to it: additional time and money have to be invested in fixing and reinspecting to ensure that it is correct before it can move on to the next step in the manufacturing process.

"In short," says Thurston, "if you don't know the quality of a component or subassembly immediately, then schedules are jeopardized, and costs go out of control."

What Thurston has discovered in the commercial ATE industry is true for all of manufacturing: there is only one way to know the quality of a component or subassembly right away, and that is by having an inspection system report it to the scheduling computer automatically and immediately. Scheduling can then take into account both good units and bad units along their different process paths.

Such integration of the QA function can be characterized by those aspects concerned with the physical functions of inspection and gaging and those aspects dealing with the management and analysis of the data thus developed. Ultimately, the objective is to provide appropriate corrective feedback to the manufacturing process and, in effect, achieve a closed-loop system that remains within specified control limits.

Feedback starts at the machine

Corrective feedback starts at the machine tool. Because CIM and, on a smaller scale, fully implemented flexible manufacturing systems (FMSs) rely heavily on untended machining, their success depends, to a large degree, on the use of automatic sensing techniques to monitor the performance of machines and processes and to compensate for the

Summary of IMS costs and benefits

Today's major cost drivers	Proposed improvements	Benefits	Implementation costs
Internal Scrap Rework Labor Capital equipment Inventory (WIP) Assembly difficulty **External** Liability/warranty	**Process** Improved preprocess Control of tools, fixtures, machinery In-process diagnostics Automated/inspection where feasible In-process/in cycle Postprocess Preprocess **System integration** Improved data gathering and information processing Improved closed-loop control	Reduced scrap Reduced rework Reduced labor Reduced capital equipment Inventory of gages Obsolescence Reduced WIP Reduced assembly difficulty Improved production-equipment utilization Reduced materials handling Reduced external failures Improved tool utilization Possible reduced precision requirement for inspection equipment	Integration of inspection technology to provide closed-loop control In-process/in-cycle Postprocess Sensor development for in-process inspection Control/communication-system development Software development Increased tooling-, fixture-, and machine-accuracy certification

Costs and benefits are predicted in a study of design and implementation of an Integrated Metrology System (IMS) carried out by Booz, Allen & Hamilton Inc for the Army Tank & Automotive Command (TACOM)

irregularities and uncertainties of the work environment (see Special Report 756, "Sensors: the eyes and ears of CIM," AM—Jul'83,p109).

Broadly speaking, these on-the-machine gaging and sensing techniques can be categorized as (1) those concerned with determining the location and size or condition of the workpiece, (2) those concerned with determining the location and condition of the cutting tool, and (3) those concerned with the condition of the machine and its functions.

A variety of techniques are available for generating the necessary signals to automatically initiate timely corrective actions or to automatically recover from failure: touch probes establish the location and size of workpieces and tools; vibration- and acoustic-analysis techniques hold the promise for predicting incipient tool failure and tool wear; power- and torque-monitoring techniques provide overload protection and, with advanced electronic techniques, can provide each tool with a sense of touch and also help detect tool wear; signature-analysis techniques can provide diagnostics to prevent catastrophic failure and pinpoint trouble areas.

Depending on the type of machine and the task to be performed, touch-sensitive probes, such as the touch-trigger probes available from Renishaw Inc (Arlington Heights, Ill), can be mounted in the machine spindle, in the tool position, on the work table, or in any other convenient machine location. They can be treated like tools by automatic toolchangers, and, when combined with software now available for most CNC systems, can be automatically cycled through significant measuring cycles for postprocess inspection.

Machine vision is also being applied at the machine tool for on-the-machine gaging. For example, the Tool Offset Sensor demonstrated on a J&L Delta 316B FMS turning center at IMTS-84 by Kennametal Inc (Latrobe, Pa) uses a machine-mounted View Engineering (Simi Valley, Calif) machine-vision system to "look" at a new cutting tool and compare it with a standard. The system then transmits any needed correction to the CNC for automatic tool-offset adjustment. Such sensors can also use pattern-recognition techniques to ensure that the correct tool is in place and detect tool contamination (such as clinging chips), fracture, or breakage.

Kennametal has also developed a tool-condition sensor that relies on three-axis force transducers mounted in the tool blocks of a turning center to measure radial, feed, and tangential forces during the cutting action. The monitor uses a computerized "tool-wear-estimator" model to continuously analyze the relationship of these forces and automatically calculates the amount of wear on a tool, independent of tool or material data. When wear reaches a predetermined threshold, the sensor signals the need for a toolchange. At that point, the Tool Offset Sensor is used to qualify the new tool and make the necessary offset adjustments. In case of tool breakage, the sensor immediately signals the CNC to retract the tool before damage can be done; then it initiates a toolchange.

Of course, it is equally important that the tools used for untended machining be properly qualified before they reach the machine tool and that the tool-setup function itself be integrated into the CIM or FMS communications network. With adoption of integrated tool-presetting systems that can interface with the machine CNC or host computer, quality assurance keeps moving upstream, and postprocess inspection gives way to process control.

Machine vision for underbody gaging*

Measurement technique	Dimension or feature measured	Gaging component(s) at each point	No. of points
Range finder	In/out	Camera and projector mounted on common base plate	14
Feature locator	Fore/aft In/out	Camera and floodlight installed in separate modules	2
Feature position	Fore/aft In/out Up/down	Camera, projector(s), floodlight, and rear diffuser installed in separate modules	7
Burr detection	Burr size	Same components used for feature position at some points	6
Crack detection	Crack size	3 cameras in addition to lighting of feature position	1

*Automatix machine-vision gaging system at Ford provides capability for 100% on-line inspection of minivan underbodies produced at St Louis assembly plant

Perceptron machine-vision system at new Chrysler Sterling Heights Assembly plant automatically checks door and decklid openings to ensure proper fit

Tool-presetting system

DeVlieg Machine Tool Co (Royal Oak, Mich) describes such a system. In it, tooling information—tool-identification number, tool length, tool diameter, type of tool, tool components, location within the tool crib, targeted-machine-tool number, and position within the toolchanger magazine—is entered by the NC programmer and stored at the manufacturing system's host computer. The host computer schedules the tool-presetting function and presents the necessary information at a CRT in the tool crib at the appropriate time.

The type of tool, the location of the tool's components within the tool crib, and a graphic display of the tool are presented on the terminal, and the information is used by a tool assembler for gathering the tool components, assembling the tool components into a complete tool, and grouping the tools. The entire group of assembled tools is then transferred to the preset machine.

The CNC-operated preset machine also receives pertinent tool data from the host computer, and this information is displayed at a CRT. The machine automatically positions the length and diameter slides to the programmed coordinates and rotates the preset turret to present the appropriate spindle. The operator then mounts the tooling in the spindle and uses a profile projector with electronic crosshairs to inspect and adjust it. Once the tooling is adjusted, a bar-code printer connected to the preset control prints a bar-code label with the tool number, targeted-machine-tool number, and the toolchanger-magazine position.

In this way, each tool is inspected, preset, labeled, and grouped with the other tools in the tool load, and the actual tool length and diameter for each tool are automatically transmitted from the preset control to the host computer. Ultimately, the preset information is used to automatically make the necessary offset adjustments at the CNC machine tool. Once the tool arrives at its destined machine, typically by wire-guided vehicle and robot, a bar-code scanner identifies it and determines its proper placement in the tool magazine.

Automation is not essential

Although the physical aspects of integrated quality assurance are most frequently achieved through the implementation of automated inspection and gaging, automation *per se* is not essential to integrated quality assurance. What is essential is that the information gained during the inspection process be integrated into the flow of data that is at the heart of CIM. In many instances, for example, manually operated digital electronic gages can be used effectively to close the manufacturing-control loop because the data they develop can be readily analyzed in real time and interfaced with the manufacturing database.

One example of such an approach is the FAN (Factory Area Network) concept recently introduced by DataMyte Corp (Minnetonka, Minn). The FAN scheme combines operator-based real-time statistical process control, through the use of a DataMyte 750 data collector, with batch-mode data collection, through the use of a DataTruck logger/analyzer.

The 750 is a multichannel data collector with statistical computational capability. It can be used to collect data from hand-held or fixtured gages, such as micrometers, calipers, and dial indicators equipped with appropriate interfaces; or readings can be entered manually. The readings (up to 2500) are stored, and statistical information is displayed at the push of a button. The unit compares readings against specifications or subgroups against process-control limits and displays an appropriate message, such as "In control" or "Range out." With the addition of a CRT, $\bar{X}$ and R control charts or histograms can be produced instantly by the operator, who uses the information to keep the process under control.

Once a day, or as often as desired, QA supervisors can use the DataTruck to "harvest" the data from clusters of DataMyte 750s. The DataTruck is a high-capacity hand-held version of the 750, designed to empty up to 20 750s and produce printed reports or transmit the data to a central computer.

DataMyte also supplies FAN software for the IBM PC and other microcomputers. The software provides data management and reporting, problem-solving tools, and long-term storage of data. Problem solving includes the ability to

Machine tool positions new cutting tool under sealed video camera of Kennametal tool-offset sensor. Sensor updates CNC offset tables within 3 sec of viewing

sort data from the 750s and compare it with indices of capability, scrap and rework estimating, and Pareto analysis to help identify the most important problems.

According to DataMyte, the FAN approach can pay for itself in less than six months, simply from the time saved in gathering and analyzing data. Not included are the saving of scrap and rework and other quality-related costs.

Microprocessor-based data loggers providing limited statistical computational capability are increasingly popular and are available from many of the traditional gage manufacturers, which have moved from the age of dial indicators to the age of digital readouts and "smart" microprocessor-based gages. The Marposs Gauges (Madison Heights, Wis) E42 microprocessor-based amplifier for use with bench gages and automatic gaging machines can store up to 32 measurements and includes statistical-quality-assurance software. A CRT is used for programming and display of data; an RS-232C port is provided to permit computer interface.

The Digital Techniques Div of GTE Valeron Corp (Brookfield, Wis) recently introduced the Dataqume/QA data accumulator. It accepts analog data from a variety of gage amplifiers, digitizes the data, and stores up to 15,000 readings. Federal Products (Providence, RI) has an output module for its Maxμm line of digital electronic indicators to provide data communication via an HP-Loop interface, making the indicators suitable for automatic data collection. Federal Products also offers appropriate data analyzers.

Operator checks shaft dimensions and records data using DataMyte Model 750 data collector. Associated CRT displays $\bar{X}$ and R charts

Supervisor uses DataTruck unit to 'harvest' readings from an individual DataMyte 750 data collector. Data is then dumped into central computer

Statistical methods essential

All of these data-gathering techniques use statistical analysis routines to develop significant information instead of just logging only raw data. Statistical quality control (SQC), probably the single most effective tool for process control, is an essential component of any integrated-quality-assurance scheme. Unfortunately, it has been long neglected by US industry, even though its principles were developed in the United States.

The underlying principles of SQC are really quite simple: set standards of quality; establish the capability of the process; monitor the process to detect any significant changes that indicate that the process is going out of control; and make corrections to keep the process in control. Thus, SQC leads to process control, and the term *statistical process control* (SPC) is beginning to dominate. The principal tools of SPC are frequency distribution, a tally of the number of times a given quality characteristic occurs, and control charts to provide a graphic means of establishing whether a process is still in control (see Special Report 762, "Statistical Quality Control," AM—Jan'84,p97).

The real key, then, to integrated quality assurance rests with the management and analysis of QA information, a task best performed through the application of computers and thus eminently compatible with the principles of CIM. According to Dr John A. Keane, president of John A Keane & Associates (Princeton, NJ), automation of the QA function is a vital part of the so-called "reindustrialization" effort now in progress, and external demands and attendant pressures in the 1980s will require computer systems to manage quality information. Some of these external pressures:

- The need to accelerate lot release to match increased manufacturing velocities.
- The requirement to centralize control of product quality to meet competition in national and international markets.
- The advantage of reducing startup-cycle time in new-product introductions.
- The pressure to improve productivity by substituting capital-intensive equipment for labor-intensive activities.
- The ability to participate in markets sensitive to product quality.

"Whether the move to computeriza-

Programmed from CAD database, Zeiss CMM checks turbine disk

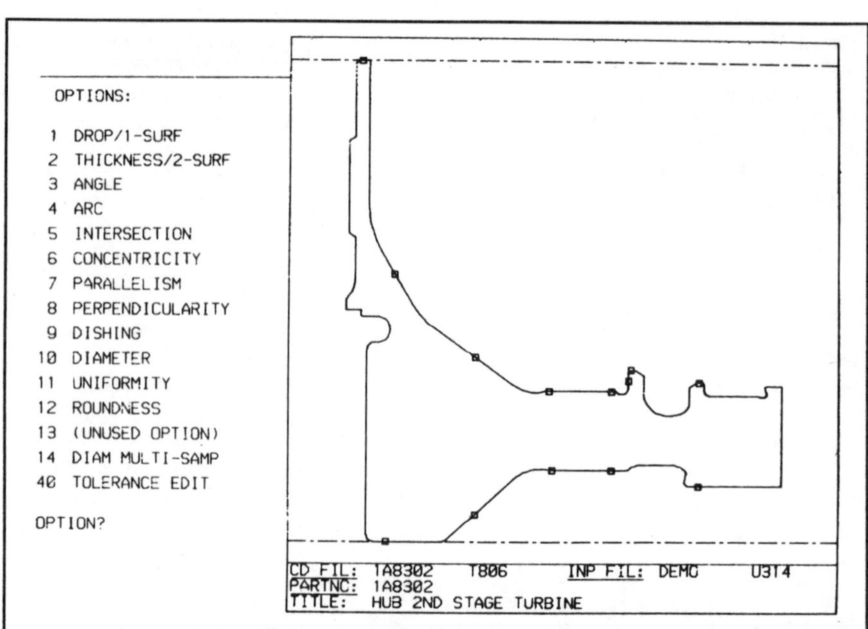

Geometry of hub, stored in cross-sectional form is called up at graphics terminal of Pratt & Whitney Aircraft system, and programmer selects appropriate inspection points

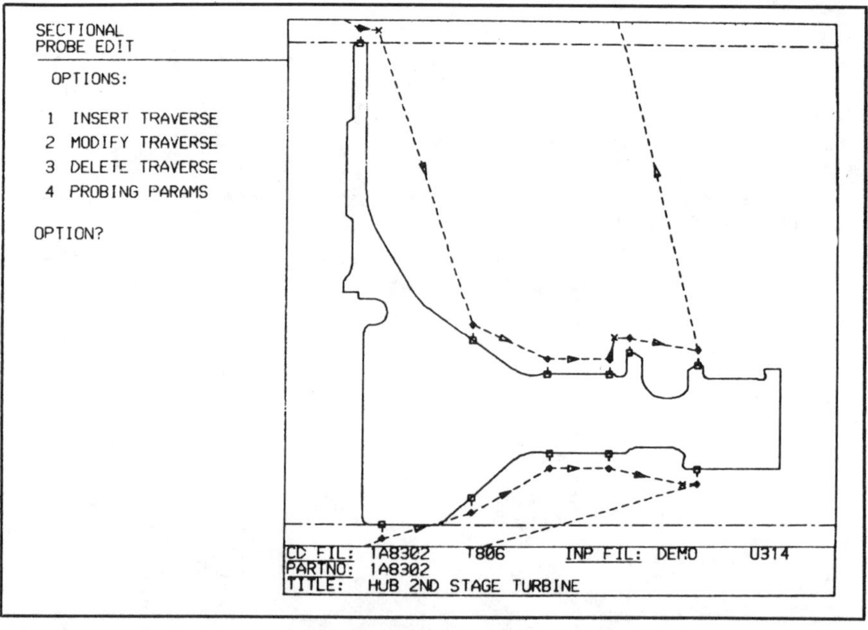

Once inspection points are selected, Pratt & Whitney Aircraft system automatically generates inspection program and displays probe path, which can be modified

tion is brought about by external pressures or is a result of internal efforts to streamline the organization, the quality-assurance function will benefit trom the change," says Keane. When data is handled manually, he points out, the cost rises proportionally to the amount of work required. With computerized data management, on the other hand, the cost is primarily up front in the initial capital expenditure, and a considerable amount of data management can be handled without any significant increase in cost.

Keane cites the example of a company that was able to justify the installation of a computerized system to collate test results from various laboratories for lot release. As it turned out, the same software/hardware configuration was able to automatically chart the necessary trends for all variable data without additional cost.

Another benefit from computerization of the quality-control function is that such analysis of the data can provide information for making decisions in a more timely fashion. "The QA function can therefore be more responsive to corporate operational needs," claims Keane. This is particularly important for new products, early detection of adverse process trends, and rapid release of lots to reduce plant inventory.

Once established, a database can be explored in many different ways to support quality and profit-improvement functions. In one instance, according to Keane, such database analysis by a firm recommended a change to one raw-material specification that produced a direct annual saving of more than $250,000. Without access to computerized data reduction and analysis, such an examination would have required the attention of already scarce personnel resources and would not have been carried out.

John Keane & Associates offers integrated software modules, called GMS programs, to handle quality information. Reports, graphs, and simple statistics are handled by core modules. Additional modules provide direct factory-floor feedback to plant- or company-wide integrated systems. Here is a brief description of some of the available modules:

- Quality Control Logistics System (QCLS) tracks the progress of lots through all stages of manufacturing and laboratory testing.
- Quality Control Operating System (QCOS), a comprehensive information system, furnishes a variety of reports and calculations, including acceptance-release reports; in-process control charts, histograms, and trend analyses; interactive access to databanks for troubleshooting; management summary reports; programmable calculations and statistical routines; labor-hour accounting, and provisions for simulation and optimizing studies.
- Quality Control Audit System (QCAS), an analytical audit package, computes confidence levels for a product's quality characteristics by quantitatively integrating prior-lot histories with release-test criteria.
- Raw Materials System permits rapid release of incoming raw material by providing such features as direct data entry, on-line acceptance-release reporting, vendor-performance rating.
- In-process Control System is a rapid-response system for detection and control of adverse process trends.
- Finished Goods Release provides timely release of finished goods into inventory by offering time-saving sampling strategies, high visibility of problems, and other performance criteria.

The appropriate modules can be selected to handle initial needs, with additional modules implemented as the scope increases. A protocol-dialogue feature allows users to customize any number of different QC applications within a single system, an important feature for flexibility.

Quality Information Systems concentrates on custom QA-information systems instead of standard products. A typical Shop Floor Quality Management System from QIS includes the following packages:

- Shop Floor Quality Control is used by the machine operator for real-time data acquisition and trend analysis.
- Production Quality Management is aimed at the quality engineer and provides data storage from multiple sources, reporting, and updating capability.
- Process Analysis provides quality engineers with detailed statistical-analysis routines, such as control charts and descriptive statistics. A true-position analysis routine is included.
- Incoming Quality Management is used for tracking materials and storing and updating vendor-performance data.
- Vendor Rating establishes perform-

The Automeasure approach to CMM programming

The first step in creating a Computervision Automeasure program is to construct a graphic model of the probe cluster to be used for the CMM inspection task. The CAD model of the cluster is built up with graphic representations of the components of standard CMM probe kits. These representations are stored in libraries of different vendors' probes and include probe heads, styluses, extensions, cubes, and angle joints (top left).

Once the probe cluster has been constructed (bottom left), it can be used to visually verify, on the CRT, that it can access all the necessary features of the part to be measured: part and probe cluster are brought up on the screen, and the probe is moved to all measuring locations. When the operator is satisfied, the probe cluster is filed for use in program generation.

To generate a CMM program, the operator first determines how the part will be placed on the bed of the CMM to ensure correct probe-path orientation relative to part geometry and then specifies the directions of the machine coordinate system, "inserting" the probe in the probe arm accordingly.

In the programming mode, a dedicated language is used for creating programs to measure features, define coordinate systems, report nominal and tolerance data on measured features, construct imaginary features from recalled features, check form and position tolerances, and execute whatever commands are necessary to satisfy the inspection requirements. The result is a graphic probe path, which also contains the part data and instructions for CMM-program execution.

Once completed, the CMM program can be simulated graphically to check for collisions of probe and workpiece (bottom right).

When the program is satisfactory, it is processed into a neutral data file. The Automeasure neutral data file is a series of generic CMM functions, which are translated by a postprocessor in the target CMM into the local language of that machine.

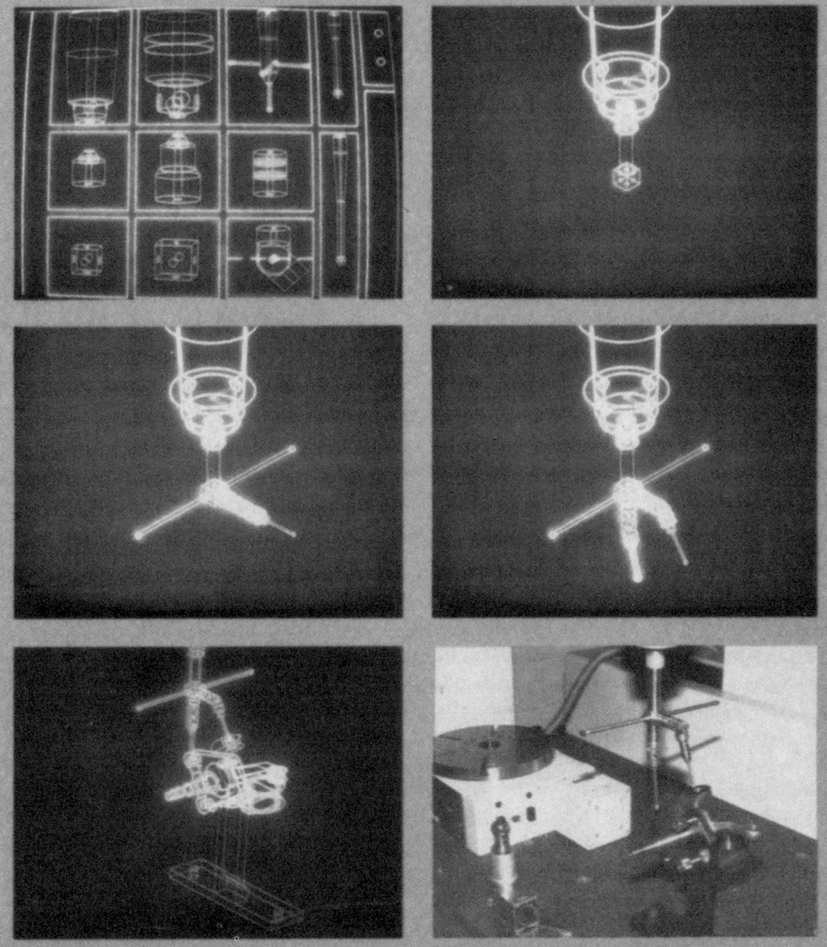

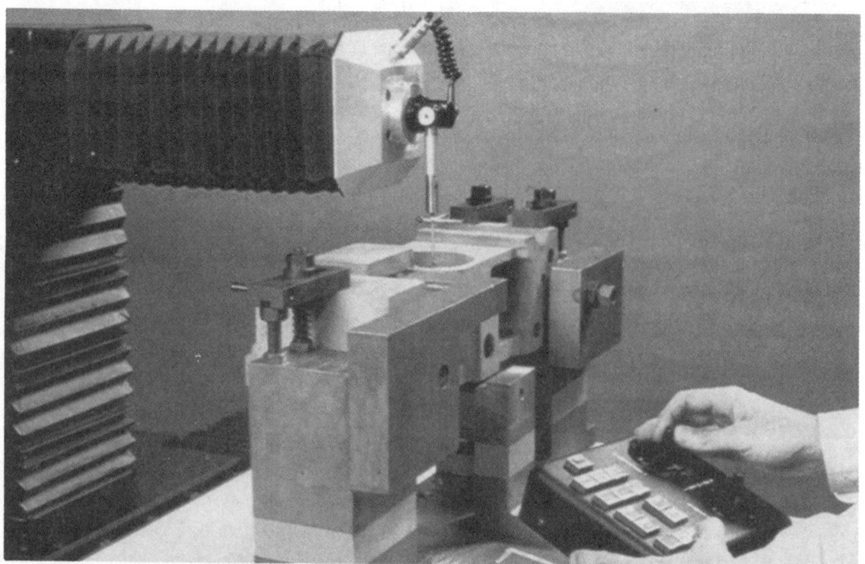

Built more like machining center than conventional CMM, Brown & Sharpe PCR has accessible horizontal design and accommodation for pallet/fixture arrangements

ance ratings and reports for all or individual vendors.

- Instruction Management, aimed at QC inspectors, provides on-line access of inspection procedures.
- Gage/Instrument Calibration, aimed at quality engineers and tool-crib personnel, provides gage/instrument control through use of a database of calibration history.

A comprehensive directory of companies furnishing QA-related software is included at the end of this report.

A combination is best

Of course, the ideal IQA approach combines automation of the inspection and gaging functions with automated data reduction and analysis. One way to combine the gage-control and data-analysis functions is through the use of programmable controllers: a PC can control the physical gaging functions and also store and manipulate the data established by the gage. "Programmable controllers are underutilized as a statistical-control device," says David Cherba, application section manager, Gould/Modicon (Andover, Mass). "Those that have used them in this manner have found them to be a very economical solution to the control/analysis problem."

Several PC builders offer special statistical software packages for their units (see QA software directory, elsewhere in this report). Gould/Modicon provides a STAC function block for its Modicon 584 Series controllers as an aid to generating statistical-process-control logic. Basically, the function block accepts a four-digit measurement value from a user-specified source register of the PC whenever the block is changed from "off" to "on." The value is then kept in a first-in/first-out (FIFO) stack of user-specified length equal to the desired sample size. When the specified sample size is reached in the FIFO stack, a new entry pushes out the oldest entry, keeping the sample size constant.

Each time the block is energized, the current value of the source register is pushed onto the stack. The block then looks at all the entries in the stack, sums them, and divides this sum by the number of entries in the stack. The result is the mean, or $\bar{x}$ value, of the entries, which is stored in a dedicated register of a table of "destination" registers specified by the user.

Along with the $\bar{x}$ calculation, the block also calculates the range of values (R) in the FIFO stack; simply the difference between the highest and lowest value in the stack. Like $\bar{x}$, R is recalculated with each new entry into the stack and is held in a dedicated holding register of the destination table.

Other data manipulations are also possible, including comparison of $\bar{x}$ and R values to upper and lower control limits (UCL and LCL) for control-chart evaluations. As with PCs in general, raw data and/or processed SPC information can be transferred to a host computer.

Allen-Bradley has developed the Series 7890 Productivity Monitoring system as a link between PC systems and a plant's central computer. A basic system includes a computer (DEC PDP-11/73, Winchester disk, and removable-disk and/or tape drive), color graphics terminals, and printers. User-configurable software monitors the status of machines and processes as governed by Allen-Bradley programmable controllers via an A-B Data Highway. User-selected production-related information is then available at the terminals. The system provides a common plant database for displays and reports, for data archiving, and for communication to a higher-level plant computer.

The 7890 System can be used to monitor such manufacturing-facility variables as production rates, scrap, downtime, machine efficiency, raw-material use, and setup time. The system also provides primary statistical-quality-control tools:

- Average and range values ($\bar{x}$ and R).
- Counts or occurrences/binomial distribution.
- Counts or occurrences/Poisson distribution.
- Pareto charts.
- Scatter plots.
- Upper- and lower-control-limit alarms.

CMM, IQA's premier tool

The one QA tool that best combines the automation of the inspection process with that of the data-processing function is today's coordinate measuring machine (CMM). Although configurations vary and some machines still rely on manual manipulation, the CMM is rapidly moving from the role of final-inspection policeman to that of process-control tool. "The CMM is used less as a standalone metrology-room device and more often as part of an integrated system," says Fred Soskel, manager of marketing and advanced planning at the XMT Div of Carl Zeiss Inc (Thornwood, NY).

Most modern CMMs use computers in one form or another and, by their very nature, provide the kind of rational digital information required by computer-oriented manufacturing processes: they talk computerese. No wonder CMMs are becoming the premier weapon in manufacturing's arsenal of integrated quality assurance.

Because most CMMs are equipped with computers to control their positioning and measuring cycles, the data thus gathered can also be readily manipulated and analyzed. Typical CMM software provides setup aids, such as automatic alignment routines; geometric-tolerance evaluation, including best-fit analysis; statistical-analysis programs, including trends analysis; and communications capability to tie in with a host computer.

Communication is vital

"As more and more data is produced in the CIM environment, the ability for discrete systems, including CMMs, to communicate with each other becomes vital," says Soskel. To meet this need, Zeiss CMMs now incorporate communications software for bidirectional communications with such support systems as host computers, materials-handling systems (robotic and transport), part-recognition systems, and CAD/CAM systems.

In fact, tying CMMs into the CIM data stream is a current preoccupation of all interested parties: CMM builders, CMM users, CAD/CAM vendors, and related professional organizations. The problem confronting them all is that there presently is no single uniform language for computer-controlled CMMs or, for that matter, other computer-aided inspection equipment: each vendor furnishes a unique computer-assist language for programming and data processing.

One effort toward CMM integration is under way as part of the Quality Assurance Project (QAP) sponsored by Computer Aided Manufacturing-International (Arlington, Tex). The project stems from the recognition by CAM-I members that rapid advances in computer-aided-design and -manufacturing (CAD/CAM) technology have placed new demands on the QA function in the manufacturing environment.

According to the QAP prospectus, technology in computer-aided quality assurance must be advanced to improve efficiency, quality, and productivity in the manufacture of durable goods. "Ideally, quality assurance should be a built-in, totally integrated function," the prospectus states, "encompassing the whole of the design and manufacturing process. The computer offers opportunity and means to bring this ideal into reality."

The QAP objective is to identify those areas in which the need for computerization of QA is most immediate. Several needs have already been identified:

- A universal high-level language for quality-assurance equipment.
- On-machine inspection.
- A study of the mathematics model vs the master model and a merging of the model and actual part.
- Standards for coordinate-measuring-machine algorithms.
- Standards for 2D and 3D graphics and graphics verification.
- Assembly quality assurance.
- Specifications and consolidated requirements for a quality database and database network.
- Dimensioning and tolerancing for surfaces.
- Automated quality-assurance testing, including tests other than dimensions.
- Image analysis via computer.
- Automated materials-review-board-type corrective action.
- Computer-aided quality-assurance administrative systems.
- Automated documentation for quality-assurance procedures.
- Quality-assurance involvement in software development.
- Computer-aided instruction for training quality personnel.
- Determination of the effect of CAD/CAM on quality assurance and its changing role.
- General quality-assurance standards.

One high-priority development project for the QAP group is a communications link between computer-graphics systems and inspection devices, particularly CMMs, to provide graphics programming and measurement feedback via CAD/CAM systems. "The project focuses on the problem of multiple languages for CMMs," says QAP Chairman David H. Miska, supervisor of CAD/CAM Systems Development, Manufacturing Div of Pratt & Whitney Aircraft (East Hartford, Conn). "It is very costly to select a measuring machine on its merits and then find out that you have to learn its language in order to make it function. The programming language should really not be a governing criterion in our choice of equipment," he points out. That is particularly true when adding new CMMs to existing ones.

Generic interface is needed

The first task undertaken in this project is the development of a Dimensional Measurement Interface Specification (DMIS) for automated dimensional-inspection equipment. When fully developed, it is intended that the DMIS will define a generic interface through which all inspection information necessary to support a CIM facility is passed. This Quality Interface (QI) is to consist of a neutral command file, a neutral data file, and protocols for transferring information. CAD systems will use these commands, data structure, and protocols to communicate with dimensional-measurement equipment.

The intelligence required to generate and decode this data will lie at either side of the interface. A postprocessor in the DAD system, quality-information system, and dimensional-measurement equipment (DME) will convert the unit's own internal data structure into the format of the DMIS. A program developed on any CAD system will, therefore, be able to run any measuring equipment, and it will be possible to send the results back to any quality-information or CAD system.

Although interfaces exist today, none are generic, and current standards do not meet all the requirements for a generic QI. For example, existing languages, such as APT, and neutral data formats, such as IGES, contain a different subset of the requirements for such an interface. The new vocabulary is to consist of terminology used in existing interfaces and current standards. The result will be a hybrid of existing standards, with each technology contributing its unique information to the overall solution.

Phase I, development of the functional specifications, has already been completed under contract to IIT Research Institute (Chicago) through a study of the inspection requirements for a variety of industries. These requirements include multidirectional communication between systems, part description, comparison, statistical and text data, and machine-position information. The position of the QI in the overall CIM computer hierarchy was identified, and the flow of QI information was defined. The QI has been identified as a vocabulary of terms recognized by the CAD system, the quality-information system, and the dimensional-measuring equipment. Phase II, which includes a demonstration model, is

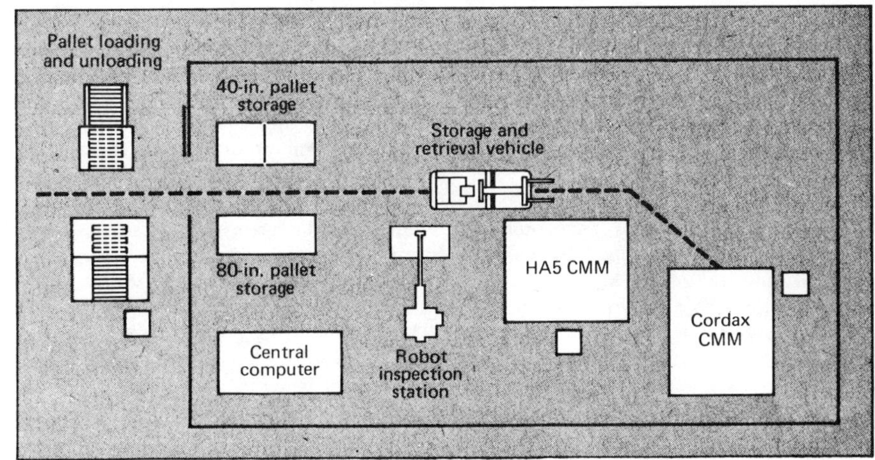

Parts to be inspected in Boeing FIS are mounted to pallets in staging area. Storage and retrieval vehicle takes pallets to machines for automatic inspection

Inspection requirements for Boeing FIS

Part or assembly inspected / Characteristic	Wing	Upper clevis	Lower clevis	Actuator	Housing	Inlet	Heat-transfer plate	Elevon housing	Antenna
Overall part size: length x width (in.)	74 x 12.21	6.2 x 6.08	6.4 x 6.2	5 x 2.9	6.2 x 3	25 x 8	9.12 x 4.06	10 x 16.2	9.13 x 8
Height (in.)	1.5	1.68	0.51	1.43	2	21.0	1.3	12.5	0.4
Linear measurement (in.)	0.010	1.010	0.010	0.010	0.010	0.005	0.010	0.010	0.005
Angular measurement (deg)	1	1	2	2	1	0.5	1	1	1
Hole diameter (in.)	0.03	0.03	NR	0.005	0.03	0.005	0.010	0.0007	0.005
Radius (in.)	0.05	0.09	0.05	0.03	0.03	0.06	0.03	0.005	0.03
Skin thickness (in.)	0.0075	NR	NR	NR	NR	NR	NR	NR	NR

NR = not required

scheduled for completion by December.

When DMIS is complete, CAM-I expects to propose it to national and international standards organizations. The objective is that DMIS be comprehensive enough to become an industry standard and inspire the vendor community to develop compatible software.

Whether CAM-I will actually succeed in developing such a standard is questioned by some industry skeptics who tend to be suspicious of committee work and prefer to go their own way. "But CAM-I is doing something that needs to be done," says David H Genest, product manager for CMM systems at Brown & Sharpe (North Kingston, RI) and a member of the QAP group. "If it does nothing else but allow people from all the major industries to sit down together and argue about what needs to be done and in that way focus on the urgency of the communications/interface problem, then it's serving a good purpose."

Some pragmatic approaches

In fact, CMM users and builders, as well as CAD/CAM vendors, have also been active in developing pragmatic solutions to the CMM communications problem. In effect, some CMM users have been working on the equivalent of a neutral data file of their own to link their disparate CMMs into an integrated planning and information system. A prime example is the CAIR (Computer Aided Inspection & Reporting) system developed by Deere & Co (Moline, Ill).

CAIR resulted from a corporate review of computer-related developments that recommended several actions:

- Develop a standard worldwide John Deere system for controlling CMMs and collecting data for them.
- Allow factories to select measuring equipment suited to their individual needs regardless of the programming requirements.
- Provide long-term retention and easy access of QA data.
- Have extensive data-analysis capabilities, including graphic presentation.

The CAIR system is a computer network comprising special software, digital-input devices, interfaces between these devices and the computer, and terminals at remote locations for programming, access to programs, collecting data, and analyzing data. Examples of digital-input devices are CMMs, the coordinate measuring head of a layout table, test-cell computers that transfer engine-performance data, and manual-input devices for entering manufacturing-process data. Any CMM, regardless of the manufacturer or the hardware/software combination, is operated by the CAIR system.

CAIR operates in some 16 North American and three European Deere facilities that can now exchange data. Most facilities can also access a larger central computer system at Deere in Moline for analysis programs not available locally. A typical local CAIR system uses a Digital Equipment PDP-11/34 or PDP-11/44 computer equipped with an RSX-11M operating system and serves as both a real-time processor to handle part dimensional data and as a multiprogramming system to perform such tasks as manufacturing-process control. The terminals provide interactive communication with the computer that stores the programs.

The software used with CAIR consists of an interpretive program called Computer-aided Layout (CAL) and several analysis programs. CAL's main function is to perform the mathematical and geometric computations necessary to convert digital data collected from measuring systems, such as CMMs, into usable data. The main purpose of the analysis programs is to perform and report statistical calculations on data stored in the computer's memory.

The latest addition to CAIR is an online-gaging system (OLGA) for use with hand and bench gages, all of which are now electronic. The gages are hooked into a data-logger and display unit that communicates with CAIR to provide instant information regarding the probable cause for measurement deviations reported by OLGA. The basic CAIR system, originally developed strictly for internal use, is now available through John Deere Technologies International on a license-fee basis.

From geometric database to CMM

Pratt & Whitney Aircraft has developed a computer-aided inspection system that captures relative data and tolerances required for manufacturing and inspection of gas-turbine disks and hubs and permits graphic, interactive programming of CMMs for the inspection task. In 1978, the company implemented a CAD/CAM system to design and manufacture turned gas-turbine-engine parts. The engineering descriptions of such parts are regularly stored in the CAD/CAM database, and the inspection system now accesses this geometric database for program preparation. Programmers at three separate locations (Southington and North Haven, Conn, and Columbus, Ga) can access the system, which is located in East Hartford.

In practice, the geometry, stored in cross-sectional form, is called up at a graphics terminal, and the programmer selects appropriate inspection points. For each point selected, the program automatically accesses the engineering database and finds the associated tolerance information, which is then displayed on the screen. The programmer can accept these tolerances or change them. A menu of measurement and inspection options is also displayed for the operator, who chooses an appropriate combination.

Once the inspection points, their tolerances, and the inspection options have been chosen, the system automatically generates an appropriate inspection program, including the inspection sequence, and displays the probe path for operator approval. The primary purpose of showing the probe path is to check for possible probe/part collisions and to optimize the probing sequence for minimum part-inspection time. The program is generated as a proprietary generic data file, and all CMMs at PWA have had their software modified to accept that file. "We don't even consider buying any software unless we get the source code," says Stevan W. Akerley, CAD/CAM-systems project leader at PWA.

In many instances, the CMM is used only to collect individual measurements, and the actual data manipulation and analysis are performed in a FORTRAN environment on a mainframe computer. "In some cases, we use the CMM measuring routines. But, in many instances, we can actually use the machines faster by offloading the results to a host and let it do the processing," says Akerley.

He would like to see vendors implement a standard such as is under preparation by CAM-I. "It would be good for vendors and users alike to have a common language for part-alignment routines and for driving the probe, but the user should have the option to use the analysis package provided by the vendor or to offload the data into a data file for a host environment. You can't integrate if you have the OEMs spit out all the answers to you," he maintains.

Akerley points out that, in an FMS or CIM environment, a variety of CMMs may be directed by a host that feeds programs as required to the local controller. Somewhere that host has to have some supervisory capability, and it's not going to be able to do that unless it gets data fed back to it from the CMM. The CMM has to be able to selectively collect information—either by continuous sampling or periodic postprocess inspection—and pass that information back to the host for further analysis. "The CMM has to communicate, or it's not going to be integrated," says Akerley.

Several advantages are claimed for computer-aided inspection of disks and hubs at PWA:

- Faster part programming.
- More-consistent, more accurate inspection.
- Recording of variable data in files for subsequent specialized analysis, engineering feedback, statistical process control, and manufacturing feedback.
- Minimum manual programming.
- Common programming for family of parts.
- Minimal programming skills required.
- Fast turnaround for changes and new parts.
- CMM freed for inspection work by off-line programming.

Storage and retrieval vehicle (in storage area) has delivered part to PUMA inspection station. Entire operation is under control of HP 1000 host computer

Robot inspection station uses Krautkramer-Branson ultrasonic gage to verify skin thickness of hollow ACLM wing. SRV is in back

CMM builders are also tackling the integration problem. For example, Sheffield Measurement Div of Warner & Swasey Co (Dayton, Ohio) recently introduced the OSCAR (Operating System for CMM Systems with Advanced Reporting) software system for its line of CMMs. Incorporating extensive data-logging and statistical-analysis capability, OSCAR is intended to integrate the total quality database by offering several capabilities:

- Part-program development and execution.
- Inspection logging.
- Delayed inspection reports from inspection logs.
- Statistical analysis from inspection logs.
- Multiple part-program and inspection-log libraries.
- Multitasking computer supporting multiple CMM and/or session-terminal users.
- Complete library-file management.
- User interface to inspection logs.

OSCAR is hosted by a Hewlett-Packard HP-1000 computer system running on the RTE real-time, multitasking operating system, which permits the support of multiple workstations, such as CMMs and ordinary session terminals. Various workstations can be executing OSCAR or other RTE programs simultaneously. For example, two CMM workstations can be executing a part program while a terminal-only workstation is being used to edit a part program in a quick-edit mode and still another workstation is running statistical-analysis routines. Also, system peripherals, such as printers or plotters, can be shared by the workstations.

At the American Motors Jeep facility (Toledo, Ohio), two OSCAR systems working together are ultimately going to integrate some 11 CMMs. Two systems are required so that multitasking capability can be provided without undue delays. The linkup of two OSCAR systems also enables programs and data to be shared among all workstations.

CAD/CAM vendors on QA track

Many CAD/CAM vendors are also beginning to turn their attention to QA integration/communications, particularly off-line graphics programming of CMMs. In concept, these efforts parallel the CAM-I effort at establishing the equivalent of a neutral data file except that, instead of being generic, it ends up as a proprietary solution.

A Unigraphics CMM interface, under development by McAuto (St Louis, Mo) consists of three modules. A generic CMM interface module, written in GRIP (McAuto's Graphics Interactive Programming Language), enables the user

The cost of quality

Contrary to some popular belief, the quest for quality does not increase the cost of a product. In fact, the lack of quality inflates the price of a product, rendering it uncompetitive.

Some estimates contend that 90% of the total cost of quality in today's manufacturing operations is spent on the appraisal of defects and the resulting scrap and rework. According to a survey made by Booz, Allen & Hamilton, scrap and rework costs are the key cost drivers for improving the overall cost of quality in today's operations. The top graph shows the total cost of quality and how the cost of control and the cost of failure contribute to the overall cost of quality in manufacturing operations.

One key factor is the rapidly declining cost of computing power compared with the relatively stable cost of inspection hardware (bottom graph). The ability to process large amounts of data inexpensively will enhance the implementation of integrated quality assurance, which depends on developing timely and meaningful information to pinpoint nonconformance as it occurs and on instituting corrective action before the process goes out of control.

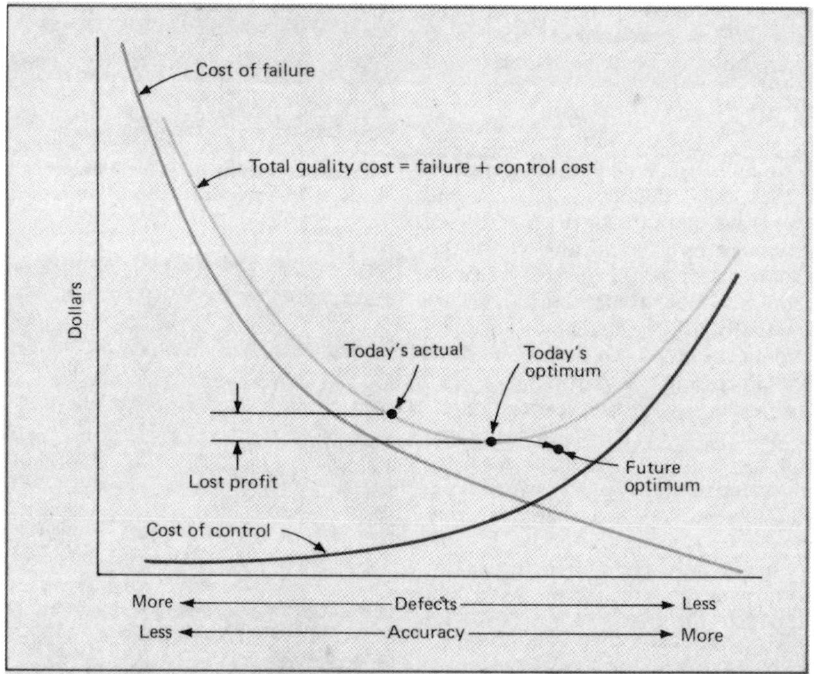

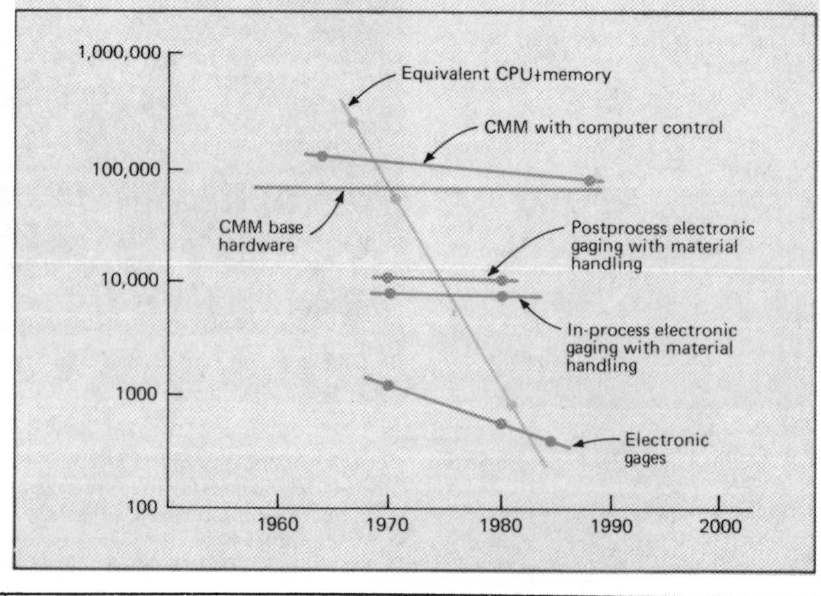

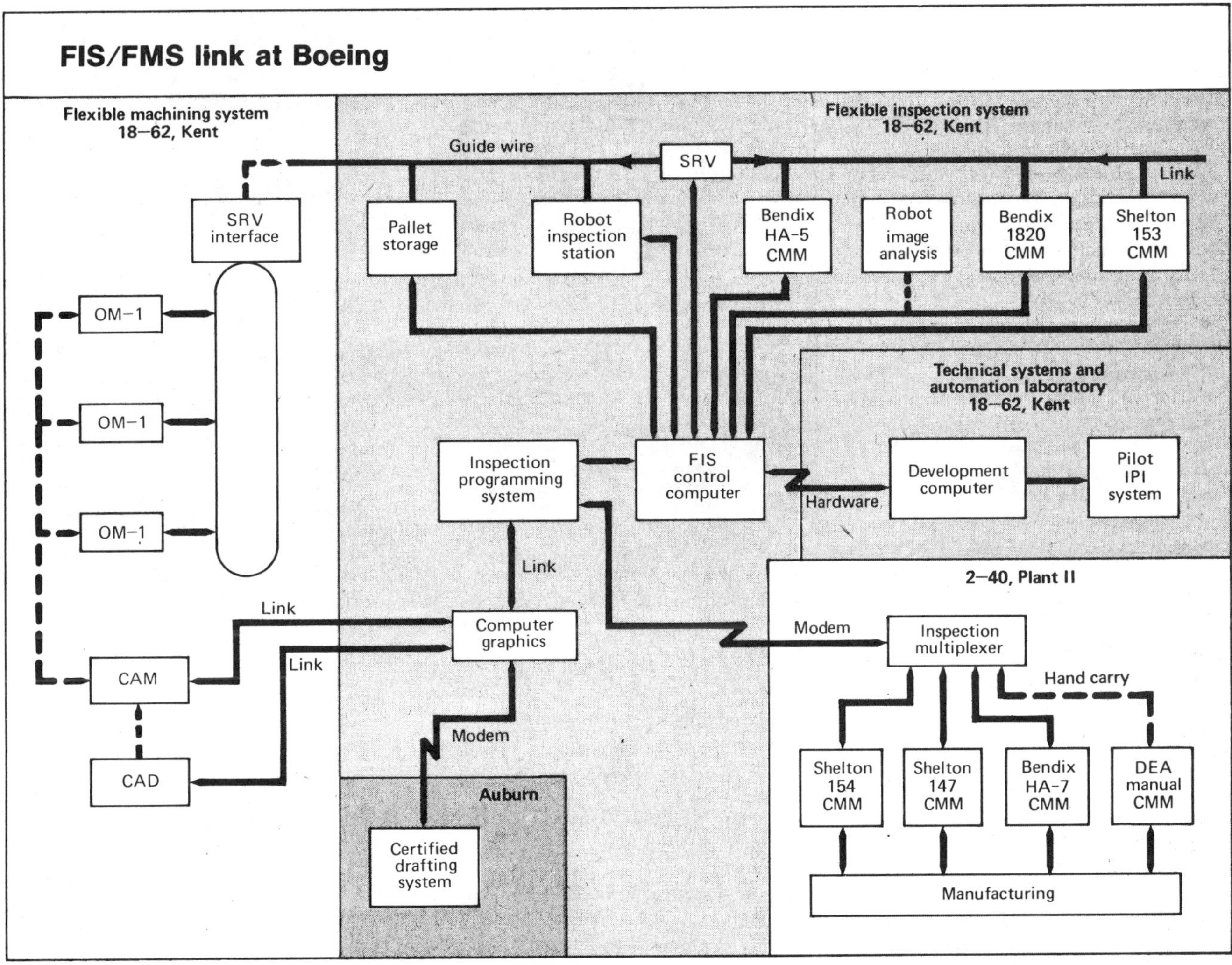

to interact with Unigraphics geometry to program a CMM; the output from this module is a CMM program in a standard language defined by McAuto. Next a pre-CMM module, or postprocessor, converts this program into the language or data structure required by the specific CMM. A post-CMM module converts the ASCII Output file of CMM measurement data into Unigraphics wire-frame geometry and surfaces for graphic display.

The output of the pre-CMM module is a standard operating-system file on the Unigraphics host system, and the CMM vendor must provide the necessary communications capability between the host and the CMM control. McAuto plans to interface CMMs from Brown & Sharpe, DEA, Sheffield (Cordax), Zeiss, and Federal Products (originally Boice). A special bidirectional interface for the video-based CMMs made by Optical Gaging Products (Rochester, NY) is also under development.

Calma has announced development of a DDM/CMM interface that works in conjunction with its Design Drafting & Manufacturing (DDM) core-system software for mechanical product development. The interface enables the user to generate inspection routines, display probe paths, and process command files directly from a DDM model database. Interfaces are currently available for Brown & Sharpe VM II command and interface files; future releases will include processors for Zeiss and DEA files.

Perhaps the most ambitious approach to graphics CMM programming is the Automeasure system under final development by Computervision (Bedford, Mass). Working with the CAD/CAM database, Automeasure simulates actual CMM inspection of machined parts by permitting an image of the actual probe cluster to be moved around an image of the part displayed on a CADDS 4X graphics terminal. The probe image is generated by combining graphic figures of standard components of available CMM probe kits stored in a computer file.

A dedicated command language for program generation handles all CMM functions, including multiple-probe-tip selection; probe motion; feature measurement of points, lines, planes, circles, ellipses, spheres, cylinders, and cones; and nominal- and tolerance-data input. Graphic machine and probe-path simulation provides visual verification to ensure a collision-free program.

The system generates a standardized neutral-data-file output, which is interpreted by a postprocessor into the local language of the CMM. Postprocessors are developed, marketed, and sold by the CMM vendors and will initially be available for Zeiss, Brown & Sharpe, Sheffield (Cordax), DEA, and Boice machines. Also, Multimetrics (Redwood City, Calif) is offering a Geovision postprocessor to link its Geomet CMM software for manual CMMs with Automeasure. The Automeasure system will be available to Computervision customers as an option on CADDS 4X, revision 4, beginning in June.

CMMs provide flexibility

In addition to combining and automating the inspection and data-processing functions, CMMs also provide the kind of flexibility needed to complement the growing use of flexible manufacturing systems, which are best suited for the increasingly batch-oriented needs of manufacturing.

FMSs are intended to operate efficiently with smaller lot sizes, shorter product cycles, and greater model proliferation. "The FMS philosophy revolves around reducing the manufacturing cycle time while maximizing machine efficiency," says Brown & Sharpe's Genest. "To accomplish this lofty goal, one key ingredient is essential: the direct computer-controlled CMM."

Genest also suggests that the CMM is not only the vehicle through which system quality is audited but also the source of such critical information as tool wear, part accuracy, rework values, and overall trends, which can be extracted on a "per-machine" basis to minimize machine downtimes and part rejection while maximizing lot-cycle efficiencies. In other words, the CMM is the key to process control.

The applications of CMMs vary. Manufacturing cells used for making relatively small parts lend themselves to integration of a small CMM serviced by a robot to handle parts between the various machine tools and the CMM. For larger parts, a horizontal-arm CMM can be placed directly in line between machines in a transfer-line arrangement.

To provide maximum flexibility for processing a wide variety of parts, flexible inspection systems (FISs) comprising CMMs, other inspection equipment, and necessary parts-handling equipment are beginning to parallel and complement the growing use of FMSs.

According to Genest, the inspection device serving an FMS should meet the following requirements:

- Be linked to the host.
- Be as flexible as the FMS it serves.
- Be fast enough to keep up with the FMS.
- Be accurate enough to provide feedback to the host so that proper machine-tool offsets can be made.
- Be self-sufficient and stable in the FMS environment.

Brown & Sharpe has specifically developed its line of Process Control Robots (PCR) to meet these requirements. First introduced at IMTS-84, the PCR is an ultrahigh-speed, totally automatic coordinate measuring machine intended for operation in a manufacturing environment. Built more like a machining center than like a conventional CMM, the Series 500 PCR features a modular, accessible horizontal design; built-in elastometric vibration isolation; accommodation for pallet/fixture arrangements; and an automatic high-speed probe-changer.

Fluorescent Penetrant Inspection Module (FPIM) of IBIS system uses helium/cadmium laser to excite penetrant phosphors and measures the resultant fluorescence

With a new high-speed multiaxis Validator controller, the PCR sports axis velocities up to 20 in./s and 0.3-g acceleration. The volumetric accuracy of a single-arm version is said to be ±0.0004 in. anywhere within its 500- x 300- x 333-mm measuring range.

Although measuring programs for the PCR can be prepared on-line in a teach mode, off-line programming is essential for an FMS linkup and can be accomplished with Validator FORTRAN or Brown & Sharpe Val Meas II geometrically based measuring programs. Direct programming from a CAD/CAM database or APT file is also possible with Brown & Sharpe's CADAM/APT postprocessor, its interface to GM's CGS, or graphics-programming software from Calma, McAuto, or Computervision. Genest reports that, every year, 20 or more customers tie their CMMs to CAD/CAM systems.

Sheffield manufactures and markets the Quadrex robotic inspection center under license from Imperial Prima SpA (Italy) for use on the production floor and in an FMS. The Quadrex features coded pallets for automatic part recognition and program selection (it can recognize up to 310 individual parts) and an optional automatic probe-changer that can handle up to six different probes.

The National Bureau of Standards is also working on integrating the inspection task in connection with its Automated Manufacturing Research Facility. Under development is a CAD-directed-inspection (CDI) system to develop inspection programs for the AMRF's Sheffield CMM inspection station. The long-range objective is to develop techniques for using the data in a CAD system to produce integrated programs for robots, machine tools, inspection machines, material-transport systems, and other devices in the AMRF.

According to Ted Hopp, of NBS's Center for Manufacturing Technology, the ultimate goal of the project is to have part-design data as a very high-level program for factory control. "Current commercial CAD systems do not keep enough information to meet this objective, being limited to providing a geometric and topological description of an ideal part," he claims. "Making use of CAD data for inspection control requires either building systems intelligent enough to understand CAD data in engineering terms (such as datum surfaces, or functional features) or finding ways to incorporate such data into part descriptions. A combination of artificial-intelligence techniques and extensions to CAD technology will be required to automate the generation and inspection procedures."

FIS integrates QA at Boeing

Flexible inspection systems typically involve more than one inspection station. A good example is operating at Boeing Aerospace Co's Kent, Wash, factory where the air-launched cruise missile (ALCM) is built. Sponsored under a Man Tech program by the Materials Laboratory of the Air Force Wright Aeronautical Laboratories (WPAFB, Ohio), the work was performed by the Technical Systems Automation Group of Boeing's Quality Assurance Dept. The FIS demonstrates the interaction of several levels of robotics, automated inspection, and automated materials handling coordinated by a central host computer. The FIS comprises two fully automatic CMMs, a multipurpose robotic inspection station, a staging area for pallet loading and unloading, a pallet-storage area, and an automated storage-and-retrieval vehicle all under the supervision and control of a Hewlett-Packard HP-1000 computer.

The specific objective of the FIS-prototype development was to demonstrate the capability of an off-line multistation automated dimensional-verification system to contribute to the productivity of a complex production line. The FIS is capable of processing a wide variety of parts (ranging in overall part size from 5 x 2.9 x 1.43 in. to 74 x 12.21 x 1.5 in.) while maintaining inspection accuracy and data-processing integrity. Demonstration parts were carefully selected to impose a broad spectrum of inspection-capability requirements on system performance.

According to Bruce Keller, manager for technical systems automation of the QA Dept at Boeing, the development was precipitated by a need to improve cost and flow factors. "Developing the necessary inspection programs and inspection aids and the actual flow time for the inspection process can be costly with manual inspection operations," says Keller. "And the time spent shuttling parts around from different parts of the factory is also costly," he adds.

Mix of parts is great

Another factor in the development of the FIS is that, although batch sizes in the ALCM program are relatively large for aircraft work, the mix of parts in any batch is also very great: the FIS may have to deal with a tail cone for the missile followed directly by an elevon housing.

The CMMs form the core of the inspection system. One CMM is a Sheffield Cordax 1820, vertical-arm three-axis machine with an inspection envelope of 78 x 35 x 27 in. It can handle large hardware, such as ALCM wings. The other CMM is a Sheffield HA-5 horizontal-arm four-axis (three linear, one rotary) inspection facility with an inspection envelope of 56 x 56 x 56 in. to allow measurement of complex parts, such as ALCM tanks. Both CMMs are controlled directly by HP 9835 microcomputers for inspection control and data analysis. They are totally automated, including setup with automatic part leveling and axis alignment.

The multipurpose robotic inspection station has been equipped for ultrasonic thickness inspection with a Krautkramer-Branson pulse-echo thickness gage. It is used to verify skin thickness of hollow wing sections formed by bonding two waffle-patterned halves together. The robot takes readings at the center of each of some 27 waffle patterns.

The materials-handling system of the FIS forms the interface between the system operator and the fully automatic inspection cell. It provides rapid, efficient response to parts staging and inspection-station loading and unloading, enabling one operator to load several pallets, which are stored for subsequent inspection. A high-rise system is used for part storage because it uses considerably less floor space than other systems and provides easy access for cleaning and maintenance.

RF-controlled lift truck

The central feature of the materials-handling system is a special wire-guided RF-controlled lift truck, a storage-and-retrieval vehicle (SRV) built to Boeing specifications by Hyster. It has a 12-ft vertical extension and 180°-rotating "turret"-type mounting for the lift forks. Because these forks provide both vertical extension and lateral motion, no special pallet-handling equipment is needed at the inspection stations.

Battery-operated and controlled by an on-board Intel SBX 8024 single-board computer, the vehicle receives commands from the host, and on-board logic converts them to vehicle motion and pick-and-place functions. Destination is identified by a combination of coded magnets embedded in the floor and an optical "final-stop" indicator at each storage location and inspection station.

A noteworthy feature of the FIS is a dynamic-statistical-sampling program, which provides automatic severity adjustment to modify the number of parts and part features to be inspected, based on inspection history. This reduces costly, time-consuming inspection of consistently good features and yet ensures thorough inspection subsequent to rejection of a part feature.

In operation, parts are delivered from the production line to the pallet load/unload area on dollies. The FIS operator loads the parts onto pallet-mounted holding fixtures, uses a bar-code reader to enter part data into the central computer, and notifies the computer that the pallet is ready for pickup. The SRV then picks up the loaded pallet and takes it directly to the appropriate inspection station if it is available or to its assigned location in the pallet-storage area.

When the inspection station is ready for that pallet, several things begin to happen: First, the host computer sends the SRV to pick up the pallet and deliver it to the inspection station. Meanwhile, the host computer selects the proper inspection program, based on part type and severity level. The actual inspection programs are arranged in severity modules and kept on floppy disks at the individual machines. The host merely instructs the machine which modules are to be used based on the continuous statistical analysis of the data. When the station has been loaded with a part, the host signals it to begin inspection.

When inspection has been completed, the inspection station signals the host, which notifies the SRV that the station is ready for unloading. The inspected part is then returned to the pallet-storage area and entered into the unloading queue. At the same time, the results of the inspection are transferred to the host for data analysis, and the accept/reject data is entered into the database for severity adjustment. In the event of a rejection, a standard reject form is printed out automatically.

At the unload area, the operator matches the part to its automatically produced inspection report and returns the part to the production flow. The host is capable of printing several reports:

- Inspection Report, identifying the part and giving inspection results.
- Inspection History Report, summarizing the inspection results of the last 15 inspections of each of the part characteristics.
- Unplanned Event Pickup, a standard rejection form.
- Collective Analysis Reports, available on an ad-hoc basis to identify hardware trends. Problem characteristics can be compiled by part number, specific feature, manufacturing machine, and location of the feature on the hardware.

Although quantitative comparisons of the FIS and manual inspection are difficult, Boeing claims that the FIS has demonstrated a 60-70% average reduction in inspection time, which is significantly better than had been expected. According to Boeing calculations, if the ALCM demonstration hardware were currently being inspected on a 100% manual basis, the potential saving with FIS implementation would amount to $2.9-million per 1000 ALCM missiles. A single inspector, running the FIS, would be able to process as many as 40 parts in an 8-hr shift.

Currently, the FIS is also used to inspect some 737-and 757-aircraft parts from Boeing Commercial Airplane Co. Ultimately, the FIS will be linked to an existing FMS. Also, a CAD/CAM-data interface is under development.

What about future FIS developments? According to John Harned, chief scientist at Autoflex Inc (Madison Heights, Wis), the concept of flexible inspection has been expanded by using robots with optical probes to inspect large assemblies, such as car and truck bodies.

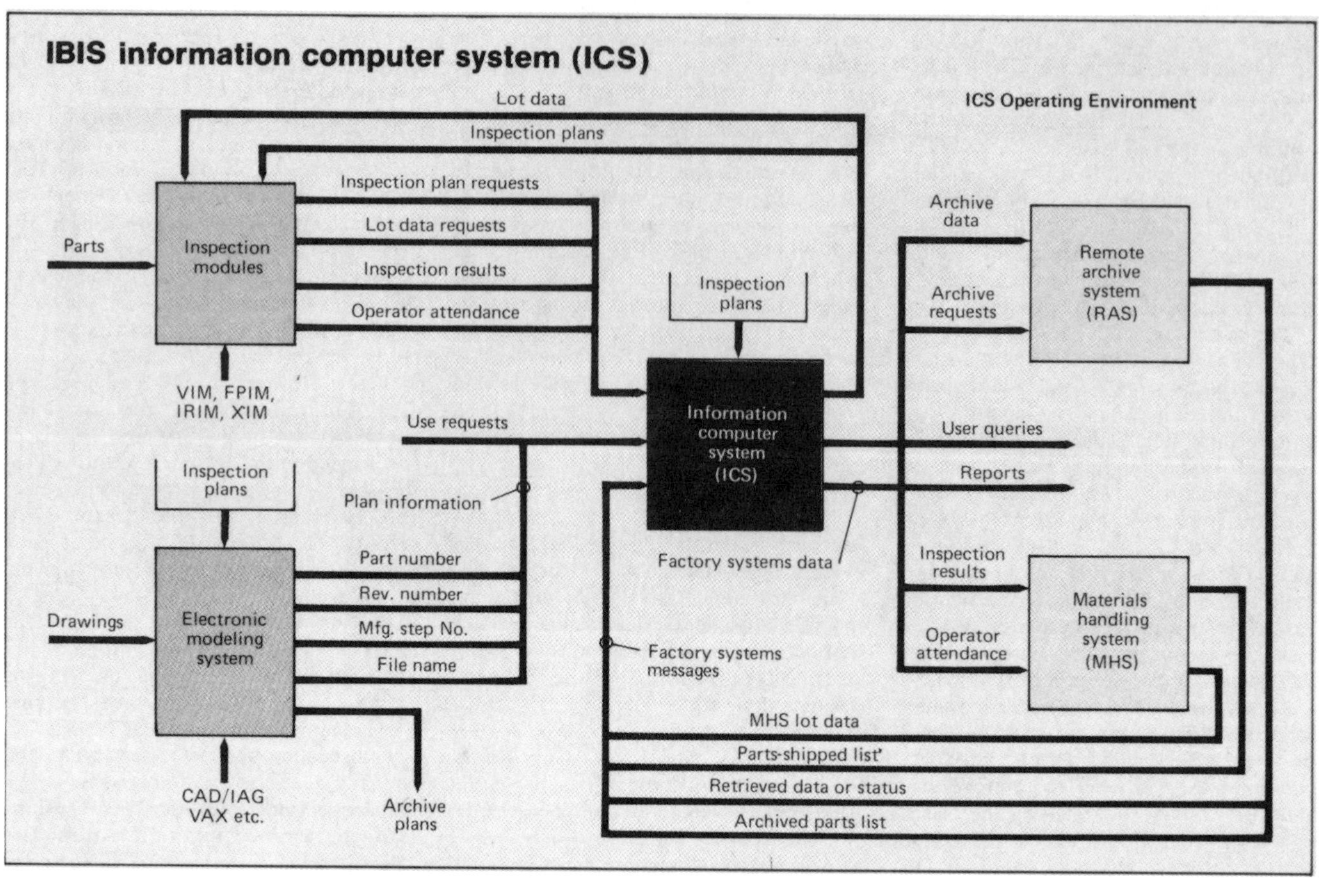

"With the recent introduction of sophisticated machine-vision systems into the workplace, it is now possible to greatly expand the role of robots in flexible inspection," says Harned. "Recent advances in CAD/CAM technology make possible the integration of CAD/CAM into robotics systems. This marriage significantly improves the productivity and economies of robot inspection systems."

But Harned also points out that robots have some limitations: they can inspect parts at moderate rates (about 100 parts/hr nominal), and their positioning accuracy is limited. He thinks vision-based robotic inspection can meet the following needs:

- Moderate throughput rates.
- Frequent part or model changes.
- Off-line part inspection requiring a large number of measurements.
- Large parts with complex geometry, such as cavities.
- Low throughput rates with parts that require many measurements.

According to Genest, the ideal QA system would use vision-based inspection to measure everything. Such a system would look at a section of a part in 3D and be able to tell everything there is to know: surface finish, hole diameters, roundness, comparison with a master, specific dimensions, color—all almost instantly. "That type of vision will be around soon," says Genest, noting that 1024- x 1024-pixel-array cameras are becoming available for vision systems and, as resolution goes up, the measuring ability increases.

A pick-and-place robot with accurate machine vision could look right into the machining area for flexible in-process inspection and control. In the meantime, Genest believes, CMM builders must start supplying their machines with interfaces to parts-handling equipment and automated guided vehicles. "CMMs must be capable of going right onto the manufacturing floor, and their controls must be hardened like current CNC systems instead of being standard minicomputers," says Genest. "We're going through the same revolution that the machine-tool industry went through with the advent of NC."

NDT also goes for integration

Although this report has, so far, concentrated on dimensional inspection, integrated quality assurance must eventually involve all quality-assurance techniques, including such diverse ones as nondestructive testing and vendor-performance tracking. One important reason for automating and integrating NDT is that most of the conventional processes involved, such as radiography and ultrasonic inspection, usually rely heavily on human judgment and, therefore, are not compatible with a computer-based factory. Not only is such inspection subject to human variability, but it is typically a qualitative judgment. CIM depends on quantitative data.

NDT is a major activity in the production and overhaul of aircraft engines, particularly turbine-engine blades and vanes. For this reason, the General Electric Aircraft Business Group (Evendale, Ohio) has been developing an Integrated Blade Inspection System (IBIS), funded under a tri-service program administrated by the Air Force Wright Aeronautical Laboratory (Dayton, Ohio).

IBIS is a computer-based, automated system for the precision inspection of individual airfoil parts. As conceived, it comprises four automated inspection modules interconnected through a computerized data-management network. Each module is designed to perform a specific type of inspection required during overhaul and/or manufacture of stationary and rotating airfoil parts. Several inspection modules have been considered under the IBIS program:

- Visual Inspection Module (VIM) for surface defects, such as dents, nicks, and scratches.
- Fluorescent Penetrant Inspection Module (FPIM) for surface-flaw indications on parts previously processed by an

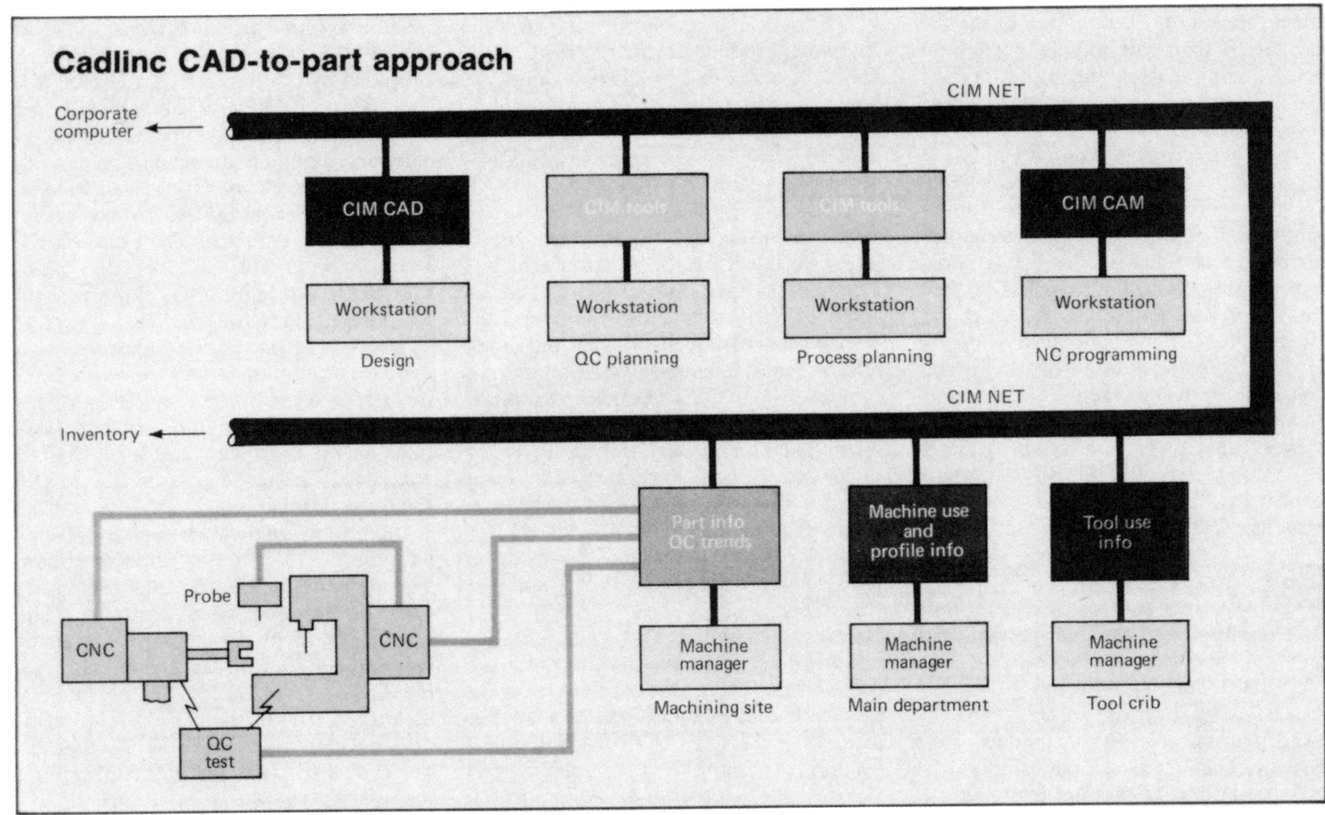

Automated Fluorescent Penetrant Preprocessing Module (AFPPM).

- X-ray Inspection Module (XIM) for internal casting flaws, such as gas porosity, and manufacturing flaws, such as internal hole position and dimensions.
- Infrared Inspection Module (IRIM) for blockages of internal cooling passages and surface cooling holes.

Each module of the IBIS system will eventually interface to an Information Computer System (ICS), which manages inspection data through an integrated airfoil database.

Although work has been accomplished on all modules, only the fluorescent-penetration, X-ray, and infrared modules are currently operating at various facilities. GE and the Air Force jointly decided to suspend work on the Visual Inspection Module because the technological development turned out to be more extensive than had been originally anticipated: part-to-part variations presented alignment problems, and the high radius of curvature at the blade tips requires many readings, which slows the process. However, the overall design and construction of the module, including the blade-transport conveyor and the blade-manipulation system, have been used for the fluorescent-penetration module.

When fully developed, the IBIS system will be programmed entirely from an electronic geometric model of the airfoil part to be inspected. This model can be created from the manufacturer's design database, which could be drawings or the CAD database. An inspection plan is then developed via the ICS and stored in an inspection-plan database. When needed by an IBIS module, the plan is downloaded into that module, which then performs the required inspection automatically. The results, including part disposition, are then passed back to the ICS for future recall and analysis.

Project plans call for IBIS to be readily interfaced with other systems for factory control for tracking and reporting the location of specific blades, off-site archiving of inspection results, analysis with engine-flight historical data, and feedback to manufacturers on the history of field overhaul and inspection results. A number of benefits are claimed:

- Improved inspection reliability and repeatability.
- Enhanced inspection thoroughness.
- Increased inspection productivity.
- Automatic information storage and retrieval.
- Feedback for process control.
- Compatibility with ICAM (Integrated Computer Aided Manufacturing program of the Air Force) concepts.

The Fluorescent Penetrant Inspection Module uses a helium/cadmium laser to excite the penetrant phosphors and a multiplier tube to measure the resultant fluorescence. Both blade and sensor head are moved to provide a full scan of the blade surface, and the intensity and location of the fluorescence is recorded and interpreted.

The X-ray Inspection Module uses filmless radiographic techniques to provide two different data representations: digital fluoroscopy, which produces a conventional film-like representation, and computerized tomography, which yields a cross-sectional image. The latter, although requiring more time because the blade must be rotated during exposure, is better able to resolve certain internal structural flaws and reveal flaws that are partially or completely masked in the fluoroscopic data format. The filmless images are produced by a General Electric-developed high-pressure xenon-gas detector array capable of imaging a 3-in.-wide scan.

In the Infrared Inspection Module, cooling flaws are detected thermographically, instead of by conventional air-flow and water-flow inspection techniques. As heated air is passed through the cooling passages, a series of thermal maps of the blade is captured and analyzed by the computer at various times for aberrant response to the temperature transients.

The Information Computer System will provide for the orderly management of inspection data by maintaining an integrated airfoil database. The goal is to provide a reservoir of various types of information to support the operation of the various inspection modules: for example, a digital description of the airfoil part, a scan plan to guide airfoil and sensor manipulation during inspection, and inspection instructions with acceptance criteria.

GE is now working on a number of programs to extend the IBIS technology to such non-airfoil jet-engine components as rotating parts and structural castings. The Evendale facility has long been moving toward a paperless factory, and IBIS is another step in that direction. "Regardless of the complexity of the product, some form of automatic inspection is going to be essential in the computer-integrated factory," says Leslie D. Meyer, manager of the IBIS program at GE. "It's hard to interface manual inspectors into an otherwise automated paperless factory," he points out.

Going beyond the shop floor

The paperless factory extends beyond the shop floor, and the quality-assurance function needs to be integrated into every level, starting with design. Unfortunately, quality assurance and the inspection criteria for a product have typically been an afterthought instead of an integral part of product planning.

"It's a major problem," says Arnold Reinhold, product-line manager at Automatix. "Lots of research is being done on how to design parts for automatic assembly, but I see almost no attention being paid to designing parts for inspectability," he complains.

Many characteristics of a part can affect its inspectability but have absolutely no effect on its ultimate use. For example, surface appearance, such as color or reflectivity, can hamper machine-vision-based inspection but may have no consequence on function or usefulness of the part. The choice of material and surface condition, established during design, can mean success or failure of in machine-vision inspection, and form and reference surfaces can also have their effect on automated inspection. It is therefore imperative that QA considerations be included at the very beginning of the product cycle.

Although computer-integrated manufacturing imposes new challenges on the QA function, it is evident that existing CIM technology also holds the key to meeting those very challenges. In fact, vendors of CIM-related equipment and systems are increasingly beginning to include QA in their programs. A good example is the CAD-to-part approach proposed by Cadlinc Inc (Elk Grove, Ill). In this approach, a product is designed with a Cadlinc workstation and CIM CAM software, and its dimensions and associated tolerances are defined and become the basis for eventual quality assurance. The CIM CAD database also becomes the basis for computer-aided process planning, tooling-requirements planning, and NC-program preparation.

The key to the Cadlinc approach is a CIM NET communications network linking workstations in the design and planning offices with machine-manager stations on the shop floor and providing all concerned parties access to essentially the same database.

Integrated quality assurance depends on timely and meaningful information. The tool for generating and processing such information is a computer. The techniques involve automation, statistical process control, and data management. The result of integrating quality assurance will be to close the process-control loop and make it possible to achieve effective computer-integrated manufacturing. ■

Directory of QA software

The following is a listing of companies offering quality-assurance software products, including calibration routines, capability studies, CMM programming and data interpretation, inspection routines, QA management, QA-cost analysis, reliability studies, sampling plans, statistical analysis, statistical process control (SPC), and supplier quality.

Not included in this listing are manufacturers of coordinate measuring machines (CMMs); most of these offer software providing measuring routines, machine programming, and associated statistical-analysis routines. For the latest directory of CMM hardware, software, and services, see Special Report 749, "Taking the measure of CMMs" (AM—Oct'82, p145).

A word of caution about compatibility: no attempt has been made to indicate the computer configurations required for running the software listed in this directory. The requirements range from handheld calculators to specific CAD/CAM systems to mainframe computers, but the majority of the packages can be implemented on microcomputers.

Many of these listings have been adapted from a "Directory of QA/QC Software" appearing in the March 1985 issue of *Quality Progress*, published by the American Society for Quality Control. The listing is alphabetical by companies, with no attempt to classify products. Not all companies offering products or services are listed, and listings of individual companies are not necessarily complete for each company.

Advanced Technology Inspection Inc
839 N Rochester Rd
Clawson, Mich 48017
313/589-3213

Qualistat: Menu-driven interactive statistical program; stores up to 85,000 measurements per job. Raw-data plot, $\bar{X}$ and R charts, frequency histogram, normal probability plot, normal curve comparison. Mean, median, mode, std dev, highest- and lowest-value occurrence, process grand mean, process mean, process total range, percent out-of-spec limits, capability calculation, kurtosis, and skewness.

Qualistat Multiple Input System Plus: Scanning multiple-user system for in-plant data collection and data processing from up to 8 remote stations. "Running window" of $\bar{X}$ and R charts display current subgroup and last 50 subgroups. Mean, std dev, highest and lowest occurrence, process grand mean, process mean range, percent out-of-spec limits.

Qualistat Smart Independent Terminal: For shop-floor use; displays $\bar{X}$ and R charts and frequency histograms. Enhancements for p, c, and u charts.

Qualistat Attributes: For attribute data only, includes c, p, and u charts and standard related statistical outputs.

Company	Product
Allen Bradley Co Systems Div 747 Alpha Dr Highland Heights, Ohio 44143 216/449-6700	Series 7890 Productivity Monitoring System: Hardware/software combination for acquiring, processing, and displaying production-process data from A-B PLC programmable controllers. Monitors status of PLC data table. Uses A-B Data Highway network for data acquisition. Statistical-quality-control tools include $\bar{x}$, R, counts or occurrences/binomial distribution, Pareto charts, scatter plots, upper- and lower-control-limit alarms (based on 1-point, 3-point, or 7-point rule).
American Machinist/Beckwith 23500 Mercantile Rd Cleveland, Ohio 44122 216/464-2181	Geo-Calc: Geometric-dimensioning and -tolerancing program verifies pattern-locating and hole-to-hole-position tolerances of hole-pattern-coordinate data. Incorporates bonus tolerance; simulated part function and functional or paper gaging. Includes programs for position-tolerance/coordinate-tolerance conversion, sine-plate gage-block buildup, and mm/in. conversion.
Applied Concepts Co 5253 S Des Moines Pl Tuscon, Ariz 85746 602/883-0664	Q Stats 1.0: 30 programs and subroutines for process-control analysis, statistics utilities and subroutines, inspection planning and analysis, analysis of experiments, sampling-plan design, and sampling-plan evaluation. Q Stats 64 adds graphics package for plotting control charts, scatter diagrams, histograms, and Pareto diagrams.
ASK Computer Systems Inc 730 Distel Dr Los Altos, Calif 94022 415/969-4442	MANMAN/Quality Management: Prepares control charts and histograms. Reports summarize failure data to highlight problems deserving prompt attention.
BBN Software Products Corp One Ailwife Pl Cambridge, Mass 02140 800/251-1717 or 617/491-8488	RS/1: Integrated software for engineers and scientists. Includes data management and analysis, statistical analysis, full graphics capabilities, modeling, and reporting.
Calian Technology Ltd 1755 Woodward Dr, Suite 305 Ottawa, Ontario, Canada K2C OP9 613/727-0606	AQOP 1000: Subcontractor-quality-management systems to provide military contractors with structured means of managing subcontractor quality on large contracts. Includes software, procedures, training, and support service to manage up to 500 subcontractors working to such standards as MIL-Q9858A and AQAP-1.
Calma Co 2901 Tasman Dr Santa Clara, Calif 95050 408/748-9600	DMM/CMM: Enables user to generate inspection routines, display probe paths, and process command files for coordinate measuring machines in conjunction with Calma's Design Drafting and Manufacturing (DDM) core CAD/CAM system.
Computervision Corp 100 Crosby Dr Bedford, Mass 01730 617/275-1800	AutoMeasure: Allows graphic off-line programming of manual and servo-driven coordinate measuring machines. Includes probe library for simulation. A CADDS 4X software option for CV CDS 4000 and Designer V-X CAD/CAM systems.
Control Data Corp 8100 34th Ave South, Box HQA030 Minneapolis, Minn 55440 800/328-1109 or 612/921-4400	SMIP: Plato curriculum for training production employees in statistical-control methods. Combines computer-based material and text to train workers in such techniques as flow diagrams, data collection, control charts, and frequency distribution. Provides hands-on experience by simulating real machinery and processes.
Crosby Co PO Box 2433 Glen Ellyn, Ill 60138 312/790-1711	/SPC: Package includes PROCESS7 for control charting, PPAPER7 for probability with chi-square test, CUSUM7 for trend analysis, p-CHART7 for p and np charts, u-CHART7 for u charts. /QCost: Tracks up to 40 individual quality-cost accounts; Pareto analysis, trend charts, custom quality-cost reports. QualityMaster: Multitask program for analysis and tracking in manufacturing. /SPC 100: $\bar{x}$- and R-chart program for briefcase computer.
Crunch Software 2547 22 St San Francisco, Calif 94116 415/564-7337	Crisp: Set of formal statistical procedures including multiple regression, analysis of variance, cross tabulation. Graphics. Interfaces to common database and spreadsheet programs.
DBi Software Products One Energy Pl Mt Pleasant, Mich 48858 517/772-5055	Trajectories SPC1: Menu-driven program written in Advanced BASIC for handling statistical process control. Control charting including $\bar{x}$, R, p, np, c, u; histograms. Program also performs pattern analysis, process capability, and percent out-of-spec.
DataMyte Corp 14960 Industrial Rd Minnetonka, Minn 55345 612/935-7704	SQC: Control charts, capability studies, histograms; particularly suited for use with DataMyte data collectors. Spreadsheet-like data editors for easy data viewing. Some versions with parts-drawing capability.

Vendor	Software
Dennison Mfg Co Corporate Quality Assurance 300 Howard St Framingham, Mass 01701 617/879-0511	StatCap: IBM PC/Lotus 1-2-3 templet program for performing process capability analysis. StatAids: IBM PC/Lotus 1-2-3 templet program comprising a collection of statistical tools for monitoring, controlling, and analyzing the performance of products and processes. StatTrac: IBM PC/Lotus 1-2-3 templet program includes FailTrac for customer-complaint/returns; InspTrac for inspection/audit data; RejTrac for reject/scrap/rework data; VendTrac for receiving-inspection/vendor-performance rating.
Digitalk Systems Inc 6633 West 28 Mile Rd Washington, Mich 48094 313/781-3095	Easy Calc: Statistical-analysis programs for variable and attribute data; copy program for backup protection and archiving. $\bar{X}$ and R charts, histograms, process-capability charts. N, np, c, and u charts.
Robert A Dovich 1200 N State St Marengo, Ill 60152 815/968-6833	Quality/Reliability Program Set: For HP 41 Series handheld calculator. Documentation for more than 35 programs, including sampling, probability, distribution test, attribute tests, $\bar{X}$ and R calculations (no plot).
GTG Microstat Ltd 2419 Drew Rd Mississauga, Ontario, Canada L5S 1a1 416/678-1001	QC-PRO Variable Data Analysis: Generates control charts, histograms, test for normality, percent out-of-spec, avg range, std dev. Predicts percent out-of-spec for process capability. Accommodates interface to measuring machine. QC-PRO Attribute Data Analysis: Generates p charts and c/u charts (defects-per-sample/defects-per-unit), Pareto analysis. QC-PRO Weibull Reliability Analysis: Generates Weibull plots with best-fit line; calculates various parameters; displays confidence bands; target testing; generates reliability plot; analysis for failed and interrupted (nonfailure) test results. QC-PRO Measurement Systems: Gage-capability program, equipment acceptance, equipment comparison, evaluating suspected deficient gage, evaluating operator's measuring abilities. Analysis includes gage accuracy, measurement repeatability, measurement reproducibility, and gage stability.
Giddings & Lewis Electronics Co PO Box 1658, 666 S Military Rd Fond du Lac, Wis 54935 414/921-7100	STAT-PAC: Menu-driven statistical-process-control package for use with G&L PIC 409 and compact PIC 49 programmable industrial computers to provide real-time monitoring of production statistics, including graphic displays for std dev, mean, range, and control charts. Gathers data continuously or at a set sample frequency specified by operator; data can be transferred to an IBM PC microcomputer.
Paul Hertzler & Co Inc 1329 E Douglas St Goshen, Ind 46526 219/533-0571	QA/S: Statistical analysis of variable and attribute data based on Ford Q101 SPC manual. Data may be entered manually or from digital gages and sensors. $\bar{X}$, R, s, and p charts; histograms. Capability index, skewness, and kurtosis. Data may be retrieved by part number, characteristic, date window, machine, tool, operator, etc.
Hewlett-Packard Co Advanced Systems Operation 11000 Wolfe Rd Cupertino, Calif 95014 408/257-7000	Quality Decision Management/1000: On-line statistical analysis provides control charts and Pareto charts. Generates scattergrams, histograms, and tabular reports.
Human Systems Dynamics 9010 Reseda Bl, Suite 222 Northridge, Calif 91324 818/993-8536	QC Plot: Attribute-data control charts (p, %p, np, c, u), variable-data control charts (mean & range, mean & std dev, individual values, subgroup sums, trend), distribution charts (hypergeometric, binomial, Poisson, normal, exponential), and database management.
Industrial Engineering & Management Press Institute of Industrial Engineers 25 Technology Park/Atlanta Norcross, Ga 30092 404/449-0460	IIE Microsoftware Statistical Analysis Group: Mann-Whitney Test analyses for set of data to determine whether two different samples come from the same population. Handles up to 200 observations per sample. Calculates several standard statistical descriptions. Histogram, polinomial curve fit, distribution fitting.
Intertek Services Corp 930 Indian Peak Rd Rolling Hills, Calif 90274 800/421-0369 or 213/377-0971	SPC Training: Statistical-process-control training program using Control Data Corp Plato format. Teaches employees the essentials of SPC and statistical sampling in four 2-hr self-paced study sessions.
JBF Associates Inc Technology Dr 1000 Technology Park Center Knoxville, Tenn 37922 615/966-5232	BACFIRE, EXCON, KITT1, KITT2, MOCUS, PREP, SAMPLE: Series of fault-analysis programs for such characteristics as common-cause failure, risk assessments, time-dependent and quantitative-dependent reliability, fault trees, and uncertainty analysis.
Kattor Research 901 40th Ave East Moline, Ill 61244	Reliability Test Plans: Generates a test plan in accordance with MIL-STD-781C. Program is menu-driven, with on-line help facility.

John A. Keane & Associates 575 Ewing St, Research Park Princeton, NJ 08540 609/924-7904	QMS: User-definable integrated modules for vendor rating; capturing and analyzing costs associated with repair, scrap, and rework (satisfies MIL-STD-1520A); analyzing process trends; analyzing receiving-inspection data; managing equipment-calibration programs; analyzing environmental data; managing product shelf-life information; and analysis of product and process defect data.
Laboratory MicroSystems Inc Hendrick Hudson Bldg, PO Box 336 Troy, NY 12181 518-274-1990	Mechanical Test: Automates collection, analysis, and storage of data from most universal mechanical-test machines. A wide variety of tensile, flexural, and compressive calculations are supported.
Jim LeDoux 3205 Lucinda Lane Santa Barbara, Calif 93105 805/687-7525	Receiving Inspection Log: Maintains receiving inspection data. Provides daily, weekly, or monthly reports by part number, supplier name, data, PO number, or inspector. Calibration: Maintains and updates calibration data, aging reports, and weekly status reports.
Logical Design Corp 33333 W Twelve Mile Rd Farmington Hills, Mich 48108 313/553-7510	Quality Assurance, Statistical Process Control: For systematic review of design, manufacturing, assembly, and quality activities. Performs capability analyses; graphical representation. Charts include $\bar{x}$, R, $\bar{x}$ and R, s, histogram, p, np, c, u, and Pareto.
J.W. Loosmore 15 E Bailey Dr Chesterton, Ind 46304 219/926-6825	SS/SPQC: Single Sample/Statistical Process and Quality Control program for batch processes. All data can be charted as individuals or in subgroups of two or ten points each.
MAR-REL Analytics Inc PO Box 9506 Rochester, NY 14604 716/265-1179	M/STAT-2000: Menu-driven interactive system for variety of statistical-analysis tasks, including variance, confidence intervals, continuous and discrete distributions, data reduction, hypothesis testing, nonparametric statistics, data management, regression, sample-size determination, time series, control charts, and acceptance sampling.
Math/Stat 45 Blaisdell Carlisle, Mass 01741 617/369-3164	inSPEC: Creates one- and two-stage inspection plans for any sampled product at any stage of production, for either product variables or percent defective. Can estimate and control recall risk or return rate subject to acceptance tests. Evaluates inspection protocol as benchmark for improvements.
McDonnell Douglas Automation Co 325 McDonnell Blvd St Louis, Mo 63166 800/325-1551	Unigraphics CMM Interface: Generic interface module, written in GRIP (McAuto's Graphics Interactive Programming language) for programming CMMs. Pre-CMM module converts generic program for specific CMM. Post-CMM module creates wire-frame geometry and surfaces from measured data.
Medallion Systems PO Box 2463 Springfield, Ohio 45503 513/399-7124	Quality Data Report: Accumulates, calculates, stores, retrieves, and prints periodic data from QC operations; for small-to-medium manufacturing operations. Reports display rejection and trend statistics.
MicroPack Inc 223 W Jackson Blvd, Suite 1110 Chicago, Ill 60606 312/787-8975	QCPack: Five statistical-analysis programs include control charts ($\bar{x}$ and R, p, u, c); and various single-sampling plans.
Minitab Inc 215 Pond Laboratory University Park, Pa 16802 814/865-1595	Minitab: General-purpose data-analysis system providing descriptive statistics; flexible plotting, including histograms; multiway tables; regression with residual analysis; diagnostics and stepwise procedures; variance; exploratory data analysis; and time-series analysis.
Multi Metrics Inc 3150 Spring St Redwood City, Calif 94063 415/367-0515	Geomet 200: Universal language for manual and joy-stick-operated coordinate measuring machines (CMMs). Can be retrofitted to any CMM. Geomite for manual CMMs. Geostat for SQC package. Geovision for link to CAD systems; initial link to Computervision CADDS 4X Automeasure. Inspector: For use with height gages, bore gages, micrometers, calipers, and a wide variety of LVDT indicators. Optimite: For optical comparators; manual, automatic, or part-programming mode. Reduces raw dimensional data to inspection reports.
Navone Engineering 2116 Waterloo Rd Stockton, Calif 95205 209/941-2669	Process Quality Simulator: A Monte Carlo simulation program that models individualized manufacturing and QC processes. Allows isolation and examination of the effects of changing many of the variables affecting the yield of a process.
Northwest Analytical Inc 520 NW Davis St Portland, Oregon 97209 503/224-7727	NWA STATPAK: Multifunction statistics package with data-management and manipulation capabilities. NWA Quality Analyst: SQC package providing standard Shewhart-Deming charting procedures. Pareto charts; general statistical computation and reporting.

Company	Products
P-Q Systems 270 Regency Ridge, PO Box 633 Dayton, Ohio 45459 800/547-1563 or 513/435-9717	SQCpack: Menu-driven general-purpose statistical-quality-control package. Includes descriptive statistics, frequency distributions, histograms, capability analysis, control charts ($\bar{X}$ and R, $\bar{X}$ and s, individual, moving avg, p, np, c, u, Pareto), cause-and-effect diagrams. Interfaces with database-management systems, such as dBase II, 1-2-3 Symphony.
Penton Software Inc 420 Lexington Ave, Suite 2846 New York, NY 10017 212/878-9600	QualityAlert: SQC package for variable and attribute data generated from samples selected from continuous manufacturing processes. Control charts ($\bar{X}$ and R, std dev, moving avg and range, p, u, c), histograms, process capability. Options for automatic data input and communication.
Perry Johnson Seminars Inc 3000 Town Center, Suite 1075 Southfield, Mich 48075 313/356-4410	SPC Analyst: Charts $\bar{X}$ and R, $\bar{X}$ and sigma, p, c, u, np, moving avg and range. Separate charting for injection-molding and stamping facilities. Determines when to start and stop tooling. Color graphics. Gaging Analyst: Indexes and records all calibrations and results. Allows attribute data to be converted to and treated like variable data. Analyzes repeatability, checks reproducibility.
Phillips Quality Consultants Inc Route 5, Box 366 Chapel Hill, NC 27514 919/933-9075	E-Z Stats: Variable-data control-chart templet for Lotus 1-2-3. Prints original data, summary of statistics, control charts with limits and interconnected points.
Porter Data Systems Inc 470 Buckeye Dr Colorado Springs, Colo 80919 303/599-5548	SQC-1/C and SQC-2/B: Graphics packages to generate standard control charts, including $\bar{X}$ and R, p, and np. Designed for microprocessor.
Q-Consultants 9300 CalleCascada Tuscon, Ariz 85715 602/298-0268	Total QA Package: Programs include Control Charts, Process Control, ANOVA and Design of Experiment, Regression Analysis, Pareto Analysis, Statistics Routine and Utilities, and Sampling Plan Design.
Quality Directions PO Box 671 Stoughton, Mass 02072 617/344-6064	qdMicroQual: Manages attribute data from vendors, incoming inspection, in-process inspection, finished-goods audits, MRB activities, and customer returns. qdQualData: Statistical analysis and reporting of in-process or finished-goods attribute data. qdVendorStat: Statistical analysis and reporting of incoming-inspection database; vendor ranking on user-selectable factor. qdReturnStat: Statistical analysis and reporting of customer-return database.
Quality Information Systems 1200-1218 Culver Rd Rochester, NY 14609 716/654-8420	Shop Floor Quality Management System: Packages include Shop Floor QC for data acquisition, geometric tolerancing, trends analysis, operator response, calibration of custom gages; Production Quality Management for customized data storage and report functions; Incoming Quality Management, Vendor Rating, Instruction Management for on-line access of inspection procedures. Process Analysis: Statistical analysis of variable data through control charts or descriptive statistics. Histogram, $\bar{X}$ and R charts, normal probability plot, normal inference control-chart reporting. Gage/Instrument Calibration: Scheduling of required calibration dates. Maintains complete calibration history; monitors delinquent equipment; searches by age, description, or location. Multiuser capability.
Quality Measurement Systems Corp 2555 Baird Rd, PO Box 578 Penfield, NY 14526 716/385-3848	Anstat: Menu-driven software allows input directly from electronic gages; allows for standardized inspection routine. SQA functions include $\bar{X}$, R, and sigma chart; frequency histogram; process capability; lot and sample statistics; attribute charts (p, np, c, u). Auto Comp II for optical comparators.
Quality Resources 840 McKinley Plymouth, Mich 48170 313/453-4616	Stat-Pak: Programs for single-sample statistics, linear correlation, F and T tests, regression analysis, Pareto analysis, chi-square contingency, and Z-scores. Chart-Pak for charting.
Quantum Co PO Box 769 Clifton Park, NY 12065 518/877-5236	Qu-All 7: Statistical analysis and simulation program. Mean, std dev, variance, skewness, kurtosis, coefficients of variation, histograms, Pareto diagrams, range, charts ($\bar{X}$ and R, $\bar{X}$ and s, p, c, np, u). Will read files from Lotus 1-2-3 and Multiplan.
QuanTech PO Box 24804 Indianapolis, Ind 46224 317/251-6161	Cal-Master I: Menu-driven program stores gage/instrument-calibration data. Searches and reports by gage/instrument description, ID number, serial number, location, calibration status. Status report includes delinquent, currently due, due within next period, in-service, removed from service. Complies with MIL-STD-45662.

Company	Product
Rapid Data 1899 Willow St San Jose, Calif 95125	Quality Pack: Menu-driven program includes descriptive statistics, normality, correlation, linear regression, X-Y plot, histogram, various distribution tests. Other programs: Descriptive Statistics, Histogram, Inspection Summary, Performance Review, Process Capability, Regression, Lotus Stat.
SQC Data Systems 29501 Greenfield Rd, Suite 116 Southfield, Mich 48076 313/557-3760	SQC System III: SPC/SQC charts for automotive suppliers (capability-analysis, distribution-analysis, process-control). Printouts for roundness (TIR), flatness, perpendicularity, squareness, dynamic balance.
SQC Systems Inc PO Box 19931 Atlanta, Ga 30325 404/881-0726	SQCS: Integrated system provides interactive capability for producing control charts, other related statistical procedures, and data management. 18 chart procedures. Pareto diagram, histogram, descriptive statistics, probability, scatter, time-sequence.
Salerno Industries Inc Computer Systems Div 4175 East Ten Mile Rd Warren, Mich 48091 313/756-9292	SPM: Integrated data-analysis system automates the mathematical and managerial functions of statistical process control in manufacturing. Supports all variable and attributes control charts, frequency distribution, capability studies, corrective-action codes, process exception, and status reports.
Software Hill 1857 Apple Tree Lane Mountain View, Calif 94040 415/969-4233	Statmate/Plus: Integrated statistical package includes multiple regression, control-chart aids, curvilinear regression, distribution tests, cross tabulations, summary statistics, correlation, ANOVA, nonparametric statistics, scatter plots, histograms. Database management.
Statistical Graphics Corp Research Park, 2 Wall St Princeton, NJ 08540 609/924-9374	StatGraphics: Integrated system for more than 350 functions for interactive data analysis, data management, and statistical graphics. Originally developed for use on large IBM mainframes, now available for variety of computers.
Stephen Computer Services Inc 347000 Grand River Ave Farmington Hills, Mich 48024 313/478-1868	SPC DataLyzer: Data entry from all major data collectors and gages. Storage for 95 analyses per disk. Variable and attribute charts, capability analysis, gage validation.
Stochos Inc 14 N College St Schenectady, NY 12305 518/372-5426	Micro/QC: Statistical-analysis module for mean, median, range, std dev, skewness, kurtosis, histogram, Pareto charts. Control-chart module for X, $\bar{X}$, R, sigma, p, np, c, u, and cumulative summary.
Synectek Inc 3933 W North Wood Lake Dr Columbus, Ind 47201 812/342-6606	DataMaster: Series of control-chart and engineering-analysis programs. Custom tailoring for small-shop application. Compatible with many networking applications.
Team PO Box 25 Tamworth, NH 03886 603/323-8843	Normal Probability Distribution Chart Plotting Routine: Generates normal probability plot from randomly input data for sample sizes up to 100. Outputs include table of ordered data, probability plots, distribution parameters. Eight choices of confidence-limit levels.
Wadsworth Professional Software Inc Statler Office Bldg, 20 Park Plaza Boston, Mass 02116 617/423-0420	Statpro XT: Integrated statistical-analysis program for IBM XT microcomputer (also runs on Apple). Accepts files originally written in various ASCII, DIF, SYLK, s-System, Lotus 1-2-3, and dBase II formats. Procedures include descriptive statistics, regression, analysis of variance. Database module allows raw data to be edited and transformed before analysis. Graphics module for customization of color graphics displays.
Paul A Ware 34 McNamara St Stoughton, Mass 02072 617/344-8204	PC(2): Menu-driven set of programs for collection and manipulation of variable and attribute data. Process-capability studies, various control charts. Control charts calculate upper and lower control limits and identify out-of-control subgroups.
Weldon Machine Tool Inc 1800 W King St York, Pa 17404 717/846-4000	Laser Machine Tool Calibration System: Menu-driven package enhances use of Hewlett-Packard laser interferometer for machine-tool calibration. Reduces labor and facilitates such measurements as pitch, yaw, straightness, and squareness. Produces statistical and graphic information on accuracy, repeatability, backlash, and lead-screw-error compensation.
Zontec Inc 1329 E Kemper Rd Cincinnati, Ohio 45246 513/671-0088	SPC TimeSaver: Integrated system for calculation, analysis, storage, and display of SPC control-chart information. Analysis functions include control charts ($\bar{X}$, median, R, p, np, u, c, run), descriptive statistics, capability, tool-wear, and economic impact (ranks problem areas).

CHAPTER 3

MEASUREMENT ISSUES

Presented at the MVA/SME Vision '85 Conference, March 1985

Induced Metrology Distortions Using Machine Vision Systems

by Richard G. Stafford
E. I. duPont de Nemours & Company, Inc.

The intent of this paper is to address some of the errors that can be induced through simplistic utilization of machine vision (MV) systems for metrology applications. The level is directed towards end users of MV who, while intimately familiar with their own manufacturing technologies, do not necessarily have extensive backgrounds in MV related disciplines. Such disciplines include, e.g., illumination technology, optical physics, optical system design, analog/digital signal processing, digital image processing and analysis, and applied statistics. The emphasis is restricted to those MV systems which use optical sensors for noncontact measurements. Such systems comprise the bulk of present MV applications in manufacturing.

In development and commercial literature, there are frequent statements to the effect that MV systems provide reliable noncontact measurements to single pixel and subpixel accuracy. While such accuracy is attainable in a laboratory environment and in certain classes of manufacturing applications controlled by highly experienced personnel, it is generally not attainable in many manufacturing environments, either due to the nature of the application or the lack of personnel experienced in MV related disciplines. A major reason for the limiting of accuracy is that most MV systems are designed for highly flexible use and are available chiefly as subsystems which require systems integration for specific applications. Generally, the degree of expertise needed to handle both flexible systems and systems integration is considerably greater than that for turnkey systems dedicated to specific applications.

A major problem in evaluating accuracy in metrology applications of MV systems is the distinction between accuracy and precision. Although measurement accuracy requires precision, the converse is not true. A given degree of precision denotes a measurement that is repeatable within a stated tolerance, assuming a unit of measure. If no biases are introduced by the measuring system, then the measurement is accurate. A simple example will clarify the problem. Take a ruler that has accurate rulings to one mil. If the ruler is then stretched uniformly, the rulings are still spaced uniformly, but no longer accurately represent one mil. While a measurement using this ruler will yield precise results based on its unit rulings, these results will not be accurate due to a bias introduced by the stretching.

This distinction between accuracy and precision is crucial in understanding the attainable accuracy in MV systems. Statements of accuracy in pixel units are valid only if the entire system introduces no distortion of the relevant parts of the original object. Although very high degrees of accuracy are possible, the end results, which are normally not pixel units, may be degraded to precise or even incorrect due to improper perception of distortion effects and calibration methods. The possible sources of bias error induced by MV systems are sufficiently numerous that time restraints will restrict coverage of the topic. However, the simple examples treated here will fundamentally illustrate the nature of the problem.

A generic MV system will be described in the next section. The response of this system to an ideal two-dimensional object, i.e., flat with zero thickness, will then be analyzed for sources of potential biases. Next to be analyzed will be the system response to a real object, i.e., nonflat with finite thickness. No attempt will be made to be rigorous in the analyses.

The thrust is to give a feel for the basic bias problem and hopefully, thereby, to encourage increased future accurate use of this technology in manufacturing.

There are, of course, applications that do not require high degrees of accuracy or precision. Much of the following content is, nevertheless, relevant to many such applications due to the complex nature of light interaction and the consequent capability for gross induced distortions.

II. MACHINE VISION: TOTAL SYSTEM

The system to be analyzed is shown pictorially in Figures 1 and 2. Figure 1 shows the optical front end of the system and includes all elements which precede an actual electronic signal. The illumination source projects light onto the object. The object is schematically shown to occupy an object volume in space. The light radiating from the object volume is transformed by the optical system into an image volume in space.

Figure 2 shows the electronic signal processing part of the system. A two-dimensional (2D) slice of the image volume is captured by a photoelectronic sensor, which is typically either a solid state array or vidicon TV camera, although other sensor types and formats are used and needed for more specialized applications. The analog camera signal is immediately digitized by an A/D converter with gray level resolution typically in the range of 6-8 bits. The 2D digital image is then stored in a memory buffer with a typical spatial resolution range of 256 X 256 to 512 X 512 pixels. Next, the image processor maps the original 2D image into secondary 2D images via spatial transforms for the purpose of enhancing the features of interest, e.g., object edges. Upon obtaining an acceptable enhanced 2D image, the image analyzer transforms the enormous amount of data in the 2D image into a result consisting of a small set of desired parameters, e.g., distances between object edges.

There are, of course, many variations on the above system architecture. Illumination schemes include back, surface, structured, and laser lighting. Available wavelength regions include infrared, visible, and ultraviolet. The optical imaging systems can be optimized for the differing illumination methods and sensors. Types of photoelectronic sensors include solid state point, linear and matrix arrays, vidicon, plumbicon, newvicon, saticon, orthicon, isocon, and photomultiplier tubes. Scanning formats include standard TV, noninterlaced scanning, variable scan rates, and single line scans. The range of variation in the electronic processing hardware is even more diverse. While each specific type of system will have a unique set of problems, there are classes of problems which are generic to many such systems. The model system chosen here is functionally similar to a significant portion of the presently available MV systems and is capable of illustrating many of the fundamental problems that arise in applying this technology to metrology in manufacturing.

III. SYSTEM RESPONSE: IDEAL OBJECT

The system response to an ideal object will now be looked at on a subsystem basis. The object, shown schematically in Figure 3, approximates a perfectly flat, transparent, rectangular hole bounded by an opaque medium with zero thickness. The object is ideal in the sense that well-behaved 2D objects are isomorphically mapped into 2D images by most optical systems.

The end result to be obtained by the MV system is an accurate measure of the width of the hole. For this case, the simplest illumination scheme is to use backlighting of the object. If the illumination system is flat and is a Lambertian radiator of white light of uniform magnitude over its entire surface, then the intensity distribution of light over the object hole can approach uniformity with proper design under laboratory conditions. Assuming the response of the remainder of the system to be perfect, then a horizontal sweep of the intensity as a function of position across the center of the object hole will be rectangular, as shown in Figure 4A. The next simple step is to do a binary image transformation using a threshold value at 50% of the peak intensity. One then determines the number of 'on' pixels in a horizontal sweep, applies a calibration correction, and the distance result is obtained to some desired degree of accuracy.

Since a perfect system can only be approached in a laboratory environment, let us now see where some problems will arise in a real system. First, near uniform illumination of an object is generally not feasible in most manufacturing environments for a multitude of reasons, e.g., due to expense or geometrical constraints. Even if such illumination were attainable, the lens system will form an image that is degraded from the object. The degree of degradation is measured by the modulation transfer function (MTF).

Any lens system has inherent aberrations which contribute to an imperfect response. For a given desired result, the lens system can be corrected through design to minimize the aberrations of importance. However, such corrections normally apply only to a fixed object and image distance with a given aperture and a characteristic frequency distribution of light. For even commercially available lenses, such data is not generally readily available. Therefore, it is rather easy to use a lens system that is ill-matched for a particular application. Typical aberrations applicable to MV are, e.g., spherical, chromatic and diffractive, and also nonuniform polarization response and additional distortions off the concentric axis. Such aberrations lead to degradation of resolution and spatial frequency response. Normally, these effects degrade radially from the concentric axis of a cylindrically symmetrical lens system.

A relatively simple and popular method to insure degraded lens response is the indiscrimate use of extension tubes to gain a desired working distance between object and lens system and an associated magnification factor between object and image. Many lenses are not designed to be used with extension tubes and, in such cases, the back focal plane distance is critical for optimum transform performance. Those lenses that are designed for variable back focal plane variation, e.g., some types of macro lenses, have well-defined regions of variation which should not be exceeded. Another method to degrade response is to have the image slightly out of focus.

Several examples of the optical degradation of light intensity profiles of the ideal object (Figure 3) are shown in Figure 4, where intensity is plotted as a function of position along the central horizontal axis of the image. Figure 4A, as mentioned previously, results from uniform object illumination and an ideal optical transform. Figure 4B results from a real lens transform, where the intensity decreases radially from the concentric axis. Figure 4C shows monotonically decreasing illumination across the horizontal direction of the object and will be discussed later. Figure 4D indicates unsharp edges which may result from, e.g., slight defocusing, chromatic aberration, spherical aberration, diffraction at high magnification, or stray scattered light. Figure 4E can result from poor radially dependent illumination or significant misfocusing.

The correction of illumination problems can range from simple to exceedingly difficult. Correcting for the gross problems due to misfocusing and misapplication of extension tubes are self-evident. However, even when working within proper lens operating regions, improvements in response are possible for highly demanding applications in metrology. For example, spherical aberration improves with decreasing lens aperature, assuming no diffraction limitation. Chromatic aberration can be minimized by using narrow band-pass light filters. Stray light can be minimized by decreasing lens aperture and proper use of light shields.

Once an image is obtained, the next step is to convert the light signal to an electronic analog signal via a photoelectronic detector. For the system in Figure 2, I will assume a solid state matrix array camera operating under standard TV format. This type of sensor is chosen chiefly since the spatial uniformity of the detector array is valuable for demanding metrology applications. In addition, the intensity response is linear over a wide dynamic range. A disadvantage is that the array of pixels generally have different gain factors. However, such a problem is correctable. Photo-electronic detectors are prone to additional error-inducing effects. Assuming that a camera manufacturer has properly dealt with these effects in camera design and the functional specifications are realistic, I will concentrate on two effects which are partially under control by the end user, i.e., signal saturation and signal noise. Saturation will cause a nonlinear response and thereby distort the spatial intensity profile of the object image, leading to inaccurate results which are difficult or impossible to correct for. Noise will cause fluctuations in the results which, if handled by proper statistical methods, can yield results that are accurately characterized by such statistical measures as, e.g., average and standard deviation.

The next step is to immediately digitize the signal since the image processing transforms and analysis algorithms, in most cases, are inherently less noisy when carried out by proper digital, rather than analog, techniques. Care should be exercised in choosing the digital resolution, i.e., gray level range, since this resolution is intimately related to the effects of the sensor noise on resultant accuracy, especially in obtaining subpixel accuracy.

Once the object image resides in a digital buffer, the following transforms and analyses necessarily assume a prior knowledge of the original object and the error-inducing effects of the preceding subsystem components. The degree of prior knowledge needed will depend on the desired results. Induced error at this point will arise from incorrect assumptions about the object, the optical system distortions of the image, and the analog/digital electronic distortions of the image. These errors will manifest themselves in faulty mathematical modeling of solutions, which will lead to applying incorrect image transforms and incorrect image analyses. Since it is impossible to estimate the capacity for human-induced error at this point, I will restrict the discussion to several examples drawn from Figure 4 for illustration.

Starting with the ideal response of Figure 4A, assume that the output of the sensor has a random noise component on the signal. Also assume the A/D converter has sufficient resolution to measure the noise component. Rather than determine the peak intensity for a given scan line, the average intensity will be calculated from those intensities exceeding a given threshold. The binary transform proceeds as before to determine the width. To characterize the accuracy of this measurement or to improve it, e.g., to subpixel levels, there are now two different ways to proceed, both statistical in nature. Method one is to perform a temporal average of the same scan line, sequentially retaking a picture of the object. The second method is to perform a spatial average over widths sequentially spaced in the vertical direction in a single picture. The measurement error can now be characterized by the average width and the sampling size. Lack of knowledge regarding the concept of random versus nonrandom signal noise, proper statistical techniques, spatial symmetry of the object, and temporal behavior of the system will certainly result in induced errors for even this simple example.

For the case in Figure 4B, by determining as before the average intensity and using a binary threshold transform, an accurate width can result. However, if the camera is shifted within the optical image plane, but off the optical concentric axis, a case approximating that of Figure 4C may arise. Proceeding as before can lead to precise but highly inaccurate results.

For the cases of Figures 4D and 4E, the location of the edge is not necessarily straightforward. The relationship between a real edge and its distorted system counterpart is a field in itself. I will again restrict the discussion to simple examples. One approach is to take a spatial derivative of the image. Due to the discrete nature of the image and the vector characteristics of the derivative, many types of derivative transforms are possible. I will take the simplest, which is a directional first derivative on the horizontal axis consisting of the difference in intensity between two adjacent points. Again, if one simply takes the derivative maxima as representing the edges, then a precise width can be determined. The accuracy of the result will depend on the relative spatial frequency composition of the edge with respect to the top of the waveform and many more factors beyond the scope of this paper. However, in many real applications, cases such as depicted in Figures 4C through 4E can have significantly improved accuracy by using spatial derivative approaches rather than simple binary thresholding. Depending on the application, there are other digital filtering techniques which may be more appropriate.

IV. SYSTEM RESPONSE: REAL OBJECT

In measuring real objects, significant sources of induced error arise in the front end of a MV system, i.e., the illumination, optical, and sensor components. Fundamentally this is due to the three-dimensional nature of real objects. Let us assume an object that is a 3D analog of that in Figure 3, except that, as the hole progresses through the object, the shape is allowed to vary. Several examples are shown in Figure 5, with cross-sectional views parallel to the concentric axis. The result to be obtained is again a distance between two points within the hole. Although the objects in Figure 5 appear to be simple, the measured distance between two given points will not necessarily be independent of the measuring system.

The choice of illumination will depend on many factors, e.g., the location of the two points, the optical characteristics of the surfaces of the object, obstructions on the object that can interfere with the light both entering and exiting the hole region, and the type of optical system. In some applications it may not be possible to properly illuminate a region of interest to obtain a measurement to a desired degree of accuracy.

Assuming that some form of practical illumination exists, an optical system will have sources of error which include those of the 2D case and additional ones due to the 3D nature. Looking back at Figure 1, remember the emphasis on an object volume and an image volume. For the objects in Figure 5, every point in the object volume is mapped via the optical system into a point in the image volume. This mapping process causes a fundamental problem which is shown in Figure 6. Points on object plane 1 (O1) are focused on image plane 1 (I1) and points on O2 are focused on I2. If the sensor is placed at I1, then a point on O1 will be focused to a point on the sensor with well-defined point-spread characteristics. However, the sensor will not only see the focused point on I1, it will also see a defocused image of the point on I2. As O2 is sequentially scanned along the optical concentric axis through the hole, it can be seen that the sensor will receive at the point on I1 an interference signal which consists of many defocused points on the hole surface. Such interference can be due to nearby edges, scattered light from a diffuse hole surface, or specular reflection from a polished hole surface. For the objects in Figure 5, it would be instructive to imagine different types of illumination and lens configurations and attempt to predict where the above-mentioned problems can occur.

Another problem area results when the two points needed for a distance measurement lie on two different focal planes as shown in Figure 5D. It is then possible that the sensor will see one point in focus and the other point out of focus. Some such cases can be solved by optimizing light collimation and optical depth-of-field and some cannot. It may also occur that, as the sensor is shifted within a given image plane or the object is rotated away from the concentric axis, the perceived distance between two points changes. All of the objects in Figure 5 exhibit this problem to varying degrees.

For a specific part and a specific measurement, the existence of a solution and the resultant measurement accuracy will depend not only on the characteristics of this specific single part, but also on variations of critical parameters of other parts of the same type. Such parameters include not only those intrinsic to the type of part but also to those dependent on how the part is presented to the system and to those dependent on the local environment. The sources of possibly induced errors by the optical front end system for 3D objects is sufficiently large, that useful discussions are limited to objects and measurements which have well-defined classes of similarity. Since the same holds true for the remainder of the MV system, further discussion of real objects is beyond the scope of this paper. Such discussions should be based on specific applications.

V. CONCLUSION

The error-inducing problems associated with utilizing optical systems in machine vision metrology applications have been clarified in a short but fundamental treatment. Such problems yield measurement results that range from accuracy limiting, to precise but inaccurate, to grossly distorted. Although the coverage has not been comprehensive in analyzing the full range of currently available optical system technology, it provides a guide to the types of bias errors that end users of machine vision systems should be aware of.

The future will certainly lead to MV systems with significantly improved sensor and processing hardware and software packaging. The development and application of such systems will be strongly dependent on related disciplines and technologies. For example, optical system technology is a mature and sophisticated field that is growing less rapidly relative to MV, yet such technology will continue to exert a major impact on the application potential of MV. Applications development engineers requiring high degrees of measurement accuracy would be well served by being cognizant of the published literature in MV related fields, where subsystem solutions may already exist and accuracy limitations may be well characterized.

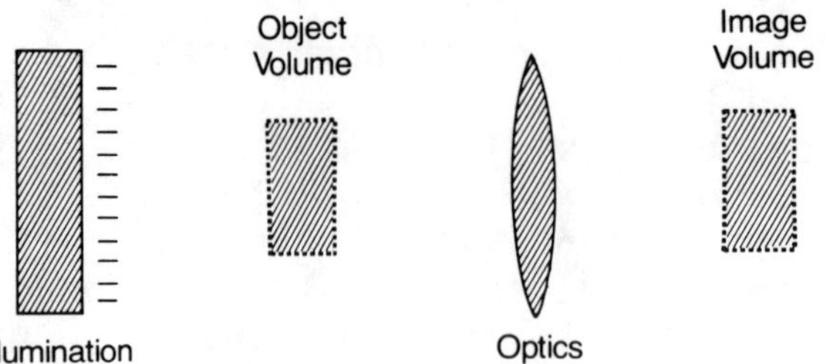

Figure 1: Optical front end of a machine vision system.

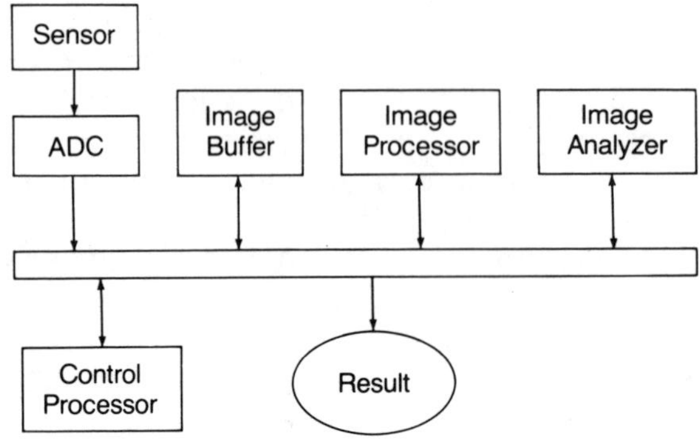

Figure 2: Sensor and processing sections of a generic machine vision system.

Figure 3: Ideal two-dimensional object - flat opaque medium with zero thickness and a transparent rectangular hole.

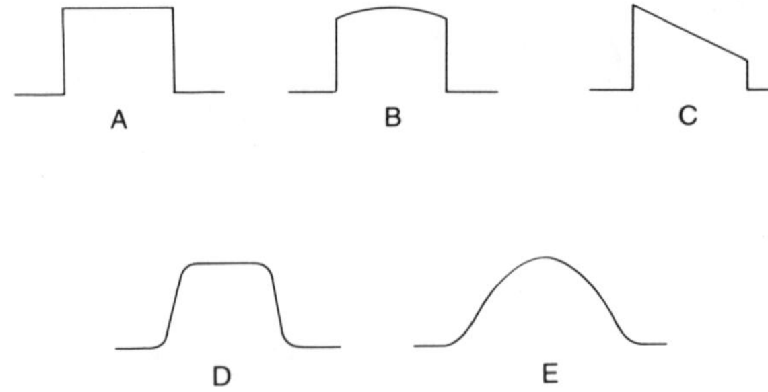

Figure 4: Horizontal intensity profile across the center of the optically produced image of the object in Figure 3 with differing optical assumptions.

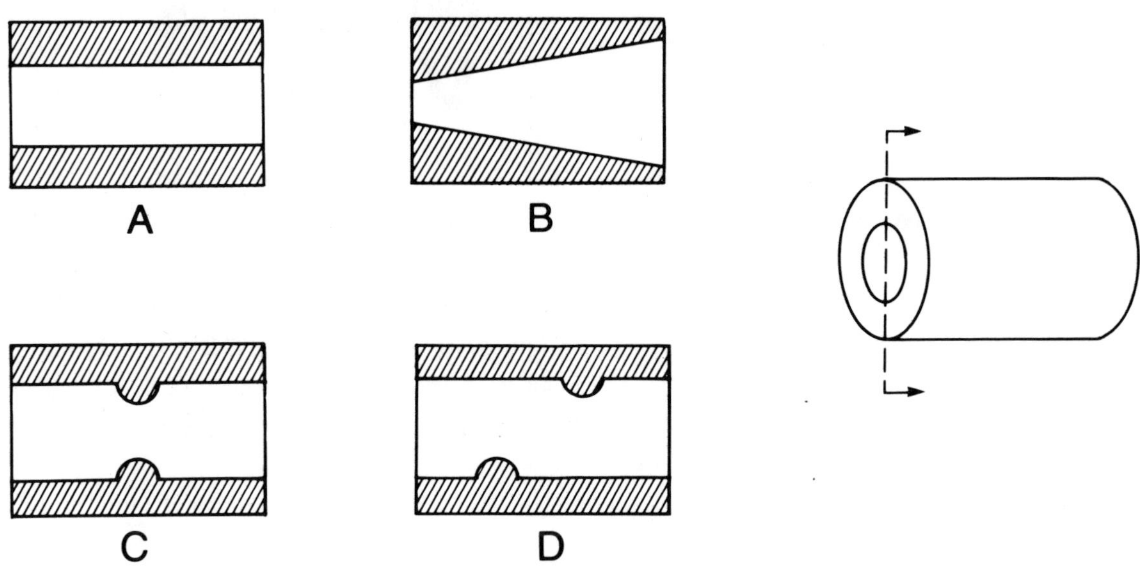

Figure 5: Real three-dimensional objects with differing holes, where cross-sectional view is parallel to the concentric axis.

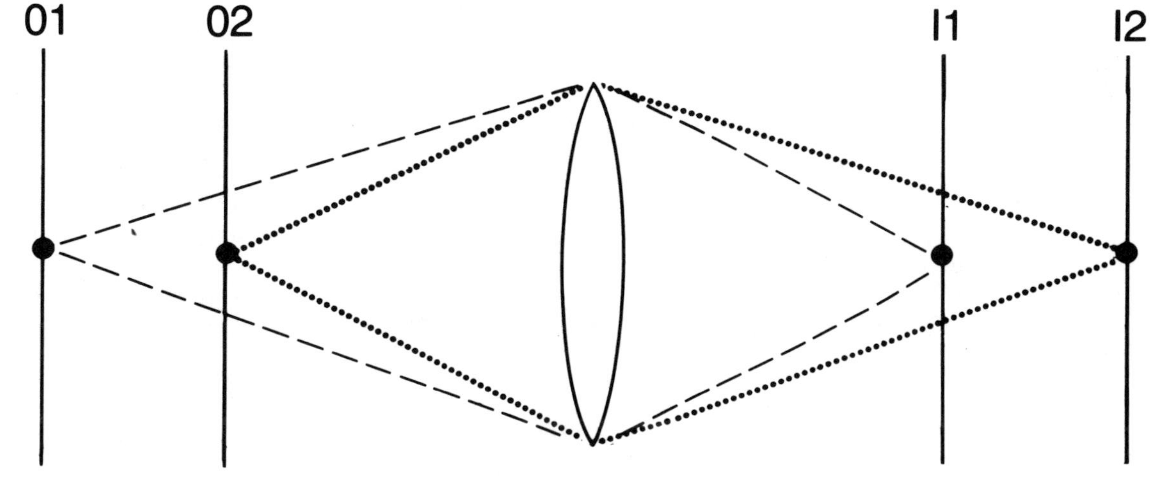

Figure 6: Differing focal properties of two points in a real object located on different focal planes.

Reprinted from *Vision*, May 1986

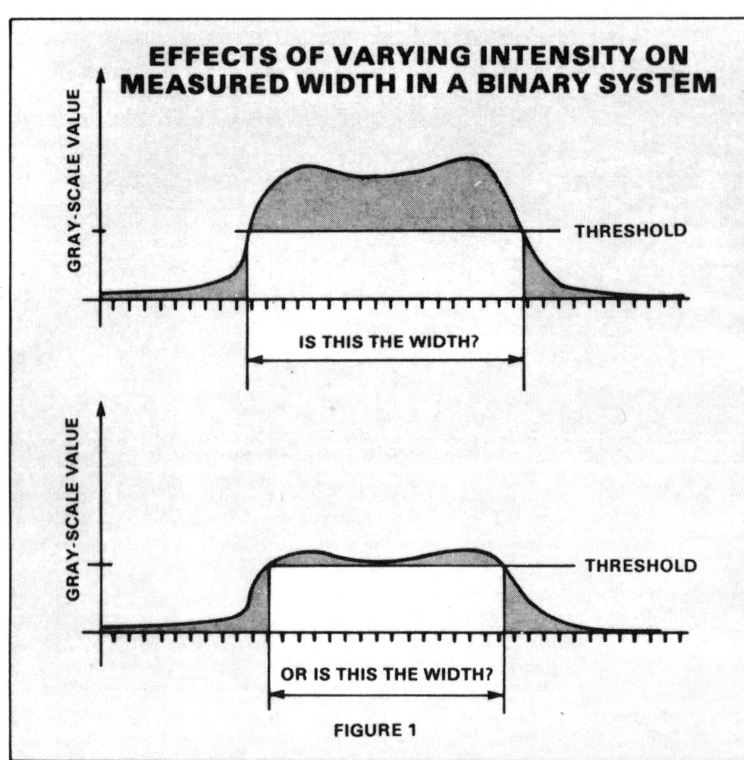

Note how measured width changes based on variations in illumination intensity.

Gaging with Machine Vision

Gaging with machine vision is an important process, yet there are a number of factors that may induce error and lower yields in actual production

BY STANLEY N. LAPIDUS

There are two sources of error in a vision system used for gaging: *systemic error* and the difficult-to-diagnose *statistical error*.

Systemic errors are characterized by a clear cause and effect relationship. Some of the common systemic errors that occur in vision systems include, in the order of likely effect, those involving lighting, staging, optical errors, camera-related errors, and front-end errors.

Lighting

Lighting errors occur when variation in lighting causes dimensions to appear to change when in fact they do not. Well designed gray-scale systems are immune to large lighting variations (up to a factor of 16 in some cases), while most binary systems are sensitive to lighting variation. The reason that binary systems are prone to lighting errors is seen in *Figure 1*.

Shown is a slice through the gray-scale image of an object. The vertical axis represents intensity (or gray-scale value) as seen by the vision system's camera. The horizontal axis represents positions along one dimension through the image of the object.

The problem at hand is to measure the width of the part. From the figure, we see that changing the lighting (which changes the amplitude of the gray-scale signal) yields different measurement results. This is caused by the change in the relative position of the threshold (the decision point between white and black in a binary system which is set at a fixed gray-scale value) on the profile of the part. A different width is picked-off for varying light intensity (and for varying thresholds, for that matter, too). Variations in illumination of the part cause measurements to fluctuate on the same part.

Vision systems with greater than 128 × 128 resolution tend to spread edges over a number of pixels. Crisp features become a little blurred, and the position of the threshold makes a difference in the measured width. The position of the threshold depends on lighting intensity, contrast, uniformity, and many other factors, but lighting-related issues are the most important in establishing a consistent, stable threshold. To remove lighting as a source of error when using a binary system, it is necessary to hold the light intensity constant and to keep lights and lenses clean. It is very important to keep in mind that individual light bulbs which have nominally the same ratings (such as a 60 W incandescent bulb) may differ by as much as ±50% in actual light output. So changing light bulbs may require a major recalibration of a binary vision system.

Staging

Staging errors are caused by part motion in unexpected degrees of freedom. Two-dimensional vision systems—that is, vision systems that use one camera per view and no structured light (structured light includes laser light stripes and other projected patterns)—can usually deal with at least two degrees of freedom: translation in *X* and translation in *Y*. Some two-dimensional vision systems can deal with rotation about the *Z* axis of the part as well. But two-dimensional vision systems cannot directly deal with translation in *Z*, as this causes the image to appear smaller or larger than it really is as it moves further from or closer to the camera. Nor can two-dimensional vision systems deal with rotation about the *X* or *Y* axis directly because this type of rotation causes dimensions to be foreshortened, as shown in *Figure 2*.

If the part can undergo these motions, use indirect measurement—which sometimes works with gray-scale systems, but is not as likely to work with binary systems—or a three-dimensional vision system, which may use multiple cameras, but typically uses structured light.

Indirect measurement can be made by comparing known dimensions in a scene, such as the pallet pegs in *Figure 3*, to their actual measured values and then computing a magnification ratio. This ratio can then be applied to the measured unknowns in the scene and consequently remove uncertainty regarding translation in *Z*. The translation in *Z* that this technique deals with should not be so great as to dramatically reduce focus, and the scene should be free from rotation about *X* and *Y* for this technique to work.

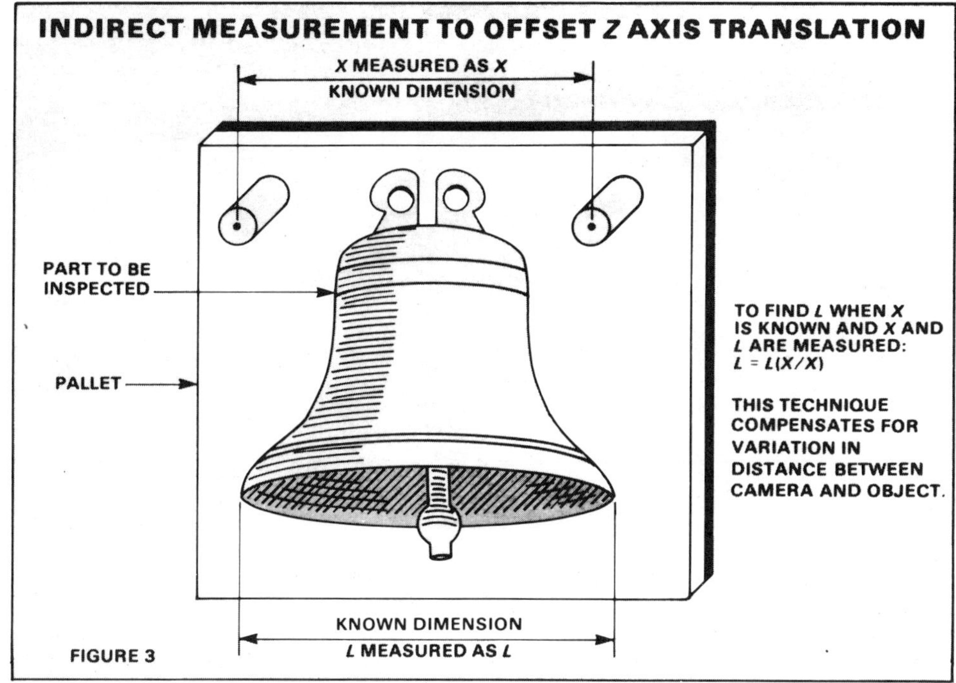

FIGURE 3

Optical errors

Optical errors are caused by lenses that do not provide perfect reproduction of a real image on the face of the camera. This usually appears as pin cushion or barrel distortion and as nonuniform reproduction of the intensity of the scene. *Figure 4* illustrates pin cushion and barrel distortions.

Pin cushion and barrel distortions usually vary with the stop of the lens and tend to be smaller in the range of *f*5.6. Further, these types of distortions aren't significant if the parts to be viewed don't move much. This is because these distortions change relatively slowly with distance, so even though a part may appear distorted, the golden part by which the system was calibrated would also be distorted by the same amount. If the part doesn't move much, the change in distortion will be small, and the measurement will change little.

There is a second condition also required for accurate measurement when faced with pin cushion and barrel distortions. This is the requirement that each measurement have a separate calibration factor. A calibration factor is established when the operator enters information into the vision system as to how many pixels per inch or millimeter there are. Calibration factors vary with the lens and the object-to-lens distance, but they also vary with distortion. In *Figure 4*, note how the distance between two lines is made smaller at the edges compared to the center in the presence of barrel distortion. If the calibration factor for each measurement can be separately entered, then the effects of the distortion can be neutralized—as long as the part doesn't move around enough to change the calibration factor.

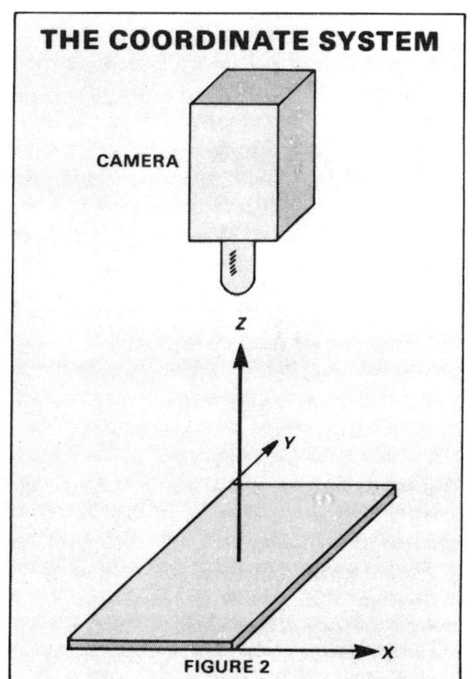

FIGURE 2

If measurements are not separately calibrated or if motion is unacceptably large, then you may be able to zero-out pin cushion and barrel distortions by adjusting the *f* stop for zero distortion and then setting the lighting by adjusting the intensity of the light source to get satisfactory illumination. Be aware that one may lose too much depth of field with this technique.

A solution for reducing distortion that always seems to work is to spend more money on a better lens or to adjust the working distance of the lens to allow the use of a focal length with inherently less distortion—anyone familiar with the extreme distortion of ultrawide-angle fisheye lenses will appreciate this. If one is using extension tubes, one may find that removing them and replacing them with a macro lens or microscope, which gives the proper magnification or minification without extension tubes, will also help—sometimes dramatically.

Incidentally, distortion should not be measured by looking at the vision system monitor, which itself is subject to its own distortions, but rather by making machine vision measurements on a grid pattern imaged by the vision system. In other words, rely on the computer, not the eye, to measure distortion.

Another important source of optical error is light fall-off, which occurs when lenses are less efficient at collecting light near the edges of the field of view than at the center. A uniformly illuminated diffuse surface, like a backlit sheet of white plexiglass, will, therefore, not produce uniform illumination on the sensor surface. As a rule of thumb, the more expensive the lens, the less severe the problem. This problem affects binary systems more than gray-scale systems because thresholds may differ in different parts of the field of view. Extension tubes exacerbate this problem.

Camera errors

Camera errors depend on the type of camera used. Solid-state cameras have one set of errors, and vacuum tube cameras—nuvicon, vidicon, orthicon, plumbicon—suffer from another set of errors. It is strongly recommended that one not use vacuum tube cameras if using the vision system for gaging. Vacuum tube cameras are subject to nonlinearity, which distorts the image, and to thermal instability, which causes the entire image to shrink and expand over time.

Nonuniformity of the photo-sensi-

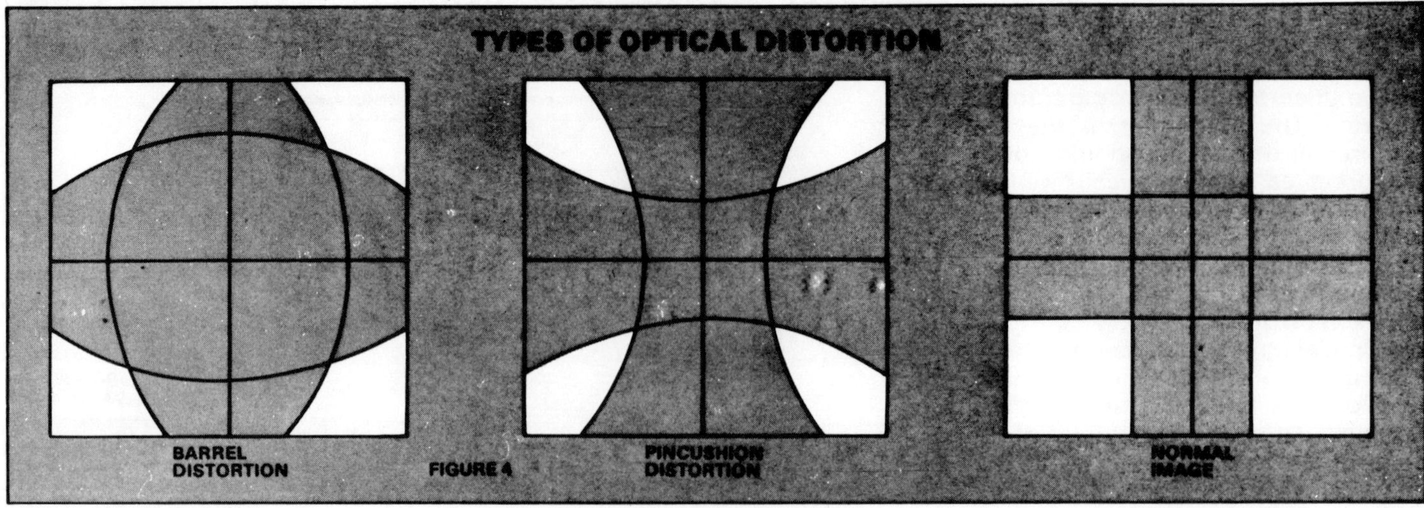

FIGURE 4

tive target on a vacuum tube can be as much as ±30%; however nonuniformity varies slowly (low spatial frequency) and tends to be severe only at the edges of the field of view. This means that true gray-scale vision systems usually can deal with nonuniformity, but binary systems may encounter difficulty. Remember that switching cameras when using a binary system may require establishing new thresholds, because nonuniformity of the vacuum tube patterns may differ.

Vacuum tube cameras are best used when lighting is weak—like microscopy. Vacuum tube cameras can work with less light than solid-state cameras—or when exceptionally high resolution is required, 1024 × 1024 pixels, for example. However, it should be noted that gaging may be unreliable; feature detection, on the other hand, will be reliable.

Solid-state cameras provide freedom from distortion and spatial drift because these parameters are determined by the very high physical stability of the camera's silicon substrate and the high electronic stability of the clock frequency—usually derived from a quartz crystal. Solid-state cameras suffer from pixel to pixel variation in sensitivity that may be so high as to render a number of pixels in the image (usually less than 10) dead. Dead pixels may cause binary systems to produce false blobs and may cause errors in some types of mathematical morphology systems as well.

True gray-scale systems, which rely on correlation and edge detection, average features over distance and thus are immune to individually dead pixels when making measurements. Some camera brands suffer from distinctly fewer dead pixels—some as low as zero.

Front-end errors

The front end of the vision system is the circuitry that receives the analog signal from a camera or cameras and converts this signal into a series of numbers for the computer to analyze. Front-end errors are much less a problem with vision systems today than they were in earlier days. The video amplifiers of many of today's vision systems are properly stabilized, and robust baseline restoration circuits along with stable integrated circuit analog-to-digital converters are used. Front-end errors manifest themselves as drift; the image appears to change in intensity with time. In binary images, this means the size of the blobs changes for no apparent reason causing measurement errors to occur.

If front-end errors are suspected, one can try connecting a high-quality TV test pattern generator instead of the camera to the vision system to see if intensity or measurements change randomly. If they do, suspect the front end. If they don't change with the pattern generator connected but the problem is present when a camera is connected, either the camera is bad or the part lighting varies. Varying light output can often be diagnosed with a photographer's light meter. Again, true gray-scale systems tend to be more robust in the presence of level and contrast variation, but gray-scale algorithms should not be used to compensate for sloppy front-end design.

Diagnostic aids

The following equipment may prove helpful in diagnosing systemic errors: a precision glass grid, available from industrial and broadcast television supply houses; a Lunar light meter, to make objective setting of light levels possible and isolation of lighting problems easy; a high-quality TV test pattern generator, helpful in isolating front-end errors; a good photographer's copy stand; and a diffuse light source, vital for studying camera uniformity.

A well thought-out vision system should have no systemic error that is not accounted for. Freedom from systemic error means fewer mistakes—calling good parts bad and bad parts good—less operator intervention, and higher yields.

Statistical errors

Statistical errors occur because of the interaction between the vision system and the manufacturing process. Statistical errors are more difficult to diagnose than systemic errors. The only way to pinpoint statistical errors is to observe what happens to production yields and to infer the cause.

The result of the finite resolution of the gaging system has impact on the yield of the process. For our purposes, we will assume that all systemic errors have either been removed or accounted for by indirect measurement techniques—that is, accuracy is ideal. We only concern ourselves with the effects of the finite resolution of the measurement system.

There are two factors which determine the yield of a manufacturing process. The first is the *intrinsic yield* that would result when using perfect gages with infinite resolution. The second factor is the *part tolerance-to-measurement ratio* (PTMR). This is the ratio of the part tolerance to the measurement resolution of the gage or the vision system. For example, a part tolerance of ±0.005" (0.13 mm) and a measurement resolution of ±0.001" (0.03 mm) in a vision system yields a PTMR of 5:1.

Good parts and bad

There are four possible results when gaging with vision: (1) A part that is actually within tolerance is reported by the vision system to be within tolerance (good part called good); (2) A part that actually is not within tolerance is reported by the vision system to be out of tolerance (bad part called bad); (3) A part which actually is not within tolerance is reported by the vision system to be within tolerance (bad part called good); and (4) A part which actually is within tolerance is reported by the vision system to be out of tolerance (good part called bad).

The first and second case represent a desired outcome of the measurement system. The third case represents what is to most manufacturers an unacceptable outcome. The fourth case at least does not allow bad parts to pass undetected, but causes lower yields and consequently costs money. In order to be sure a vision system will never call a bad part good, it is necessary to program the limits of the vision inspection measurement to be tighter than the part tolerance by at least the measurement resolution of the vision system.

In the previous example of measuring a ±0.005" (0.13 mm) tolerance with a ±0.001" (0.03 mm) resolution in the vision system, the vision system should be programmed to a tolerance of ±0.004" (0.01 mm). Adding the programmed tolerance of ±0.004" to the ±0.001" resolution uncertainty of the vision system yields a worst case of ±0.005", which satisfies the requirement that no bad parts will be called good. But the result has been to let some good parts perhaps be called bad. For example, a part that is measured with a precision gage at a tolerance of +0.0045" (0.114 mm) could be measured by the vision systems as +0.0055" (0.140 mm)—keep in mind the ±0.001 uncertainty of the vision system—and consequently rejected. Assuming uniform distribution on the vision system's measurement in the ±0.001 range, we can assume that half of the parts that the vision system gages in the range of ±0.004 to ±0.005 are, in fact, good parts.

It's important to keep in mind that the probability that a part will be good is the product of the probabilities that all of the individual, and independent, manufacturing operations are good. This means that if 10 holes are to be made in a part and each hole (made by a different tool) has 99.5% yield, the probability that parts are made with all 10 holes within tolerance is 95%. This says that individual process steps must be under extremely good control to achieve good results in completed parts.

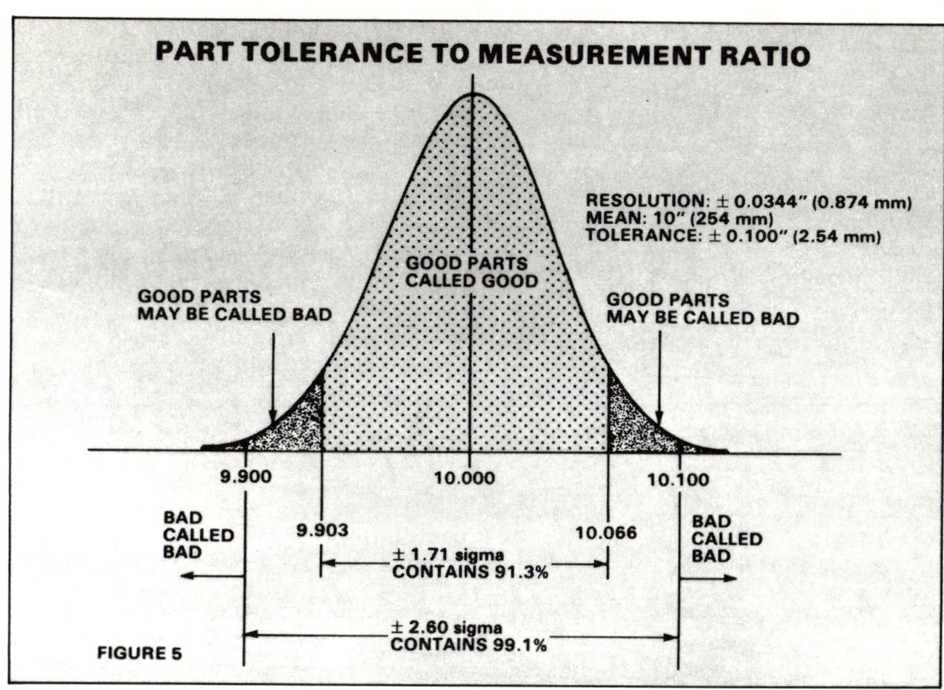

FIGURE 5

Let's examine the effect of PTMR on yield in two cases, one in which the process is extremely well controlled, so the defect rate on a particular measurement is, say, 0.2% when measured by high-precision gages (parts made within ±3.15 sigma of the mean are good). The other case is one in which the process is still good, but yield is somewhat lower—say 99.1% of the parts are good as measured by high-precision gages (this corresponds to good parts within ±2.60 sigma of the mean).

For example, using real numbers, let's make the following assumptions. First, the vision system has none of the systemic errors discussed previously, or they have been accounted for. Second, the nominal diameter of the part is 10" (254 mm). Third, the tolerance on the diameter is ±0.100" (2.54 mm). Fourth, the inspection must catch all out-of-tolerance parts, even if this is at the expense of calling some good parts bad. Fifth, a single-camera vision system is used. To allow for a 1" (25 mm) positioning error of the part, the field of view is 11" (280 mm). The resolution of the vision system is 320 × 240 pixels. The pixel size is therefore 11" divided by 320, which equals 0.0344" (0.874 mm). This implies a PTMR of approximately 3:1. Note that the axis of the vision system, which contains the maximum number of pixels, is aligned with the diameter of the part. And, finally, let's assume that the measurement repeatability of the vision system is ±1 pixel. This means that if the measurement is repeated many times at varying part positions and under varying lighting conditions with the same part, all of the measurements will be within 2 pixels of each other.

Figures 5 illustrates one example of the process. Assuming that there is an equal likelihood that a measurement will fall anywhere in the ±1 resolution unit zone, the important data is summarized in the following chart. A 10:1 PTMR is included for reference. This implies approximately 1000 × 1000 resolution on the part of the vision system.

Conclusions

The better the process—that is, the higher the intrinsic process yield—the less the yield suffers because of measurement resolution. This is because the slope at the edges of the bell-shaped curve is flatter

INTRINSIC PROCESS YIELD	10:1 (APPROX.) INSPECTION YIELD	3:1 (APPROX.) INSPECTION YIELD
99.8%	99.65%	97.98%
99.1%	98.60%	94.70%

than in central sections.

Further, textbooks written on gaging often make the point that the gage resolution (and accuracy) should be 10 times finer than the tolerance to be measured. Even this is no panacea—the reduction in the yield can be significant especially if it is assumed that this reduction in yield can occur on a number of dimensions on the same part—remember the multiplicative effect of probabilities. Reinspecting the bad parts with super-high resolution tooling (by hand, presumably) may yield a surprising number of parts which are good.

Increasing the PTMR for a well-controlled process may not be of significant value, but increasing it for a less well-controlled process pays.

Spending more money to improve a process without a high PTMR in the vision system may yield disappointing results.

Programming the tolerances of a vision system tighter by ±1 resolution unit will help assure that bad parts will not be passed.

Last, understanding the causes of statistical error is an important step to getting yields up and in making informed purchases of vision equipment. ◍

Stanley N. Lapidus is president, Itran Corp. (Manchester, NH).

Presented at the MVA/SME Vision '86 Conference, June 1986

Resolution and Accuracy in Gray Level Machine Vision

by Yen C. Chu
Santa Clara, California

Gray level machine vision is being considered by manufacturing engineers to meet the demand for better quality and higher productivity. Most machine vision systems installed in the factory are performing measuring functions.

The successful use of the vision system in inspection tasks depends greatly upon knowledgeable analysis of the measurement error of each component for the selected vision system. Last year Lapidus and Englander [1] discussed the danger of attempting to get more information out of the image, Stafford [2] talked about the error-inducing problems associated with utilizing optical system in machine vision metrology applications.

This paper examines the optics processing, image sensor, and computer image acquisition functions of the machine vision system. It discusses how and where the errors affect system accuracy, which is extremely important to metrologist and inspector.

OPTICS

The optical front end of the vision system must be designed with care. Otherwise, a precise measurement might hide a significant error caused by limitations in the optics, or more sophistication and expense will have to be put into the vision system to compensate for effects which could have been designed out of the front end more cheaply. Prull [3] discussed the effects of the optics in detail.

Image quality is limited by the optical aberrations and diffraction of the lens. Aberrations are lens defects which cause the lens to reproduce increasingly fine detail with increasingly poor contrast. On other words, as the detail in the object becomes finer, the contrast in the image becomes worse. The aberration can be reduced by closing down the lens aperture, if enough illumination is available. In most cases, the resolution of camera lens is limited by aberrations. However, because of the diffraction, even a perfect lens with no aberrations, could not focus on an infinitesimally small point.

For a perfect lens, the image of a light-to-dark transition (the light intensity profile of an edge) is shown in Figure 1. The diffraction blurs a sharp thin edge even at the best lens focus. Contrary to aberration, the diffraction-limited resolution increases with increasing lens aperture. High quality lenses are often aberration-limited at wide apertures and diffraction-limited at small apertures.

For reasons of cost and convenience, most vision systems employ general photographic lenses which have not necessarily been designed with high measurement accuracy in mind. The contrast requirement for vision inspection systems is generally much higher than the requirement for photography. The designer should always try to choose a lens which will be working near its designed optimum object distance and image format size.

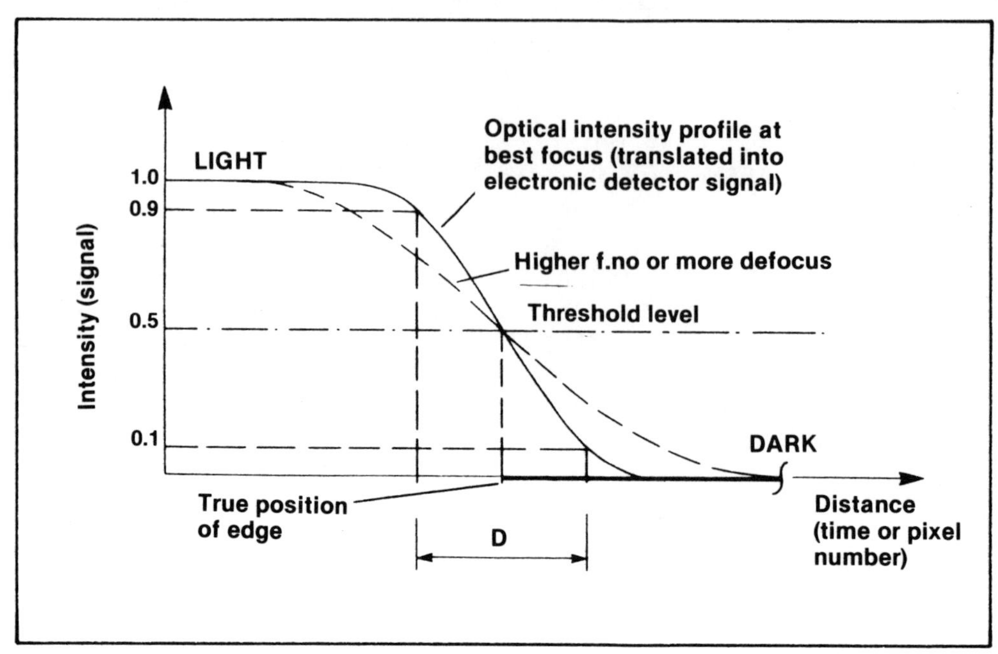

Figure 1. Edge Optical Intensity Profile [3]

Image Sensor	Sensitivity μA/1m	Gamma	Lag
Sb_2S_3 Vidicon	40-1200	0.6	20%
Pb0 Plumbicon	300-400	0.95	2-5%
Silicon Vidicon	4000	1	8%
CCD Camera	3500	1	1%

Table 1. Sensitivity, Gamma and Lag for Image Sensor [13]

The best camera lenses may achieve 100 line pairs per millimeter resolution at the best aperture setting and on-axis condition. With the higher contrast, larger aperture constraints of the vision inspection system, it would be very difficult for the best camera lenses to match the performance of a solid-state sensor capable of sensing 70 pixels per millimeter resolution. Therefore, the lens is the limiting factor for system accuracy.

IMAGE SENSOR ACCURACY

An image sensor is an electro-optical device which translates lignt to electric voltage or current. There are two types of image sensors commonly used in the industry; the image pickup tube (vidicon tube) and solid-state image sensor. Generally speaking, the solid-state image sensor is superior to the vidicon tube in most gauging applications. The disadvantages of the vidicon tube are image burn, lag, blooming, geometrical distortion, and non-linear sensitivity (Table 1). Solid-state image sensors are not perfect either, especially when it is designed into a camera for T.V. broadcasting applications.

When solid-state cameras became commercially available, they were made for the T.V. broadcasting industry. To be compatible with existing broadcasting hardwares (vidicon tube cameras) solid-state cameras were required to generate the standard RS-170 output signal regardless of image sensor arrangement.

In every T.V. camera design, there are two hidden operations which may affect accuracy of the vision system. A hardware low pass filter is built-in to the T.V. camera to smooth sharp edges for better T.V. viewing effect. For video applications, the gamma of the solid-state T.V. camera has been adjusted at the factory to 0.45, instead of 1 which is linear sensitivity. The gamma is a gradiant measurement of the curve of the image sensor sensitivity. It is defined by the following:

$$\text{gamma} = \frac{\log \text{(output video magnitude)}}{\log \text{(input illumination magnitude)}}$$

The solid-state sensor has approximately linear transfer characteristics (gamma = 1), and most vidicons range from 0.65 to 0.75. This modification of gamma in the camera is intended to be comparable to existing vidicon cameras and to compensate for the non-linearity in the T.V. monitor.

A "blemish" is defined as any pixel output voltage varying over 10% from the theoretical standard fluctuation of the video output of each pixel in a solid-state sensor. In other words, a blemish is a defective part of an image cell which can not provide the video signal linearly to the input light. Some vendors do not perform any blemish compensation, others compensate it with the pixel value in front of the blemished pixel. Any form of blemish compensation should improve the picture image presented to the T.V. viewer, but might be misleading if the camera is used as part of a vision inspection system.

Pixels with video output controlled within 10% of the standard fluctuation are considered good pixels. The fluctuation, or noise, may be caused by dark current, non-uniformity, random noise, cross-talk, or charge transfer losses of the sensor. CCDs (Charge Couple Device) have fewer random noise than CIDs (Charge Injection Device), but CIDs have no charge transfer losses (Table 2).

Most solid-state area image sensors used in vision systems are in the range of 128 X 128 to 512 X 512 pixels. Few sensors with 1024 X 1024 pixels are still very expensive. Most CID and CCD arrays have rectangular pixel spacing, partly due to their structure and partly due to the rectangular TV video frame format.

Figure 2 shows three popular solid-state image sensor design concepts in transfering and storing video data. The photon charges on the CCD array are transferred to an output storage either directly one line at a time (line transfer) or via a temporary frame storage area (frame transfer).

The frame transfer device has a section for photon integration and another section for storage of a full frame of video data. The frame transfer device tends to present a contiguous surface for collection of photon source data. This is an advantage which vidicon sensors share, in that all image data can be collected. This is not true of other types of transfer devices.

Type of Array	CCD	CID
Resolution Spacing (μm)	488 × 380 23 × 13.4	248 × 244 46 × 36
Random noise	Low	High
Dark current	High	Low
Non-uniformity	High	Low
Susceptibility to blooming	High	Low
Change transfer losses	Yes	No

Table 2. Comparison of Properties for CCD and CID Types of Array [4]

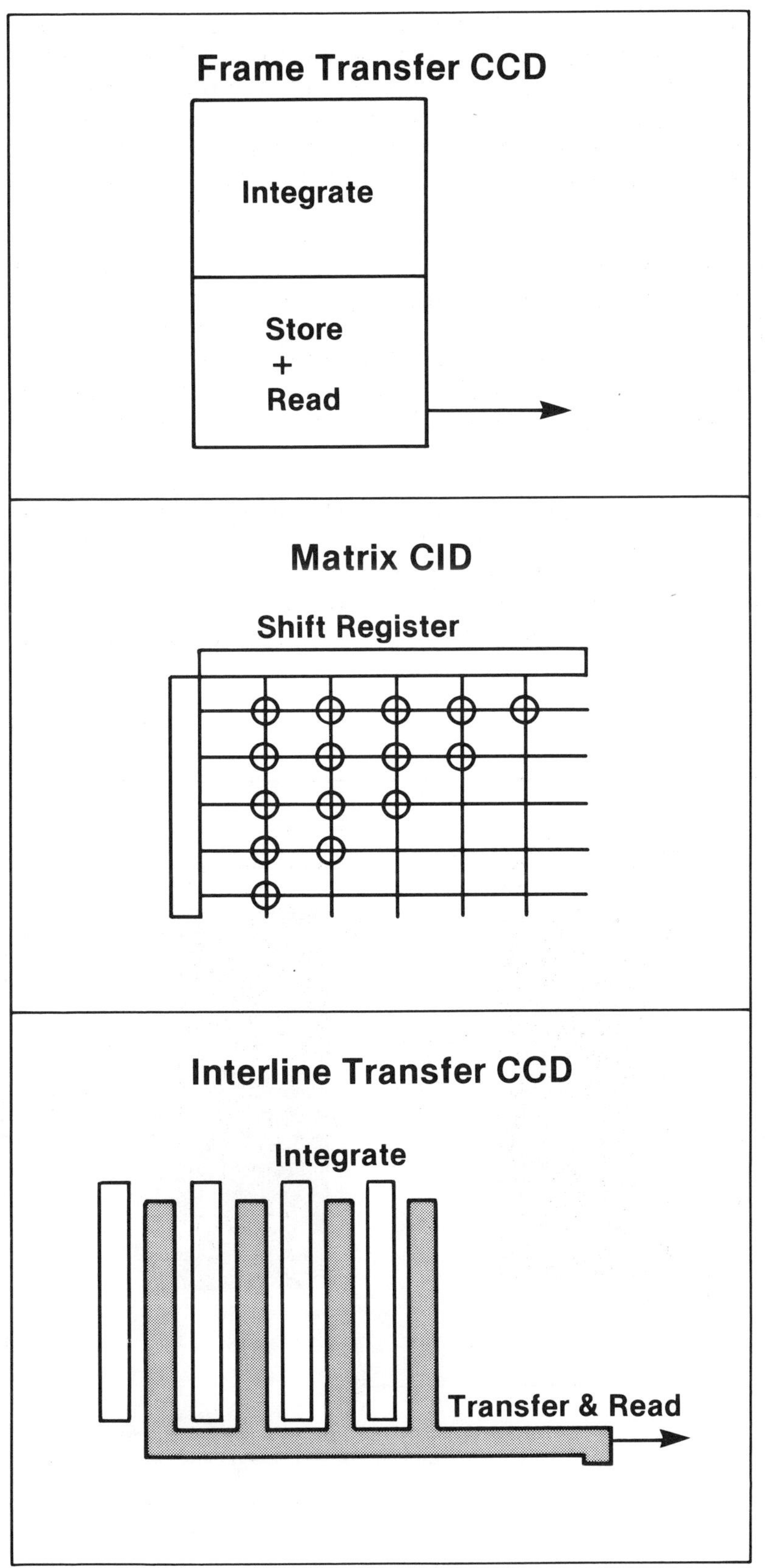

Figure 2. Solid-state Image Sensor Design Concepts [5]

The line transfer device interleaves the storage and transfer registers with the registers used for photon integration. One advantage of the interline device is that "smear" is virtually avoided. Because the distance from the photon integration registers to the optically shielded storage and transfer registers is very short, the charge transfer can be very rapid. On the other hand, this approach presents a significant obscureness to the photon image.

Greater than 50% of the horizontal pitch of each pixel is obscured (Figure 3). This also presents a very interesting challenge if single or sub-pixel accuracy is required for the inspection system. For the matrix-addressed device concept, such as CID and photodiode, each pixel can be read sequentially or randomly. They all tend to be contiguous sensors, with varing degrees of obscuration. Less than 10% of obscuration occurs in current CID design.

CCDs are particularly susceptible to blooming if a given pixel receives a light level which is far in excess of the level needed to saturate the pixel. The overcharged pixel will simply spill over to adjacent pixels. CIDs are less susceptible to blooming than CCDs because the basic cross section structure of the chip includes a layer which is capable of soaking up a large number of excess charge carriers.

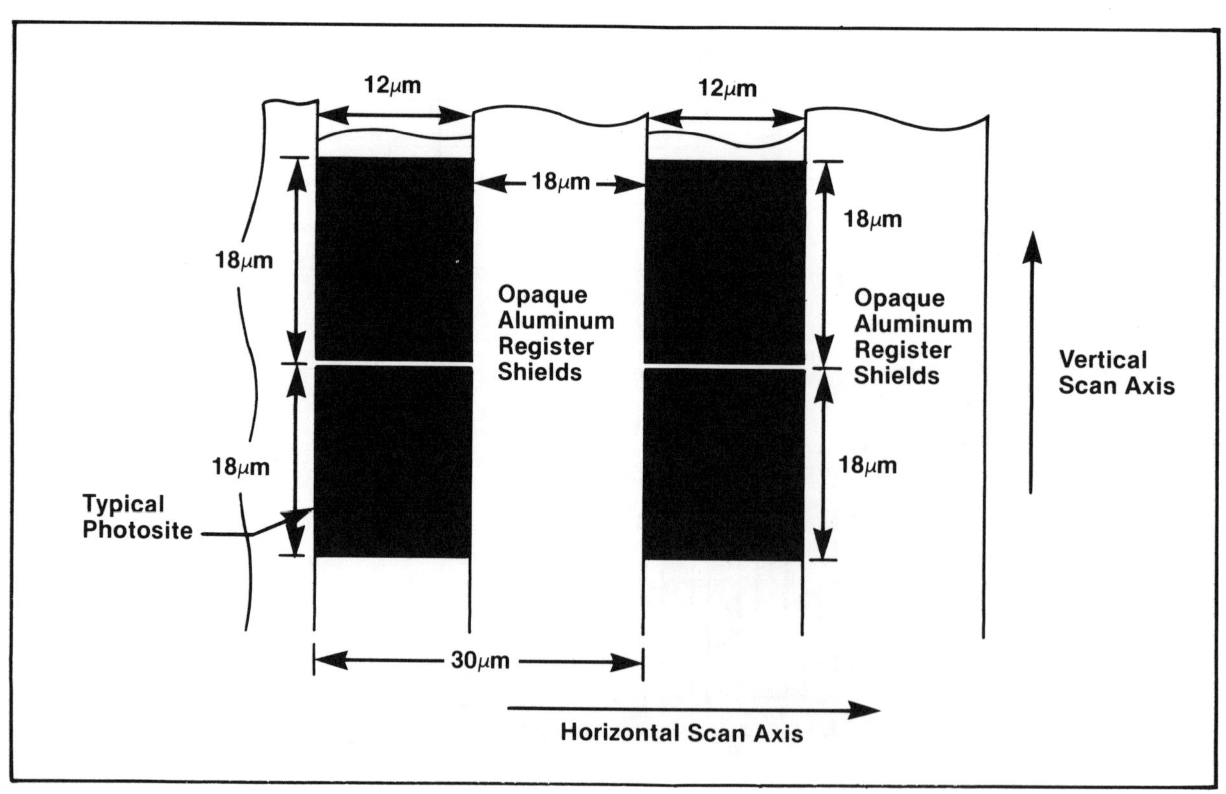

Figure 3. The Line Transfer Device Structural Detail [15]

IMAGE GRABBER RESOLUTION

The process of storing video data into computer memory is another important source of error. Most vision systems digitize the standard RS-170 video signal.

The RS-170 video signal is interlaced with two fields to form one image frame. There are 525 horizontal scanning lines within one frame. The frame repeats at a rate of 30 frames per second. The aspect ratio is 4:3. All solid-state cameras have to map their pixels according to RS-170 format regardless of pixel organization. For a 384(H) X 491(V) CCD image sensor, only 378 pixels is actually used to generate one horizontal line of video signal. The leading and trailing blanks period fill up unused horizontal scanning time (Figure 4). Each field of the frame starts with 20 blank horizontal lines, followed with 242.5 lines of video signal. The total useful pixel area is 378(H) X 485(V).

Some cameras, such as GE TN-2500 CID camera, do provide both analog (RS-170) and digital (8-bit wide) outputs. Each pixel charge is digitized at video frame rate. But the actual useful pixels are less than 244(V) X 248(H).

If the application demands single pixel accuracy or better, an accurate image digitizer is essential to ensure direct digital translation of each pixel. Camera calibration is extremely important to the inspector to insure the same interpretation of the image data.

CONCLUSION

The effects of optics, image seneor, and illumination cause the actual edge image to degrade from the ideal edge profile. The vision inspection system must push machine vision technology to the limit to achieve single or sub-pixel measurement accuracy.

A better resolution vision system is always demanded by industry. What industrial users should really demand is to have better resolution and accuracy cameras designed for machine vision industry. A new video output interface standard should be established. It would be impossible to get 1024 X 1024 pixel resolution out of the old RS-170 video signal. Larger image size is needed to improve resolution rather than shrinking pixel size. The average pixel size (20 um) is about the limit of camera lens resolution. The camera should generate true pictures rather than modified pictures for television. The camera should minimize the error during photon integration and video transfer process. Digitized video data output and random access of each pixel would be ideal features.

BIBLIOGRAPHY

1. S. Lapidus and A. Englander, "Understanding How Image Are Digitized", VISION '85 Conference Proceedings, March 1985.

2. R. Stafford, "Induced Metrology Distortions Using Machine Vision Systems", VISION '85 Conference Proceedings, March 1985.

3. D. J. Purll, "Optics for Image Sensors", Chapter Five, Automated Visual Inspection, IFS (Publications) Ltd., UK, 1985.

4. B. G. Batchelor, D. A. Hill and D. C. Hodbson, Automated Visual Inspection, IFS (Publications) Ltd. UK, 1985.

5. F. Sachs, "Sensors and Cameras for Machine Vision", Laser Focus/Electro-Optics, July, 1985.

6. K. Yamagata, F. Nagumo, H. Kosaka and H. Kawamoto, "CCD Camera for Machine Vision", Electric Imaging, November 1984.

7. T. Hori, "Solid-State Miniature Camera for Vision Systems", Pulnix America, Inc.

8. A. Wilson, "Solid-State Camera Design and Applications", Electric Imaging, April 1984.

9. M. Schardt, D. Weisner, R. Miller, R. Bredthauer, R.smith, R. Ginaven, R. Zenick and E. Odeen, "Design and development of a 1024 X 1024 visible image", SPIE Proceedings Volume 501, August 1984.

10. S. Sternberg, "Grayscale Sub-pixel Gauging", Machine Vision International, September 1984.

11. N. S. Chang, "SMV - A Computer Program for Loading Surface Mount Components", SPIE Proceedings Volume 557, August 1985.

12. R. Freeling, "The Significance of Lighting in Industrial Inspection Tasks", 1985 IEEE International Conference on Robotic and Automation.

13. J. Bretschi, Automated Inspection Systems for Industry, IFS (Publications) Ltd, UK. 1981.

14. E. T. Spencer, "Configuring a Binary Vision System with a CCD 3000 Camera", Fairchild Camera Co.

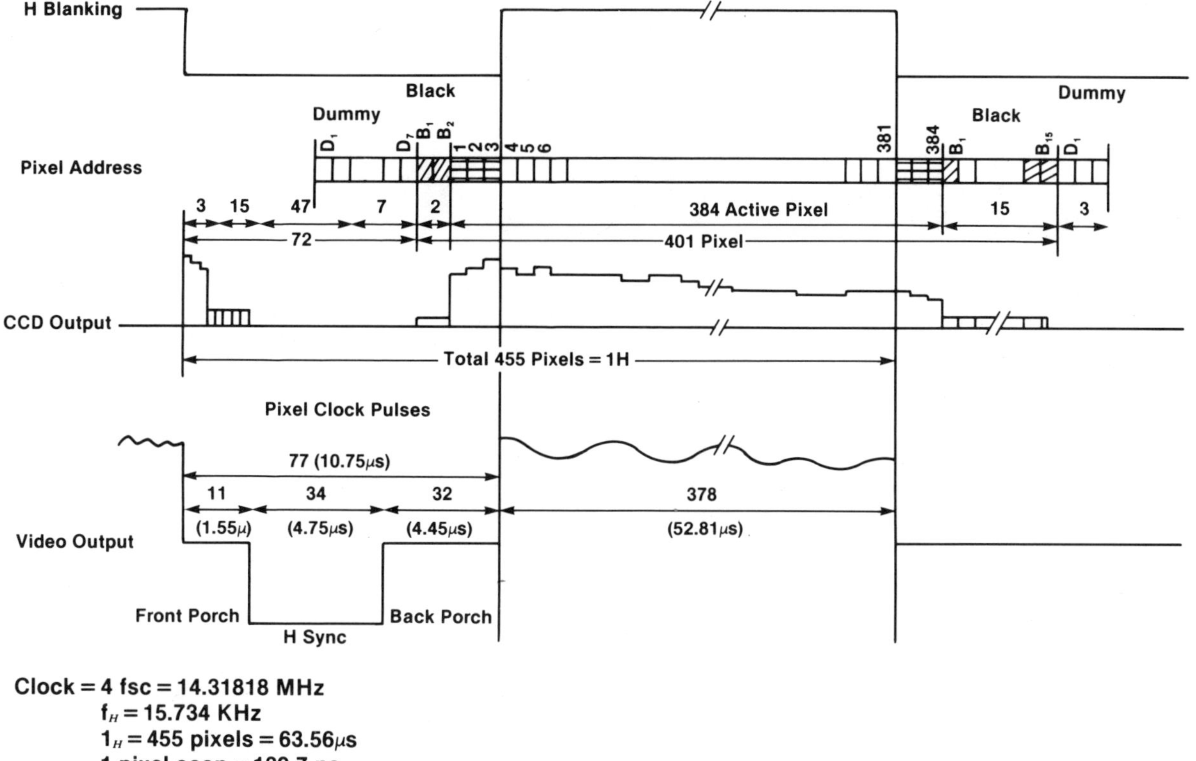

Clock = 4 fsc = 14.31818 MHz
f_H = 15.734 KHz
1_H = 455 pixels = 63.56μs
1 pixel scan = 139.7 ns

Figure 4a. Pixel Allocation and Timing Chart in One Horizontal Line [7]

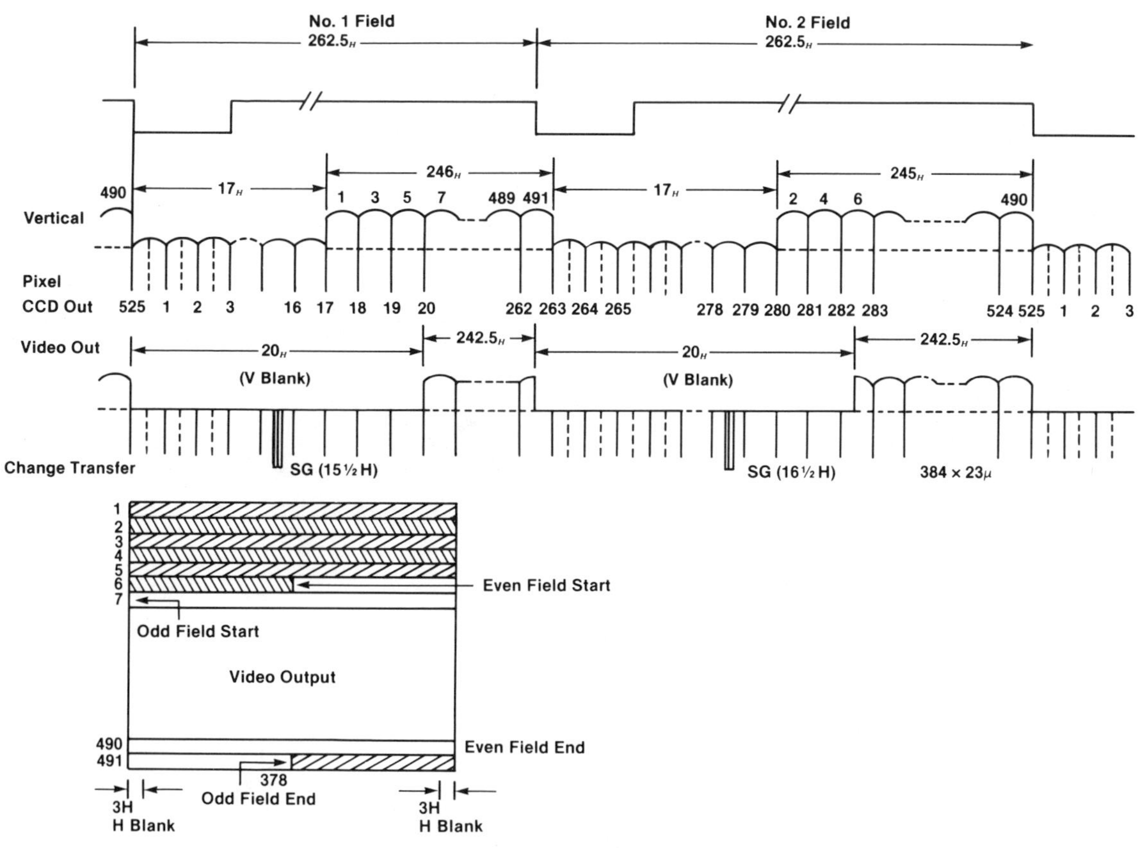

Figure 4b. Timing Chart and Video Frame Output [7]

Reprinted from *Robotics World*, April 1986

MACHINE VISION MEETS THE CHALLENGE FOR SMALL MEASURES

By Arman Garakani and Judy Cobb

The fudicial mark on the left is used by a vision system to determine positioning. A vision system using center-of-mass techniques would fail to locate the damaged mark on the right.

In order for automatic manufacturing cells to be able to adapt to changing conditions, they must receive sensor feedback. In small-scale or very precise manufacturing operations such as those found in electronics, adaptive assembly machines usually must be able to see.

"What does it mean, to see? The plain man's answer (and Aristotle's too) would be to know what is where by looking. In other words, vision is the process of discovering from images what is present in the world and where it is," according to David Marr in his book, *Vision*.

To see why small scale measurements require specialized machine vision systems, consider the following questions. How tall is this magazine? If you pull out a ruler you can easily determine that it's eleven inches, give or take about an eighth of an inch. Now, what is the diameter of the dot at the bottom of this question mark? The ruler that worked perfectly well for inch-scale measurements is no longer adequate. Clearly, accurate measurement requires tools designed specifically for the size scale involved.

For accurate mil-scale measurement, a machine vision system must have a superior technique for finding objects. We'll examine two possible techniques, center-of-mass searching and model-based searching, and will see that model-based searching is superior. Then, after a vision system finds an ob-

Arman Garakani is applications engineering manager, and Judy Cobb is marketing communications manager for Cognex Corp., Needham, Massachusetts.

A grey-scale system is capable of seeing an edge as several small steps of intensity change across a row of pixels. The system performs statistical analysis on the transition in this row, and on those rows above and below it along the same edge.

An example of SMDs with equivalent lead-through devices.

In order to make accurate and consistent small-scale measurements, a machine vision system must employ model-based searching and grey-scale analysis. The best way to determine which vision system performs your job best is to actually test systems in factory environments using your part samples.

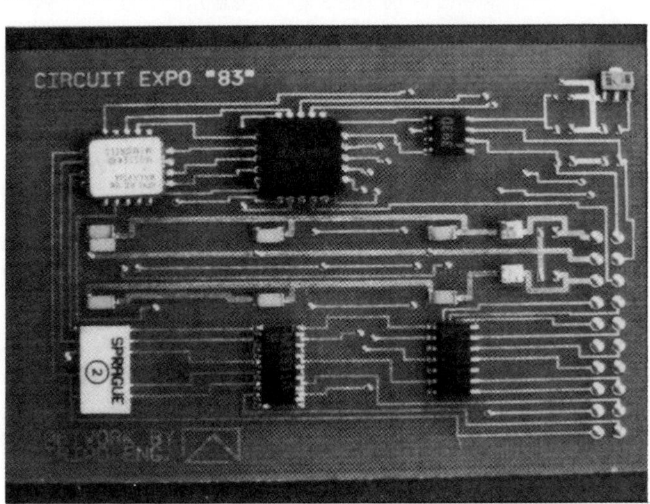

A printed circuit board containing SMDs.

Depicted above is a mispositioned SMD.

ject, it must have a superior technique for analyzing that object's image. We'll examine both binary and grey-scale analysis, and will see that for accurate small scale measurement, a vision system must employ both model-based searching and grey-scale processing.

Searching techniques

In center-of-mass searching, the system looks for regions of the image where several adjacent pixels (image segments) have an intensity very different from that of the background. When the system finds such a blob, it measures characteristics such as area and perimeter to determine if the blob is the object in question. If so, the system reports that the object's position is wherever its mathematical center is.

In model-based searching, on the other hand, the vision system looks not at a blob's measurements, but at its distinguishing shape elements. The system looks for blobs that have the same unique features as an ideal image stored in memory. When it thus locates a valid object, it reports that the object's position is wherever it would be if the object were a perfect duplicate of the model.

To see how the two searching techniques can produce different results, let's apply each to the sample task of locating a fiduciary mark on a printed circuit board. The mark is a simple + measuring 250 mils squared. The vision system must determine the position of the mark accurate to +/− one mil, so that the automatic component placement head will know exactly where to seat the 10-mil, surface-mount chips. That requirement may seem tough enough, but to add to the challenge, marks are not always formed properly. In fact, in routine manufacturing the marks may be missing up to 50 percent of their shape.

Consider the example of a mark that is missing its right arm. If a vision system using center-of-mass techniques succeeded in identifying the mark, it would fail to accurately locate it. For such a system would calculate the center of mass as falling to the left of the crosshair intersection, and would report this spot as the mark location. On the other hand, a model-based vision system would correctly identify the mark and correctly report its location as the location of the intersection feature.

Grey-scale analysis

In addition to using model-based searching techniques, a mil-scale measurement system must also employ grey-scale analysis. To understand grey-scale analysis and its alterna-

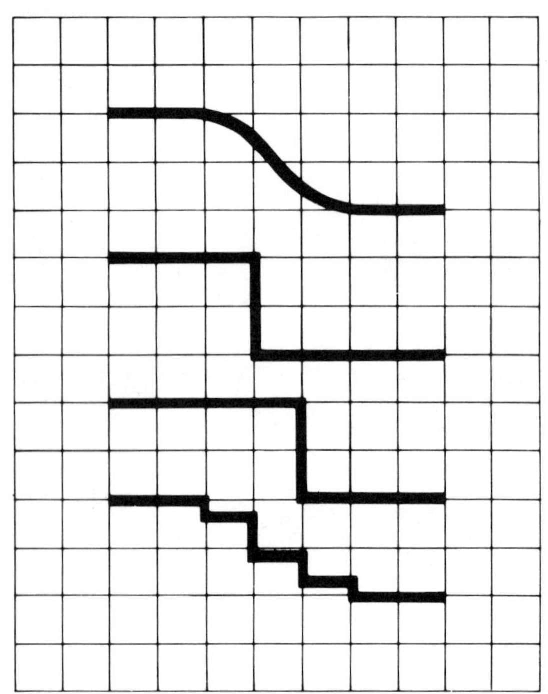

Consider the edge between a white object on the left and the black background on the right. The top curve shows how the edge actually looks; high intensity on the left, low intensity on the right. The middle two curves show two possible binary interpretations of where the edge lies. A more accurate and consistent representation of the edge appears in the grey-scale curve at the bottom. In the graph below is a two-dimensional transition of the same edge above.

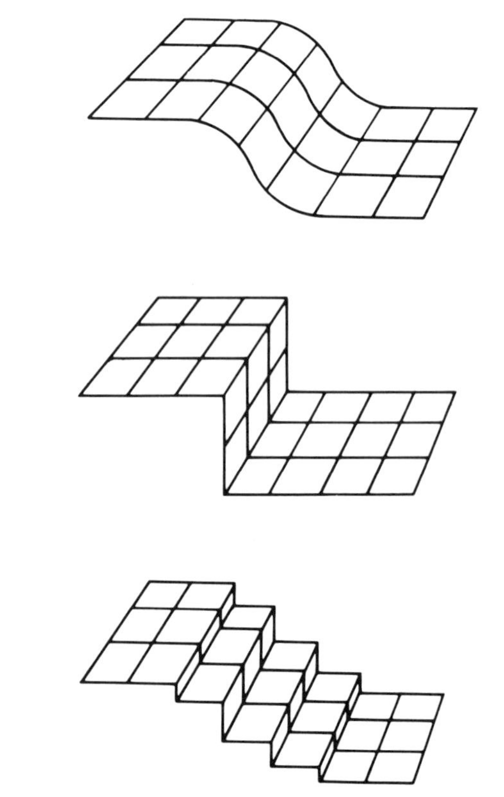

tive, binary analysis, we must first briefly review a few fundamentals of how images are captured.

When a vision system camera senses an image, it records the level of light intensity at each pixel as an analog value. Then an analog-to-digital converter assigns to each analog signal level the nearest available digital value. The precision of the A/D converter determines the number of shades of grey contained in the image sent to the processor. For example, a six-bit converter could assign any of 2^6 or 64 grey levels, whereas an eight-bit converter could assign any of 2^8 or 256 grey levels.

Most vision processors today use all of the shades of grey provided by the converter during preliminary image processing. However, in many machine vision systems, the crucial decision-making software operates on only two shades of grey. Consequently, the system must set a threshold, calling all shades of grey above that level black, and all shades of grey below that level white. These thresholded or binary systems discard much important information contained within the subtle shade changes in an image.

A true grey-scale processing system never sets a threshold, but rather operates on all shades of grey throughout decision-making. As a result, grey-scale analysis provides superior performance. To see why, let's return again to our sample task of locating the fiducial mark, considering the issues of accuracy, field-of-view, and consistency.

Accuracy

First, we examine accuracy. In an ideal situation a very sharp black/white object edge would occur exactly along the dividing line between two rows or columns of pixels. In this case, some pixels clearly would be object pixels, others clearly would be background pixels, and the edges would be easy to locate precisely.

However in the real world, edges usually fall within pixels. Moreover, since we are talking about examining small objects at high resolution, the edge may not be sharp at all. Rather it may be a more gradual change in intensity across a few pixels. You can see that many edge pixels in real images will contain part object and part background.

A binary system must call each of these transitional pixels either all object or all background. Thus it artificially produces sharp object/background jumps at dividing lines between pixels. In effect, the binary system "rounds off' its edge location to the nearest whole pixel, necessarily introducing some error. Consequently, its measurements are accurate to only +/− one pixel at best.

A grey-scale system, on the other hand, is capable of seeing an edge as several small steps of intensity change across a row of pixels. The system performs statistical analysis on the transition in this row, and on those rows above and below it along the same edge. In this way, it produces an extremely accurate measurement of the edge's location.

Using this grey-scale analysis, a vision system can calculate edge locations accurate to 1/16 pixel. However, due to video noise and other uncertainties, systems can practically measure with accuracy to about 1/3 or 1/4 pixel.

Field of view

Secondly, we examine field-of-view. Say that the materials handling equipment positions circuit boards in the assembly cell with an accuracy of +/− 1/2 inch. Then to find the mark, the system must search a one-inch-square (1000 by 1000 mils) area.

Recall that we must determine the mark's position accurate to +/− one mil. Since a binary system's measurements are accurate to the next whole pixel, such a system would have to be configured so that each pixel contains no more than one square mil of area. Thus to obtain a field of view of 1000 x 1000 pixels, the system would require a camera with a pixel grid larger than 1000 by 1000 pixels. Though cameras with

Machine vision systems today are automatically measuring everything from automobile parts to food jar labels. Most of these objects are on the scale of inches. Yet, in the electronics industry, machine vision systems must measure things on the scale of mils—things like component leads on surface-mount chips. In order to handle such tasks, machine vision systems must be equipped with the proper software tools.

such large arrays have recently become available, they add significantly to the cost of a system.

Grey-scale processing systems can avoid this problem because of their sub-pixel measurement capability. A grey-scale system can measure to one-mil accuracy when each pixel contains an area three to four mils squared. Therefore a grey-scale system with a standard 512 x 512-pixel camera can have a large enough field of view to find the mark and still locate it with the required accuracy.

Consistency

Finally, we examine consistency. Some degree of variation exists in the images captured by any machine vision system due to video noise and changes in ambient lighting. In a binary system such minor fluctuations can cause pixel values to make the radical change from black to white or vice versa. Consequently, binary processing systems tend to amplify image inconsistencies and produce even more inconsistent measurements.

In a grey-scale processing system, image inconsistencies will likewise produce changes in pixel intensity values. However, since those changes are between values that are very close together, the actual image processed does not change radically. The system will produce the same results in repeated runs dispite the presence of video noise and lighting changes.

We've seen that in order to make accurate and consistent small-scale measurements, a machine vision system must employ model-based searching and grey-scale analysis. Yet the best way to determine which vision system performs your job best is to actually test systems in factory environments using your part samples. The system that's right for you is the one that knows what's where by looking—the right way.

Reprinted courtesy of Dr. Veeder South III
Westlake Village, California

Automatic Precision Video Measurement for Quality Inspection

by Dr. Veeder South III
Matrix Videometrix, Inc.

INTRODUCTION

The word precise, as defined by a dictionary, has several words used to describe it such as: "exact"; "strict"; "adhering to rule"; and "accurate". Precision is defined simply as "the quality of being precise". Accuracy, on the other hand, is defined by words such as: "in exact conformity with truth"; "free from error"; and "precise". Even though each definition contains the other, in referring to measurements, the terms "precision" and "accuracy" may be understood to have different meanings. That is, the precision of a measurement is referred to as the degree of repeatability (or non-repeatability as it is usually specified) while accuracy is understood to be how close the measurement relates to an agreed upon standard ("truth"). This distinction is quite important when designing, manufacturing, and utilizing automatic video measuring equipment (or any measuring device for that matter) since it is entirely possible to have precision without accuracy while it is not possible to have repeatable accuracy without precision. While the ultimate goal of a measuring instrument is accuracy, the primary design goal is to ensure precision. Once precision is obtained, accuracy can be calibrated into the instrument.

FACTORS AFFECTING PRECISION

Mechanical Effects

In order to measure parts larger than the optical field-of-view most video inspection systems utilize some sort of precision staging mechanism. This varies from a simple X-Y-Z stage of small travel to a six-axis robot with large motion capacity. In order to utilize such a staging system, it must have a precision of its own which will affect the overall repeatability of the measuring instrument. I want to categorize two primary non-repeatability effects in staging. The first is associated with the position encoding system such as digital readouts or steps on a stepper motor. Let's call this the basic stage encoder resolution and define it as the limit to which the position of the stage is known, R_s. The second category of stage non-repeatability is to be a catch-all which includes such detriments to precision as hysteresis, loose coupling, backlash, poor workmanship, and other such culprits. For the sake of simplicity let's call these the Mechanical Non-repeatability, δ_m.

Video Digital Resolution

The basic digital resolution of a video system is dependent upon the camera and digitizing electronics as well as the optics used to project the image on the video sensor. Assuming that the optical resolution is not the limiting factory (as is the case for most video inspection systems) the achievable video resolution is obtained by dividing the optical field-of-view by a number which corresponds to the capabilities of the sensor and digitizing electronics. This number varies but is typically a number between 256 and 1000 for most systems. The Rank Videometrix System uses a digital resolution number of 500. Thus, the digital video resolution (sometimes referred to as the size of a pixel) is:

$$R_{\sqrt{}} = \frac{F.O.V.}{500}$$

Typical values of optical magnification used to produce digital video resolution are shown in Table I.

Optical Field-of-View	Approximate Magnification	Approximate Depth of Focus	Digital Resolution
3"	0.25X	0.200"	0.006"
1"	0.75X	0.100"	0.112"
0.5"	1.5X	0.050"	0.001"
0.1"	7.5X	0.010"	0.0002"
0.05"	15X	0.005"	0.0001"
0.01"	75X	0.001"	0.00002"
0.005"	150X	0.0002"	0.00001"
0.002"	375X	0.0001"	0.000004"

TABLE I. Some typical optical/digital parameters used in precision video inspection systems.

The effect of digital resolution on a video measuring system is to set the basic building block upon which all field-of-view video measurements are derived. This is not the limit of precision of the video, as is often stated, but the basis for which to compute the limit of precision (non-repeatability) for video measurements.

Environmental Factors

Environmental factors which affect precision are extremely important to consider since they are usually under the direct control or influence of the user of a video inspection system. These precision inhibitors include: temperature variations, vibrations, variation in lighting due to line voltage, different position of lighting, variations in part presentation to the axis (fixturing), dirt on the part, and several other inhibitors that can be lumped into the category: "others". Giving each of the environmental factors a symbol (for later reference):

δ_t = Thermal Non-Repeatability

δ_v = Vibrational Non-Repeatability

δ_e = Lighting Non-Repeatability

δ_a = Alignment and Fixturing Non-Repeatability

δ_d = Dirt Non-Repeatability

δ_o = "Other" Environmental Non-Repeatabilities

Statistics

It is convenient to digress a little at this point and discuss a very useful topic - statistics. I realize that while it sometimes seems a bit like black magic, there exists a powerful tool here. The fundamental power of statistics is to be able to make conclusions from a limited sample of data. For example, by making a number of measurements (say 30) on a sample part, one could possibly predict the precision of measurement expected on any other part with better than 99% confidence. There are, of course, several questions about the sample to be considered such as: Was the part representative of all parts? Was there sufficient time taken between measurements to allow all variables (such as temperature) to be represented in the sample? Were there any biases in the sample? These factors should be considered but should not deter one from making use of this powerful tool.

Utilizing Statistics to Improve Precision

The general rule of thumb in measurements is that the more data taken and utilized, the more precise the answer. As an example, suppose we wish to measure the length of a small stick known to be exactly four inches long with a six inch ruler. Let's say the finest increment of the ruler is 0.001" (about 1/4 the size of a human hair.) This value represents the basic resolution of the measuring device. Is it the precision? The answer is most likely yes for any single measurement. However, suppose the task of measuring the stick was given to several different people and their answers averaged to produce a single answer. Would this answer have a better precision than the single measurement? The answer again is yes since it is quite probable that there will be several answers slightly on the high side and several answers slightly on the low side and that their average will have a precision better than 0.001"! From a statistical point of view, the results of such an experiment follow the equation that the error associated with the averaged result is equal to the standard deviation divided by the square root of the number of samples used in the experiment. (See Figure I) Thus, the effective resolution of measuring the stick would depend not only on the basic resolution of the ruler but also the number of people (samples) used in determining the final answer. Thus, we would expect a 10-fold improvement in effective resolution by averaging the results of 100 people rather than accepting the measurement of one person.

For a normal distribution of sample measurements, the ruler resolution is a good estimate of the standard deviation (σ) and the laws of statistics state that 99.7% of all samples will fall within $\pm 3\sigma$ of the mean value. Thus, it is possible to estimate the Resolution Limited Non-Repeatability (δ_r) of this measurement as:

$$\delta_r = \frac{3(R)}{\sqrt{N}}$$

Where: R is the basic digital resolution of the measuring instrument and N is the number of samples used in the measurement.

Naturally, this same argument holds whether the measuring instrument is a small ruler, a glass scale encoder, or a video measurement system, i.e., the basic resolution of a digital system can be statistically improved by using several samples for each measurement being taken.

ESTIMATING PRECISION

All precision measurements in a video inspection system depend upon the answers to two simple questions: 1) Where is the camera?; and 2) Where is the point of interest in the field-of-view? The combination of this information using the appropriate mathematical equations produces the answer required. The degree of repeatability of this answer is the measure of precision of the instrument. And since there is usually a data processing computer included in an automatic video measuring system, it is a relatively simple matter to measure the degree of precision that the system can produce by repeating measurements and printing out the results. From a quantitative standpoint, the standard deviation of these measurements from their arithmetic mean (average value) is an excellent way of reporting precision.

One method of estimating non-repeatability is to simply add up all of the non-repeatabilities of the measuring system. While this would definitely give the worst possible case of the measuring instrument, it is highly unlikely that all of the uncertainties would stack up in the same direction. A more reasonable estimate would be to compute an RMS value. In equation form,

$$\delta_{rms} = \sqrt{\delta^2_{R,stage} + \delta^2_M + \delta^2_{R,video} + \delta^2_t + \delta^2_v + \delta^2_e + \delta^2_a + \delta^2_d + \delta^2_o}$$

Where δ_{rms} represents the best estimate of the non-repeatability of a measuring instrument taking all of the previously discussed non-repeatabilities into account.

It is the mutual goal of both the manufacturer and the user of precision measuring equipment to specify the precision required for a particular measuring application and then control all variables contributing to the non-repeatability of the instrument to meet this specification.

MINIMIZING NON-REPEATABILITY

There are several trade-offs in achieving precision in a video inspection system. Mostly these are between cost of the equipment and speed of measurements. In general, to achieve high precision requires more money and more time, neither of which is endearing to the efficient designer nor the prudent customer. Thus, it is in both of their best interest to minimize all non-repeatabilities which cost the least to implement and have a negligible effect on speed of the system. Included in this category are:

δ_t; δ_v; δ_d; δ_a; δ_1; and δ_o

The other category on non-repeatabilities involves those that are entirely the responsibility of the manufacturer. These are:

$\delta_{r,}$ stage; δ_m ; and $\delta_{r,}$ video.

These three non-repeatabilities are the basics upon which a precision video inspection machine is founded. It is the responsibility of the manufacturer to ensure that these (along with all of the others) can be minimized such that his precision measuring system has the capability to honestly meet or exceed its advertised specifications.

Thermal Non-Repeatability (δ_t)

Temperature affects everything in a video inspection system. Thermal expansion plays havoc with staging, contributing to errors involving straightness of stage travel. Temperature affects the electronics in cameras, both digital and analog, and can cause variations in the linearities of video measurements. The time effect of temperature plays a role in non-repeatabilities from day to day in a non-thermally controlled environment. In general, the more constant the environmental temperature the more repeatable system in a reasonably controlled environment is recommended. It is also possible to characterize the thermal response of a given system and take this into account in the system software. Mostly, this is experimental in nature and must be accomplished on a system by system basis.

Vibrational Non-Repeatability (δ_v)

Vibration isolation is a relatively easy thing to accomplish in most environments for inspection systems specifying non-repeatability in the ±0.0001" range. Vibration becomes a more severe problem as the spec approaches ±0.000010" or less. In the former case, a solid construction between the camera and stage is usually sufficient. In the latter, additional preventive measures are recommended such as air shock absorbers between the measuring system and its surroundings.

An additional level of repeatability when vibration is present is achieved in the video by taking many data points per measurement. The same statistical laws as previously discussed apply here since vibrations typically cause the errors to be random about a mean value.

Non-Repeatability Due to Dirt on the Part (δ_d)

Dirt on parts to be measured is a common problem. In a manual system, it is easy to handle since the operator is able to brush the part off or use his/her built in "pattern recognization" capability to ignore it. However, unless the automatic vision system analyzing the part has been "taught" this useful skill, non-repeatabilities due to dust particles and dirt (up to several mils) will show up in the measurements. Again, using multiple data points in a video measurment with built-in software filtering routines is imperative in the design of an automatic video inspection system if precision is to be expected. Of course, it is recommended that the user of such a system clean parts to be measured as well as possible but it is not reasonable to require that all parts be perfectly clean in order to maintain precision measurements.

Alignment and Fixturing Non-Repeatability (δ_a)

The primary function of the user of a video inspection system is to ensure that the part (or camera if that is moved) does not wobble or slop around while measurements are being taken. Sloppy fixturing or not fixing the part securely to the (moving) X-Y stage is an instant non-repeatability. For the manufacturers contribution, software alignment routines should be provided such that the fixturing requirements are not elaborate and expensive. Compensation for part misalignment using a "cosine correction" should be standard and easy to use. Systems should correct for stage motion to allow parts to be presented at almost any angle with respect to the axis of the measuring system.

Lighting Non-Repeatability (δ_L)

Again the principle of "repeated conditions yields repeatable results" applies. The subject of lighting, especially when surface illumination is present, is a very complex one involving the direction and spectral distribution of incident and reflected light, type of surface being measured, type of light, sensitivity of the camera, and other technical considerations. Thus, a change in lighting conditions may result in a different picture as seen by the video measuring instrument. It is the responsibility of the manufacturer to provide the means to allow the user to be able to repeat lighting conditions from part-to-part. This includes light positioning fixtures which can easily be set to previous conditions and intensity controls and meters to allow repeatable incident illumination. Additionally, the vision system should allow for a reasonable variation and still repeat measurements. The user must provide a uniform voltage to the light source in order to avoid extreme variations in light intensity. Many times lights such as room lights should be shielded so as not to present unwanted reflections on parts being measured.

Other non-Repeatabilities (δ_o)

Included here is everything not previously mentioned such as electronic noise in the image, poor handling of the equipment, etc. Suffice is to say that common sense and proper treatment of the equipment should minimize any external event causing problems.

RESOLUTION LIMITED NON-REPEATABILITY (δ_r)

Stage Non-Repeatability (δ_r, Stage)

Digital position encoders (such as glass scales) typically have a basic resolution (R_s) of from .0005" to as low as .000005". Unfortunately, the cost goes up as the resolution goes down. Another less costly resolution approach is a stepper motor system driven open loop. Stepper motor technology has improved such that step resolution (and hence stage position) can be known to within 1/50,000 of a revolution. Thus, combining this capability with a fine pitch leadscrew yields a stage resolution near .000001". For example, a 32 pitch leadscrew with 25,000 steps per revolution gives a stage digital resolution of .00000125".

For measurements requiring extreme precision, the same statistical improvements can be accomplished by using several stage moves for a single measurement as in the previous example of measuring the 4" stick. However, stage moves take the longest time and are usually minimized in order to maintain throughput on a system. Nonetheless, the non-repeatability contribution of the position resolution of the stage can be minimized by taking multiple measurements and averaging results:

$$\delta_{r;\ \text{Stage}} = \frac{3R_s}{\sqrt{N_s}}$$

Where N_s = number of stage positions used in computing the desired measurement.

Video Non-Repeatability (δ_R, Video)

The primary advantage of precision measurement using video (other than the non-contact aspect) is its speed of measurement. Today's standard cameras are capable of scanning up to 60 pictures per second. In each picture up to 1,000,000 pixels may be present with up to 256 gray scale intensities per pixel. These values give the astronomical figure of 15,360,000,000 bits of information per second. While most systems don't process information this fast, numbers like 640,000,000 bits per second is not uncommon. Thus, it is quite possible to play the statistics game with very little cost in speed (as is not the case in reducing effective stage resolution.) In general, a precision video system must utilize this capability in order to achieve its optimum performance. The reduction in video non-repeatability follows the general resolution formula:

$$* \quad \delta_{R,\ \text{video}} = \frac{3R_v}{\sqrt{N_v}}$$

Where N_v = number of pixels used in computing the desired measurement.

* Figure two is an actual example of this effect.

Mechanical Non-Repeatability (δ_m)

Mechanical non-repeatability is perhaps the most difficult and costly to minimize. It involves the size of stages, quality of staging, coupling leadscrews, assembly, materials, and anything else that may produce slop in the system stiffness. It is usually the limiting factor in precision (and hence accuracy) of measurements involving parts larger than the optical field-of-view. The range of mechanical non-repeatability is from a few millionths of an inch for extremely tight systems to a few thousandths of an inch for loose ones. Periodic maintenance and calibration is recommended to minimize this value.

SYSTEM ACCURACY

In many applications, precision alone is all that is required to satisfy the requirements. Included are applications in production control where a satisfactory part exists and the only requirement is to measure deviations from the acceptable part. No one cares what the size of the part is but they are vitally interested if it deviates from the standard. Another application is in automatic alignment or registration of robotics. The concern here is the part location and precision is important in order to guide the robot to its assigned task. However, a large category of automatic inspection problems involves comparing manufactured parts with their associated drawings and prints. This usually involves absolute dimensions requiring extremely accurate measurements with little room for error. The usual rule of thumb is that the measuring instrument should be repeatably accurate to within one tenth of the tolerance of the dimensions to be measured. In today's manufacturing world it is not uncommon to find tolerances of a few ten thousandths of an inch or even less. This requires not only the system non-repeatability be extremely low, but the answer produced by the measuring device be "correct". In the case of a measuring device, the "correct" answer means that it can somehow be traced to a standard that represents "truth."

It stands to follow that if the precision characteristics of a measuring system are known, then a method can be devised to translate its repeatable (though not necessarily accurate) results into accurate ones. This is possible with video measuring systems by utilizing the software capabilities of the system along with a calibration standard or certified master part. By simply measuring the master part, comparing the measured results with those of "truth" according to the standard, enough information is obtained to correct any measurement subsequently made. Note that this calibration procedure can be performed as often as desired since in an automatic system it takes little time. The added benefit to frequent calibration is that non-repeatable effects which are the result of time (such as temperature effects) can be eliminated. Also, confidence in results is obtained by measuring the standard before and after measurements in question are performed.

CONCLUSIONS

It is possible to evaluate and estimate the precision of an automatic video inspection system using the principles previously discussed. Accuracy can then be calibrated into the instrument via software correction factors as determined by measuring a standard part that represents "truth". It is the responsibility of the manufacturer of precision video inspection and measuring equipment to honestly evaluate and specify repeatability and accuracy of the systems he sells. It is the duty of the potential buyer to thoroughly evaluate the system before he buys with regard to its stated capabilities. It is the responsiblity of both the supplier and user to minimize all possible non-repeatabilities that affect the quality of measurements used to evaluate the quality of parts being inspected.

CHAPTER 4

MEASUREMENT TECHNIQUES

Reprinted from *Machine and Tool Blue Book*, December 1984

Vision System Senses Needs of Sheetmetal Fabrication

Robotics, controllers, microprocessors . . . they're all extending the capabilities of laser-based inspection systems. Now, positioning, guidance and calibration are part of the package.

By VICTOR J. WOLANSKI, Manager, Diffracto Ltd., Grosse Pointe Woods, Mich.

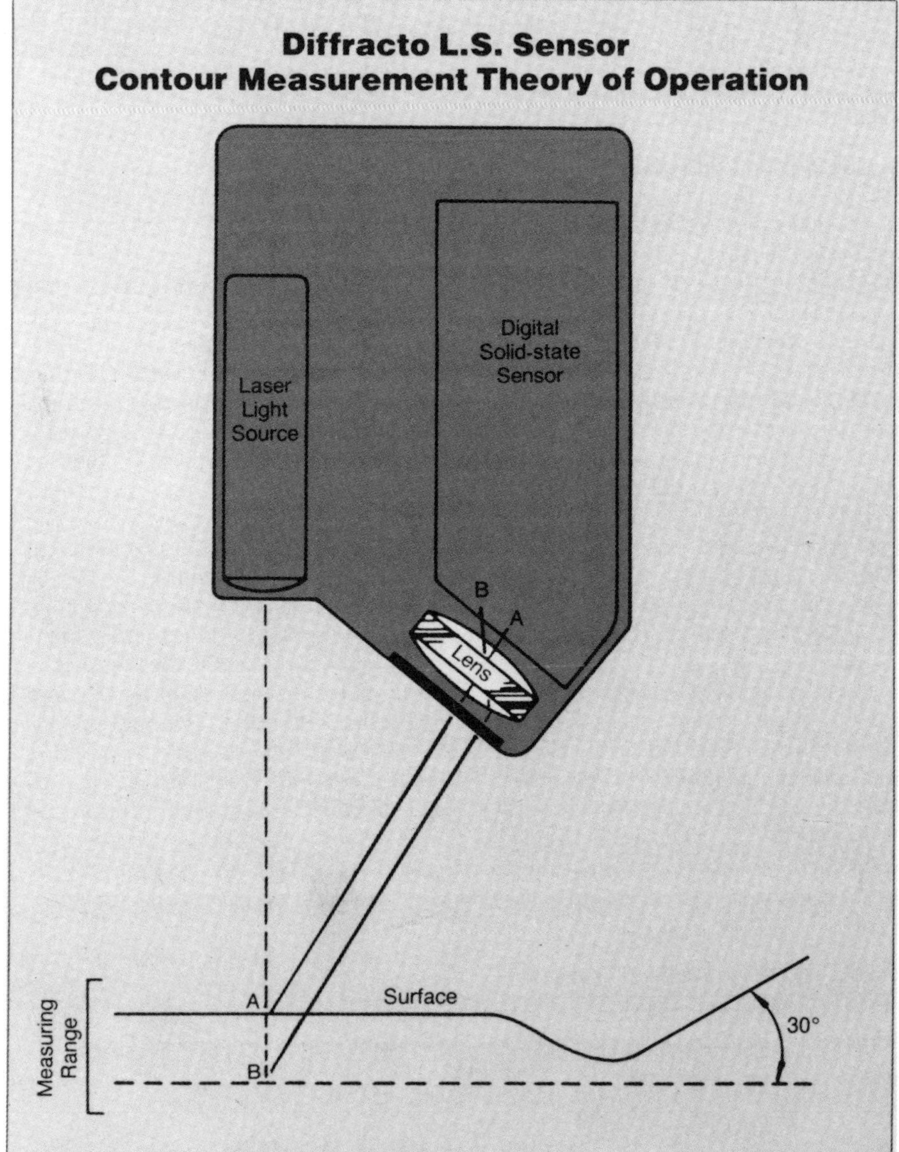

Vision sensors employ a triangulation method of measurement by sending a laser spot of light out to the part surface. As light strikes the surface, a lens images the point of intersection onto a photodetector.

Today, the use of vision techniques for precision measurement of parts is growing at a rapid pace and in a variety of applications. The method, be it laser-based, white light, or continuous or strobed operation, depends on the part to be inspected. Correspondingly different detection methods are employed to suit the application requirements, whether single element photodetectors or two-dimensional matrix array cameras are used.

New applications demand a high level of measurement capability. One such area, for example, is sheetmetal fabrication where alignment and fit of body parts is critical.

Vision sensors mounted on a measuring robot can now perform automatic calculations on sheetmetal stampings for gap and flushness, hole diameters and edge locations in a fraction of the time it took in the past with manual or ring gage methods. In order to know which sensors to implement for a given application, an understanding of how state-of-the-art sensors operate would be beneficial.

The vision sensor itself employs a laser-based triangulation method of measurement for ranging. This is accomplished by sending a finely focused laser spot of light out to the part surface at a nominal standoff. As light strikes this surface, a lens in the sensor images this point of intersection onto a solid-state, position-sensitive photodetector.

Any subsequent deviation from the initial mastered point can be measured accurately. The degree of accuracy varies depending on other performance requirements such as standoff and range. Typically, as range requirements increase, accuracy tends to decrease, therefore multiple-sensor types are designed to perform within various application tolerances.

In many instances, a hole diameter or slot is present on the part and needs to be measured, or an edge location for referencing must be determined. To extend the performance capability further, feature measurements can also be accomplished with the same sensor.

The internal light source portion

of the sensor illuminates an area surrounding the feature to be measured. Imaging then projects that hole diameter or edge onto the photodetector for measurement, allowing the sensor to perform both the ranging and imaging measurements at typical data rates of 200 readings per second.

Main advantages of these vision sensors are:

- Noncontact measurement allows no distortion of the part nor wear on the sensor.
- High-speed operation reduces conventional inspection time.
- Dual measurements are accomplished by one sensor.
- Optical magnification allows typical tolerances in sheetmetal applications to be easily achieved.
- Self-contained sensor includes light sources and detector in a small lightweight package.

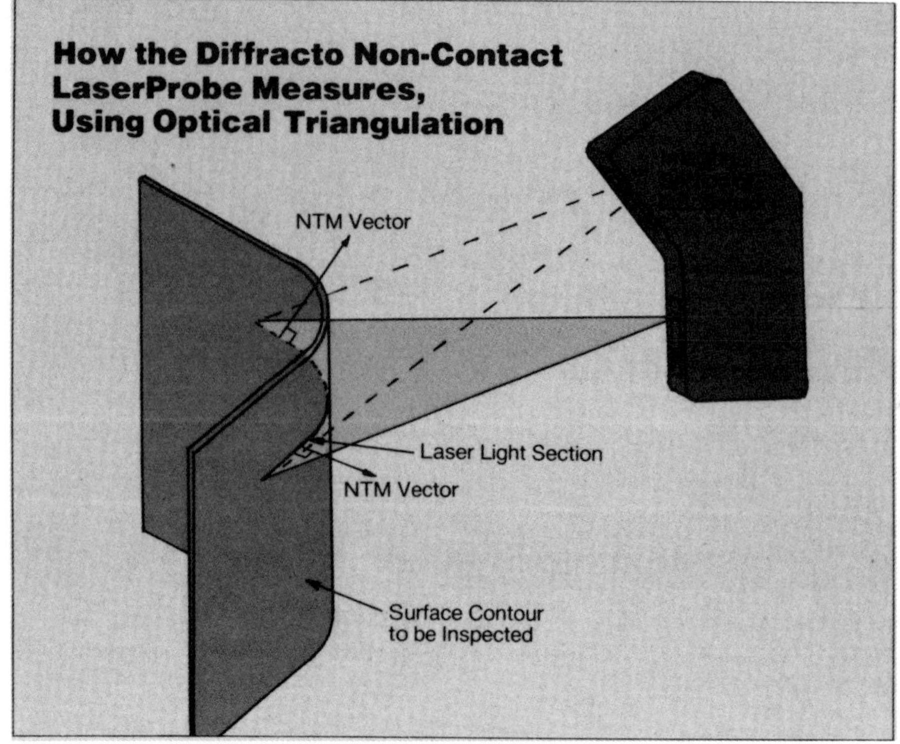

In contour measuring of sheetmetal surfaces, the sensor projects a line of light onto the contoured part. When viewed by the system, the image of the line takes on the shape of the surface.

Adapting the Technology to Production

Under laboratory conditions, the workpiece, surface, ambient light and many other factors are ideal. This invariably does not extend to the real-life situations that exist in an industrial manufacturing or assembly environment. Special features, therefore, must be incorporated into the sensor to compensate for the harsh factory environment for variable surface conditions.

The LaserProbe series of sensors manufactured by Diffracto for industrial use have built-in devices which allow for variations in surface quality and ambient light level. Two factors which affect the light intensity include surface reflectivity and color. If the workpiece or surface to be inspected is reflective, there is little light for the imaging system. The sensor will determine whether or not the intensity is sufficient and automatically implement a gain control feedback, allowing parts having varying surface conditions, roughness and color to be brought into the gaging station and be automatically inspected without operator adjustments.

In order to inspect a large volume part such as a sheetmetal panel, fender, door or hood, programmable motion must be introduced either to the part or the sensor. RoboGage 642, a measuring robot system developed by the same firm, includes a four-axis robot arm to manipulate the LaserProbe sensors about the fixtured workpiece. A large volume measurement up to 6 by 4 by 2 feet is achieved by stepping through preprogrammed points about the part and determining at each location whether or not the part is within design tolerances. In addition, the edge location, slot presence or hole diameter can be determined at any or all of the gaging points.

All of the axes of the robot are encoded so that the position of the sensors are always known. In addition, at any point throughout its gaging path, a self-calibrating position can be achieved by programming travel to a known master station where recalibration is performed, if necessary. This allows continuous updating of accuracy and repeatability.

Flexibility: Key to Long-Term Use

As systems become more sophisticated with a multitude of parameters for each application, costs soar! Justification for expenditures on such systems becomes more difficult and flexibility becomes crucial. However, with the aid of microprocessors and programmable controllers, a new age of "smart" machines will allow diversification within part families and total flexibility for reprogrammable measurements due to engineering or design changeover. Instead of a two-year payback on a dedicated unit, a flexible system can extend the return over a four- to six-year term, even longer in other cases.

State-of-the-art systems can now perform multiple parameter measurements alternately on symmetrically opposite parts such as left and right fenders, doors or hinge pillars. A single fixture capable of holding both a left and right part can be indexed into the gaging station by means of a rotary table. An operator or automatic transfer system will be able to load one part while the other is being measured.

In the case of the programmable RoboGage, the sensors and main measurement system are reusable on different part types. The fixture is

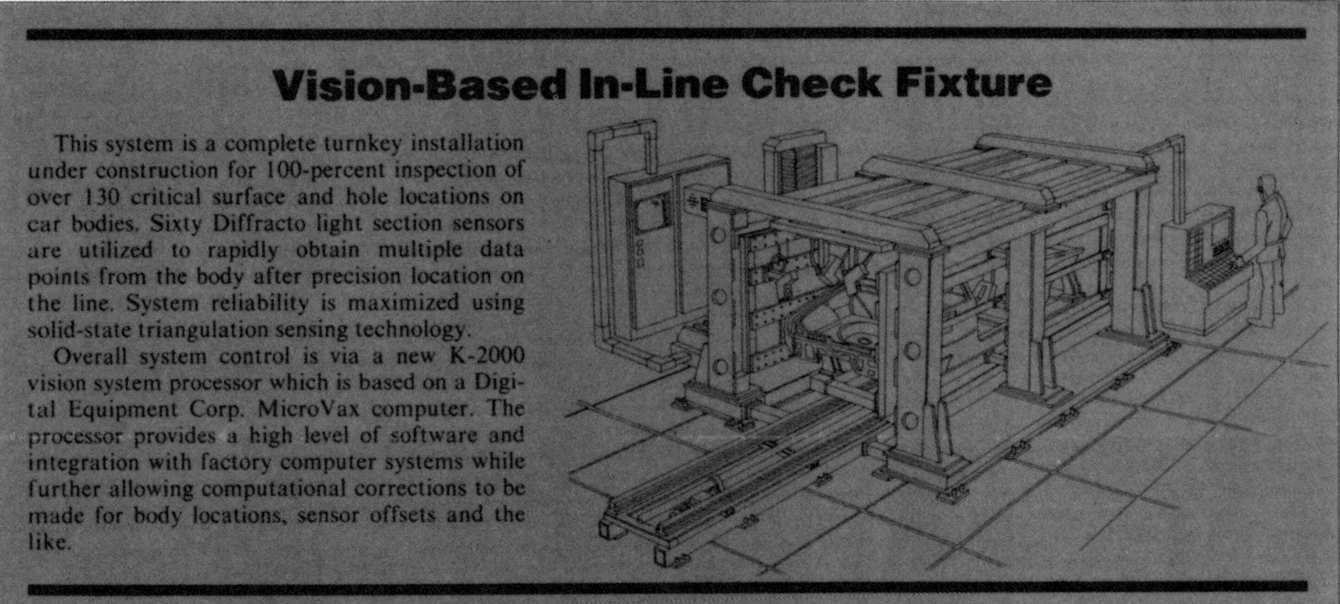

Vision-Based In-Line Check Fixture

This system is a complete turnkey installation under construction for 100-percent inspection of over 130 critical surface and hole locations on car bodies. Sixty Diffracto light section sensors are utilized to rapidly obtain multiple data points from the body after precision location on the line. System reliability is maximized using solid-state triangulation sensing technology.

Overall system control is via a new K-2000 vision system processor which is based on a Digital Equipment Corp. MicroVax computer. The processor provides a high level of software and integration with factory computer systems while further allowing computational corrections to be made for body locations, sensor offsets and the like.

replaced to accommodate the new workpieces but at only a fraction of the cost of the total system.

The teach mode of the robot allows the operator to program all desired measurement points into the controller, or a theoretical math database may be downloaded into the microprocessor from a host computer. Measurements are then performed automatically. Inspection rates are typically three to four minutes per part including fixturing, sensing, processing and reporting. This time is a major reduction from the traditional methods of ring gage measurements which take up to four hours per part. This also changes the quality control process to a much higher percentage audit capability.

There are other advantages as a result of this in-line gaging. One advantage is the increased volume of piece part inspection. In addition, earlier detection of die wear or malfunction is observed, resulting in fewer scrap or rework pieces. This is a cost factor not always attributed to the justification of the measurement installation.

Contouring

Previous discussion included applications where motion was introduced to the sensor by means of a robot to measure integrity on a stationary part in a point-to-point inspection mode. In the sheetmetal area, another application requires the measurement of features and contours on larger subassemblies such as a car body shell. For this situation, special laser-based sensors are installed into a fixture which surrounds the entire structure. Multiple-sensors are grouped and oriented so that many individual checkpoints over the complete body are measured in one stationary position.

To measure the contours, L.S. (light section) sensors extend the use of optical triangulation from a single point to a structured light section measurement. Illumination and imaging for both hole and edge identification and contour inspection is packaged into a single industrial unit. Automatic compensation for varying surface conditions and dual measurement of both range and hole, slot or edge gaging are maintained.

Instead of a single point, the sensor now projects a line of light onto an angled or contoured surface. When viewed, the image of the line takes on the shape of the surface and a measurement of that contour can be made. Determination of the normal-to-metal (NTM) vectors, the radius of curvature and the distance from the apex to the sensor (range) are achieved in one measurement. Optionally, the capability of locating edges, slots or hole diameters is accomplished by strobed white light front-surface illumination contained in the same sensor. The microprocessor translates the digital data from the sensor into engineering units and can generate a measurement report, provide statistical analysis or interface with a host computer.

When integrated into multiples, these L.S. sensors now provide in-line measurement capabilities of sheetmetal alignment, mating and fit of welded parts, and correct locations and diameters of holes and slots for further assembly. Quality control is no longer an audit of one assembly per 100, but rather 100-percent inspection.

Over the Long Term

These sheetmetal applications are only a few of the many new opportunities of improving quality by means of vision. The noncontact sensor advantage aided by the microprocessor for reprogrammability make vision systems flexible for a variety of measurements over a long term. As in past years, where parts were designed to accommodate hard tool jigs and fixtures for gaging, the new part families can be engineered to incorporate vision sensing techniques, with only repositioning and reprogramming, instead of new tooling and machining. Flexibility and 100-percent quality control are becoming the state-of-the-art trademarks in manufacturing, today and in the future. **BB**

Reprinted from *Lasers & Applications*, August 1984

Laser Gauging Goes Three-Dimensional

Lasers Flourish in Industrial Gauging as Interest in Depth Measurement Grows

by Lawrence A. Murray and Cuong Chu

Much of the excitement in robotics a few years ago concerned machine vision — the use of "intelligent" digital logic and image processing to size and sort parts on a production line. Excluding a few specialized holographic measurement tasks, that early flurry of activity in machine vision was mostly two-dimensional and didn't, for the most part, require laser illumination. But recently, the general question of 3D viewing of industrial parts and the more specific application of lasers to that task have received a second look.

Renewed Interest in 3D

The great majority of machine-vision companies and prospective machine-vision users agree that the measurement of the third dimension of targets, objects, and scenes is a prime near-term objective for the vision industry. Researchers in leading machine-vision companies believe that, once a fast and precise 3D method is at hand, the demand for such capability will avalanche.

The market thrust is not so clear. Thus the field appears technology-driven rather than market-driven. Historical wisdom suggests that the road to commercial success must be rocky and filled with detours and meanders. Also, technology-driven market segments tend to arrive at market success somewhat later than market-driven segments. While it's too early for definitive statements about this market's driving forces, some of the detours, successes, and meanders are included in this article. History will decide which are which.

Today's machine-vision modules, in the main, extract information about the objects placed in front of them by examining 2D silhouettes. This simply is not adequate for many applications unless object orientation is controlled in both ϕ and θ directions. Even then, such objects may not touch or overlap. For present-day production lines, which only recently have faced up to programmable automata, this situation is acceptable and tolerable. But as the experience and expectation levels of industry grow, so must the demands on vision technology.

This overview, besides addressing laser-based 3D, looks at some of the market areas calling for 3D vision and the other technologies picked to meet their needs. Table 1 orders

Larry Murray is director of advanced technology and Cuong Chu is an engineering assistant at Vuebotics Corp., Carlsbad CA. This article is extracted from *Machine Vision for Industrial Control and Robotics* to be published by Marcel Dekker in late 1984.

Table 1 — Markets and Technologies of 3D Measurements

Market	Technology
Arc Welding	Adaptive Optics
Automotive	Algorithmic
Bin Picking	Gray Level
Gauging/Control	Phase Modulation (Laser)
Military	Stereo
Mobile Robot	Structured Light (Laser)
Timber	Time of Flight (Laser)

markets and technologies alphabetically. Any good matchups among markets and technologies on the same line are purely accidental.

Driving the Demand for 3D

The automotive market, at present, gets the most attention. This is not surprising, because it was the automotive industry (as represented by General Motors in its CONSIGHT program) that selected structured light as the correct application-oriented technique. Thus automotive needs have unconsciously placed a market-driven boundary condition on 3D vision.

Conversely, the technology-driven boundary condition fell out of university research on algorithms to simulate artificial intelligence. But the bin picking application most often used to justify 3D research seemingly stems more from university supposition than from in-place industrial need.

A third factor driving 3D development comes from military support, where a prime application is enemy tank detection. Time-of-flight ranging is a chosen technology. Many observers believe that this technique and variations on it offer the speed and accuracy required in industrial applications. But before describing it, an overview of other techniques, their successes, limitations, and potential, is in order.

Structured Light

As noted earlier, the impetus for this technological development came from the automotive industry. It did not originate from any need to measure the third dimension but rather stemmed from the need to find an ambient-light-independent measurement system. The auto industry offers an enormous diversity of inspection tasks and environments for performing them. The measurement tasks cover dimensions of body panels (Figure 1), transmission labyrinths, door hinges, spot welds, pings on the body, hybrid circuit patterns, threads on bolts, ignition contactor placement, molded pipe burrs, and serial numbers on radiators. Ambient lighting is extremely diverse in the auto industry, ranging from the dim blue haze of a stamping plant to the wild brightness excursions in an assembly/arc welding plant.

Early structured light systems aimed merely at illuminating the edges of the object to produce high contrast. The bulk of all machine-vision systems are binary, so that it is only the edges that are seen, anyway.

A typical structured light system aims a line (plane) of laser light across a conveyor belt. The plane is generated either with a mechanical scan or by spreading the laser spot with a cylindrical lens. As the part to be inspected moves down this belt, it intercepts the laser line. The reasons for using laser illumination are generally twofold: it is much more intense than ambient lighting, and its monochromaticity allows broadband background filtering.

A line-scan camera is positioned to stare at the laser line. The movement of a 3D object through the laser line displaces the line toward the rear, since the line sits above the reference plane. The displacement distance is a function of the height of the target at the point of interception and the angle of incidence of the beam. As a result, a sharp discontinuity in brightness occurs at the edge of the part and is seen by the staring line-scan camera.

Most vision systems are binary: they see any light level above a preset brightness threshold as white and any level below it as black. In addition, binary vision looks only to locate transitions between black and white (i.e. edges). The combination of laser structured light with edge-detection vision processing (as in the casting inspection system shown in Figure 2) overpowers any ambient lighting conditions that can be reasonably postulated, static or dynamic.

Clearly, measurement of the third dimension was not a driving force behind this development. Note, too, that 3D information may be extracted if the edges *and* the locus of the entire laser line are analyzed as the target passes by. Figure 3 depicts the White Scanner produced by Technical Arts. The distance between the intercepted and the expected position is a measure of the height of the target above the reference plane. Resolution is determined by the width of the laser line and how much it spreads (due to focusing optics) as it climbs up the target. A second limitation to the technique is that the 3D information cannot be extracted and processed until the entire object has passed through the light plane. Thus, large quantities of pixel data must be temporarily buffered or compressed.

Many variations of this structured light concept have been advanced, using different structures such as dual lines, grids, a matrix of points, or holographic projections. None of these fully address the need to make measurements in real time.

Table 2, which lists the various machine-vision companies involved in 3D measurements, shows how structured light dominates implementation planning. Of the companies involved in structured light, Applied Scanning Technology and Technical Arts service the timber industry, while RVSI and Sensit serve the military community. The others all look toward the automotive market. At this point, laser ranging by time of flight (listed in Table 1) is not being pursued industrially, only militarily.

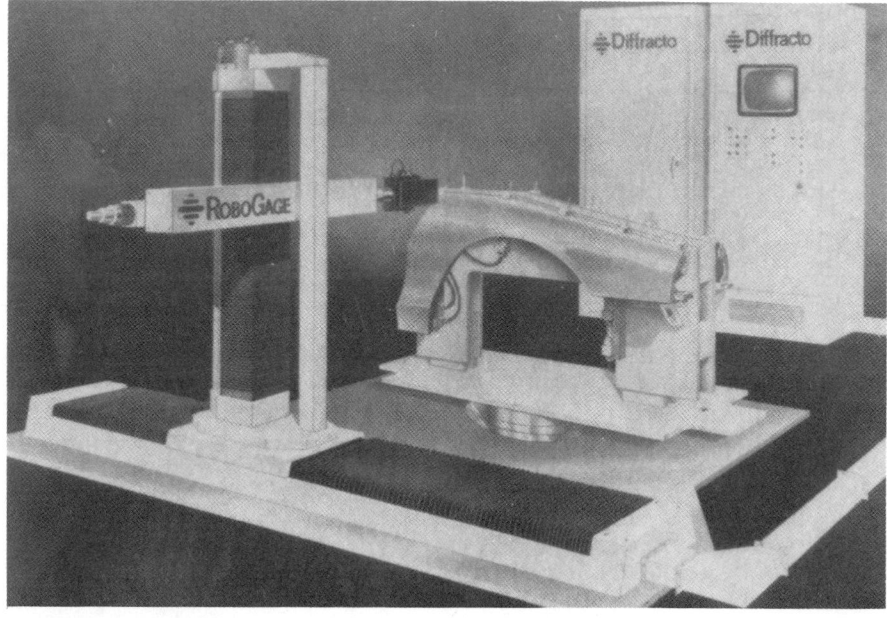

Figure 1. Laser-based robotic system developed by Diffracto Ltd., Windsor, Ontario, inspects auto fenders on an assembly line.

Table 2 does not take into consideration all the 3D research being performed by leading universities or government labs (e.g., Carnegie-Mellon University, the Jet Propulsion Laboratory, University of Rhode Island, MIT, Stanford University, University of Maryland, and New York University) or by manufacturers not competing in the machine-vision market directly (e.g., IBM and Westinghouse). However, a new structured light variation from NYU merits mention. Here, Ratio In-Depth Sensing (RID), which does not necessarily use a laser, views the entire scene in two frame times. White light is structured by passing it through a graded filter whose absorption characteristic varies monotonically from edge to edge. A camera, staring perpendicular to the incident light, sees an illumination field that varies monotonically in intensity down the target. By measuring the ratio of intensities reflected from the target, with and without the filter, depths can be determined.

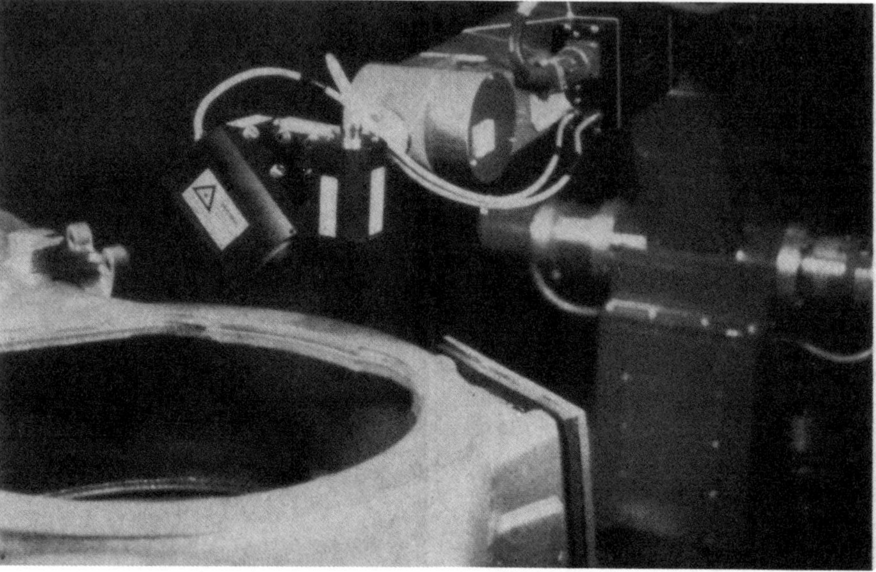

Figure 2. This robotic inspection system manufactured by Selcom, Valdese NC, uses a laser probe to monitor machine castings in a steel factory.

Stereo Vision

This obvious approach stems from our human-based method of depth perception. It views the target from two directions using two closely spaced cameras. The position of each of the two target pixels of each point, referenced to the charge-coupled device image chips, is then used to calculate the distance of those target points from the camera. The technique, so easily stated, involves complex binocular matching algorithms which are quite robust. Even so, problems with occlusion, shadowing, optical illusions, out-of-focus smearing, and specular reflections abound. The time required to process the information is usually quite long.

But for programmable robotic applications where cycle times exceed 6 seconds and where production volumes are low, image processing speed is not a driving force. International Robomation/Intelligence (IR/I), which uses stereo vision, is a robot manufacturer; in this case stereo fits speed requirements. In an application-specific assignment, Machine Vision International uses 16 pairs of stereo cameras for robot vision gauging.

Enormous efforts in algorithmic research have been undertaken to understand the workings of human vision. After all, the eye/brain combination uses a bagful of tricks to reduce the amount of information it has to process for viewing even simple objects. The system is iterative, adaptive, intercommunicative, slow, and error-prone.

The latter two attributes of the human vision system raise a pragmatic question: to what extent *should* machine vision emulate man? On the one hand, structuring machine-vision systems as analogs to human vision suggests, ultimately, systems with fairly common hardware and some sort of near-universal software, interchangeable from task to task. This would require massive software development efforts, with the expectation of ultimately being able to sell many, many systems. On the other hand, the industry could develop application-oriented hardware and techniques that avoid human limitations. But the sales volume for any given hardware system would be much smaller. Various mixtures of the two approaches can also be readily envisioned. (Those who struggle with the architectures of systems exhibiting artificial intelligence face a similar predicament.)

Gray Level 3D

Since the intensity of light reflected from an object into a camera is a function of the distance from the object to the camera, the amount of light received could be a measure of this distance. But it generally can be considered only as a rough measurement, because it requires that lighting be uniform and the target's reflectivity be invariant. If these conditions are met, the user can infer depth data and differentiate one plane of a target from another. Other variations on gray-level extraction deduce shape from shading or shadowing.

The algorithms used to implement gray-scale 3D are based on complex software. Hence, inspection of parts using the technique tends to be slow (Figure 4). The speed limitation of the software is only one of the reasons that R&D attention is now being directed more toward tray picking than bin picking in some application areas. There, the electronic separation/recognition and differentiation of touching and overlapped objects within the field of view receives primary scrutiny. Early approaches using gray scale to preprocess the image before extracting the data have been encouraging.

Dynamic Focus

The technique of dynamic focus is sometimes called adaptive optics to be consistent with, but perhaps more broadly based than, the thrust of several government programs toward adaptive control. It takes many forms. View Engineering extracts depth information from movement of its x-y-z stage. In the Honeywell-Visitronic approach, the scene is imaged onto a set of 24 microlensed photodetector pairs. When the scene is in focus, the intensity signatures from the two sets of 24 photodetectors match. As the lens focus is moved, the focal distance at which the match takes place may be used to determine the target's distance.

Generally, this technique provides global information. For more specific applications, such as in microscopic inspection, the sensor may scan an integrated circuit chip to measure step heights of metal lines

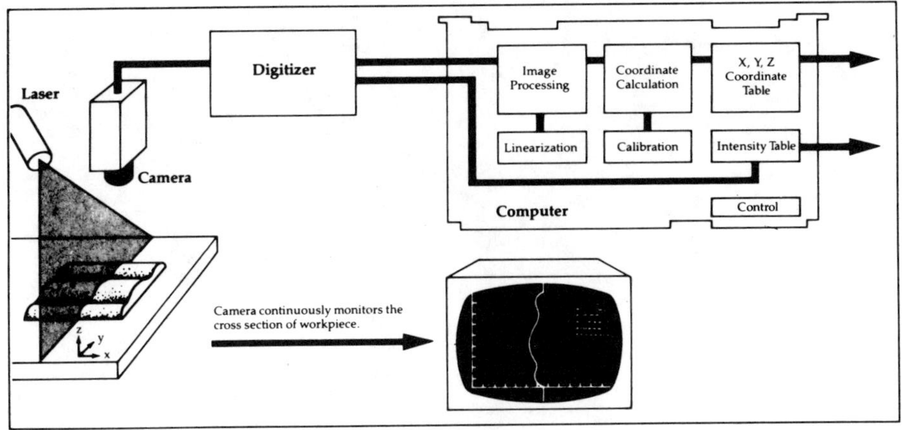

Figure 3. The White Scanner from Technical Arts, Seattle WA, calculates from 1,000 to 3,000 x-y-z coordinates per second as the object under test traverses the production line.

or contact hole depths, provided they encompass a major portion of the viewed field. In this form of adaptive focus, the scene itself determines the point of focus.

An innovative optics-based method developed by Associated & Ferran provides a lens system with an extremely narrow depth of field. At each focal length, the scene is sliced up into sharply focused 2D images. These images may then be individually analyzed, electronically reassembled, and recognized as individual 3D objects. Locations may then be transferred to the associated automata for whatever sort of processing is needed.

Machine Vision is Imperative

Three of the aforementioned techniques — range in depth structured light, gray scale, and adaptive optics — slice the 3D scene into a quasi-continuum of unconnected 2D images. Each sensor system generates a slice per frame time, which approaches the requirement for real-time sensing. However, in most software-intensive computer-based vision systems, image analysis time is prohibitive. High-speed, real-time image analyses need to be developed. These would be able to take all the annular silhouette objects in a 2D frame, connect them electronically into 3D objects, and then recognize or inspect them against the stored base of memorized objects.

The next and more complex step would occur if the objects in view had not been placed down in some standard orientation. Then it would be necessary to rotate the stored data until a memorized object conforms with the target image. This is impossible, at present, with pure 2D technology. For simpler tasks, such as collision avoidance in mobile robot situations, where only location data are needed, the real-time image processing problem is not so severe — at least not until the objects (e.g., humans) start to move.

In all three of these techniques, resolution is determined by the camera's resolvable increment of gray levels, by the number of gray levels, or by the narrowness of the depth of field. The number of resolvable steps is limited. In the discussion of the next method, sideband-modulated laser ranging, the resolution limitation is not so strict. But the same real-time image processing considerations still pertain.

Laser Ranging

Laser ranging is usually done on a time-of-flight basis, which implies that the speed of light is the measure of the distance of the detector/source and the object. Since the speed of light is approximately 3×10^{10} centimeters per second, the roundtrip time to cover a single centimeter would be 67 picoseconds. This time interval is much shorter than can be measured with most available photodetectors, which typically have risetimes on the order of a nanosecond. At that, an accuracy of 1 cm does not meet the needs of the industrial inspection/location world. Moreover, 15 cm is actually a more realistic resolution limit in pure time-of-flight measurements.

If, however, the light source were modulated at, say, 100 megahertz, the cycle length would be 300 cm. Phase detection schemes can improve resolution to perhaps 1 part in 1,000, so the limit becomes 0.3 cm. Even though a hundred times better, this accuracy still does not come to grips with industrial application needs.

A further refinement based on electronic theodolite manufacture was recently suggested. If the 100 MHz signal were heterodyned with a 10 kHz single sideband, a time expansion factor (TEF) of 10,000 would be obtained. Now the 67 ps of the first example becomes 670 ns. Using a *pin* photodiode with a 1-ns risetime, the accuracy limit (z_{min}) becomes

$$\begin{aligned} z_{min} &= c \times t_r/TEF \\ &= 3 \times 10^{10} \times 10^{-9}/10^{4} \\ &= 30 \text{ micrometers.} \end{aligned}$$

where c is the speed of light [cm/s] and t_r is the risetime of the photodiode [s]. This distance resolution begins to make industrial sense.

Alas, with only one photodetector involved, we have only gathered range information for a single pixel. For an industrial image, as viewed by a CCD, many tens of thousands of pixels are worthy of targeting. They each need their distance dimension extracted for image analysis prior to locating, gauging, inspecting, or identifying the target.

Table 2 — The Chosen 3D Technologies of Machine-Vision Companies

Structured Light
Applied Intelligence Systems
Applied Scanning Technology
Autoflex
Automatix
Diffracto
Perceptron
Robot Vision Systems
Sensit
Technical Arts
Dynamic Focus
Associated & Ferran
Honeywell
View Engineering
Stereo
Automatix
Int'l Robomation/Intelligence
Machine Vision Int'l
Gray Level
General Electric
Object Recognition Systems
Single-Point Sensing
Selcom/ASEA
Autech

In this table, single-point sensing is defined as a system that sends out a split beam of light over two different paths and triangulates distance from spot separations.

Even with vector scanning and data compression, thousands of pixels require study. A laser scanner would direct the spot to the appropriate points, but to gather good signal-to-noise ratios, the dwell time of the laser would have to be a substantial fraction of a millisecond. Thousands of milliseconds inexorably become seconds, which is not real time. The system conceivably could be speeded up by using linear arrays of photodetectors and lasers.

As an interim solution, the ranger can be aimed to look only at particular windowed areas for inspecting and gauging critical areas. This can be accomplished by using the initial 2D image to determine part orientation and place windows. Alternatively, the part can be placed in front of a preset electronic window using hard fixturing.

The best solution would be to develop a solidstate imager with individually addressable, high-speed pixels. These would be capable of datalogging the phase delays of the crossings of the modulated packets of charge during a single frame time (16 ms). Each x,y pixel in the array would then have its z coordinate known. Development of such an imager is at best futuristic.

Actual laser ranging work is still in the laboratory stage, as at the Environmental Research Institute of Michigan. Prototype units use mode-locked helium-neon and solidstate lasers acousto-optically modulated at 3 MHz, which produces ambiguity intervals in the 2 to 8 meter range. The laserbeams are scanned over the scene by two rotating polygon mirrors. Future developments include a 3D scanner for a walking machine and plans for decreasing the ambiguity interval to make the technique more applicable to industrial needs. Similar technology is also being applied to aerospace programs by General Dynamics, McDonnell- Douglas, and Perkin-Elmer.

Figure 4. An Automated Propeller Optical Measurement System (APOMS), shown here with a mock up propeller, uses lasers and detector arrays to measure large parts in three dimensions. (Photo courtesy of Robotic Vision Systems Inc., Hauppauge NY.)

Overview

A great body of research, development, and product work concerning 3D technology has surfaced in recent years. But the drives of the various kinds of R&D organizations are very different. Universities aim their efforts at artificial intelligence and at understanding the workings of the human body. The military looks to gather offensive weapon placement and verification data. The government funds both enterprises.

The young, resource-limited machine-vision industry looks to established R&D results to service the needs of its industrial customer base. Because of the diversity of R&D sources and the diversity of the machine-vision industry's entrepreneurial engineer base, approaches to 3D technology abound. At best, each approach serves a specific need of a specific industrial customer. For a new 3D job, then, another technological approach may be selected or generated. Only laser structured light has found general use, but it is used more for edge contrast than for extracting depth data.

The successful industrial 3D technology of the future will probably have to satisfy three constraints simultaneously: real-time 3D image analysis, high accuracy, and capacity to study large numbers of pixels.

The first constraint calls for improvements in the architecture of machine-vision systems. The quickest route to advancement along these lines will likely come from a hardwired, modular-board approach. The second constraint calls for the development of a new sensor system, preferably one that can transmit its data off-chip for real-time image analysis. This does not exclude sensors that acquire data through acoustic or tactile means, but optical sensing is generally a cleaner approach. The third constraint refers not only to sensor chips but to the number of pixels in the target whose z dimension will be measured.

If machine-vision developers who believe in the marketability of 3D vision are correct, 3D will become a reality in the next few years despite the obstacles and plethora of approaches now being put forth. □

Reprinted from *Electro-Optical Systems Design*, May 1982

Linear Measurement with Photodiode Arrays

Non-contact (electro-optical) measurements in industrial processes have become accepted techniques in everything from sorting objects on a high-speed conveyor belt to making a detailed high-precision inspection of a large diesel engine. The humble photodiode, arranged in a linear array with many of its twins, is the key to some of these techniques.

Albert R. Tebo
Associate Editor

Industry is always looking for more efficient, faster, and cheaper ways of testing and inspecting its products for acceptability and quality control. Any means that reduces human handling of the products or eliminates the touching of damageable surfaces with mechanical instruments will find favor with industry. So it is that electro-optical techniques have become popular and are being developed in many directions. This is attested to by the clinic on "Electro-Optical Gaging and Inspection Techniques", sponsored by the Society of Manufacturing Engineers in Detroit on March 9-11, 1982.

In any electro-optical technique of testing or measurement, light from a source—whether it be a laser, a light-emitting diode, or a tungsten-halogen lamp—must be reflected from, or interrupted by, the test object. The detection of this reflection or interruption is conveniently done by a photodiode. By using several or many photodiodes in an array, the detection process can be extended in space to give a length measurement, or extended in time to give a sequence measurement, or both.

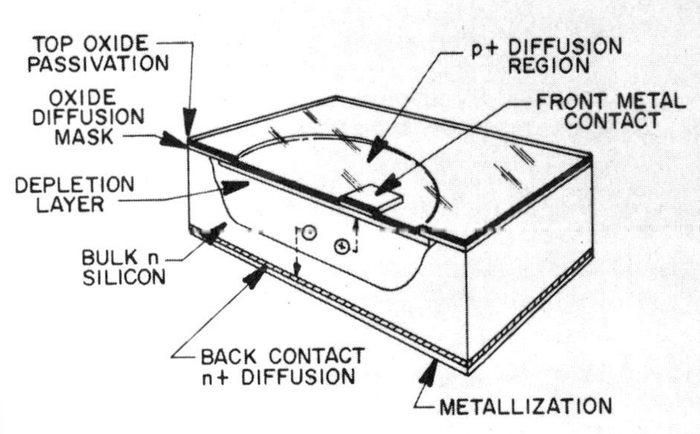

How Does a Photodiode Operate?

To set the stage, let's look briefly at how a photodiode functions, using a silicon pn type as our example. See **Fig. 1.** The diode is fabricated from an n-type wafer with n^+ Si diffused on the back surface (bottom in the figure) for attaching a metal contact. A p^+ diffusion on the front surface (top in the figure) defines the light-sensitive area. Metal contacts are evaporated on front and back surfaces for lead attachments.

Fig. 1. Construction of a pn photodiode, fabricated from an n-type wafer with a p^+ diffusion on the front surface (top in the figure). (Adapted from "The fine points of sensing light". by Tony Green and John M. Tait, Centronic, Inc., in Machine Design, Sept. 11, 1980)

The depletion layer is very thin if no voltage is applied to the diode. Free electrons and holes attracted out of the region by the n and p^+ regions create a strong electric field across the region. A reverse bias voltage (positive on the n-type Si) applied to the diode widens the depletion region.

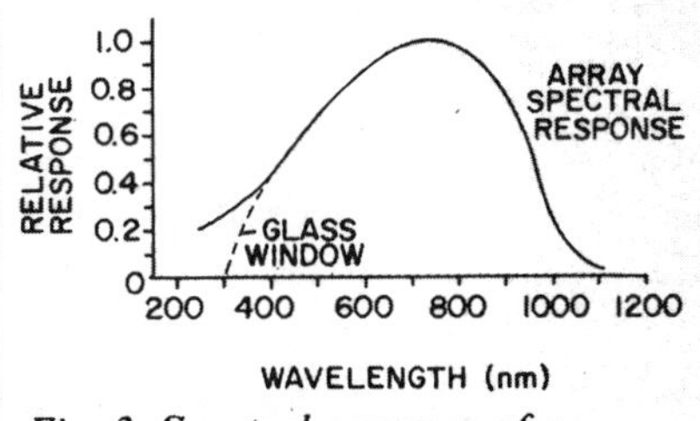

Fig. 2. Spectral response of a quartz-windowed silicon photodiode array. A glass window has the same response except at the uv end, shown by the dashed line. (Courtesy of EG&G Reticon)

Light of wavelength shorter than 1150 nm striking the diode surface creates electron-hole pairs. Such charge carriers formed in the depletion region are separated by the electric field and drift to their respectively oppositely-charged sides of the device, producing a photo-current.

If a sufficient reverse bias voltage is applied to the device, the depletion region will extend to the n^+ diffusion region, enabling a much greater current to reach the electrodes, and increasing the speed of response of the photodiode to the incident light.

Photons of light of wavelength longer than 1150 nm do not have enough energy to overcome the bandgap of silicon, and thus will not create electron-hole pairs. However, photons of wavelength shorter than 1150 nm can do so. In fact, photons of wavelengths throughout the visible and into the uv will be effective in contributing to the photodiode cur-

Fig. 3. Photodiode array designed as a bar-code analyzer. (Courtesy of Silicon Sensors, Inc.)

rent, although not all with the same effectiveness. The peak of the spectral response occurs at about 850 nm.

When the photodiode is enclosed in a window material the overall spectral response can shift somewhat. **Fig. 2** shows the spectral response of an array of Si photodiodes. Quartz and glass windows have similar responses, but the glass window cuts off in the uv at about 300 nm.

In setting up an electro-optical test or measurement system it is necessary to select a light source whose spectral output matches reasonably well the response of the detector. Since silicon covers the whole visible and the very near ir, selection of a light source is not generally a problem. Some photodiodes are made of other materials having spectral responses quite different from that of Si, such as GaAsP, but these are only a small minority, so we shall not consider them further.

The sensitivity of a photodiode chip is proportional to the active area of the chip—a feature that argues for choosing as large a chip as possible for a given application. However, the dark (leakage) current—that which flows in the absence of light when reverse bias voltage is applied—increases with the area of the chip. So a tradeoff must be made. Additionally, if the application requires the discrimination of minute spatial features, by using an array of photodiodes, small chips are likely to be necessary.

Who Uses Photodiode Arrays?

In trying to gather material for this article, I found that many manufacturers of electro-optical measuring equipment were reluctant to divulge details of the techniques used, because of proprietary rights. The brochures that they publish simply mention that their system uses a laser (or whatever light source) and an electro-optical sensor.

Furthermore, many manufacturers of photodiode arrays do not indicate the specific applications in which their customers use them. Most of their work is custom service, and details are just not available.

So, I shall talk about arrays and applications somewhat separately, in an attempt to cover the wide range of applications that do exist.

A punch-card reader or tape reader is a common application of photodiode arrays, using just a few diodes. **Fig. 3** shows an array used as a bar code analyzer, and **Fig. 4** shows an area array that reads a whole 12 × 80 position computer card at one time. None of these is a measurement application, but they do indicate some of the configurations of arrays.

Who Makes Arrays?

EG&G Reticon seems to be one of the largest producers of photodiode arrays, having a catalog of several categories of standard products. The family of linear photodiode arrays (line scanners) made by EG&G Reticon includes devices having from 64 to 2048 elements. These devices are enclosed in an integrated-circuit type package with a ground and polished window. On a single silicon chip they contain the row of photodiodes each with an associated capacitance and an MOS multiplex switch. Also contained on the chip is either a digital shift register or an analog CCD shift register for serial readout. In addition, some devices

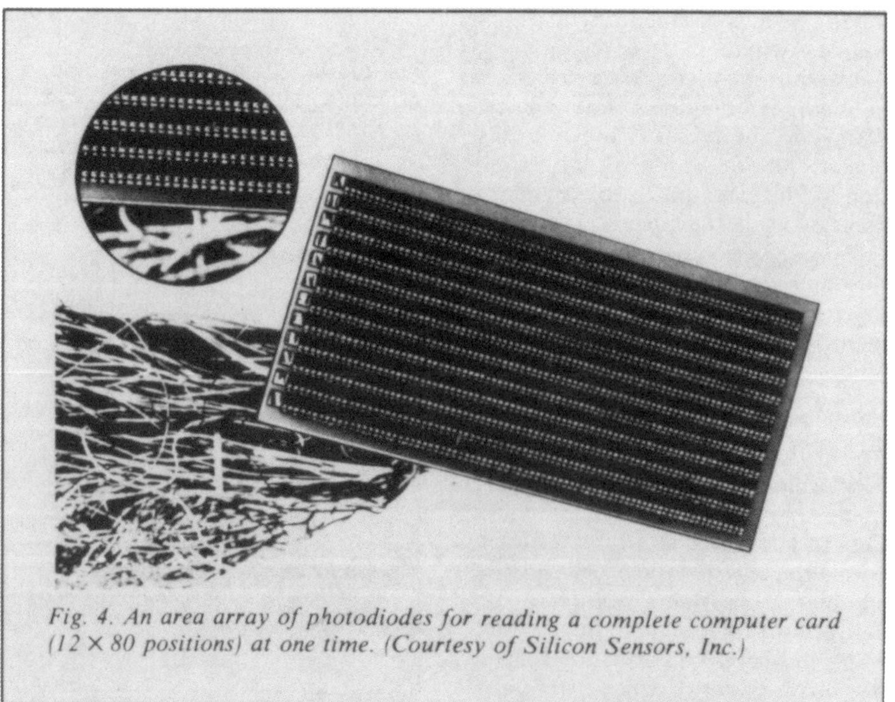

Fig. 4. An area array of photodiodes for reading a complete computer card (12 × 80 positions) at one time. (Courtesy of Silicon Sensors, Inc.)

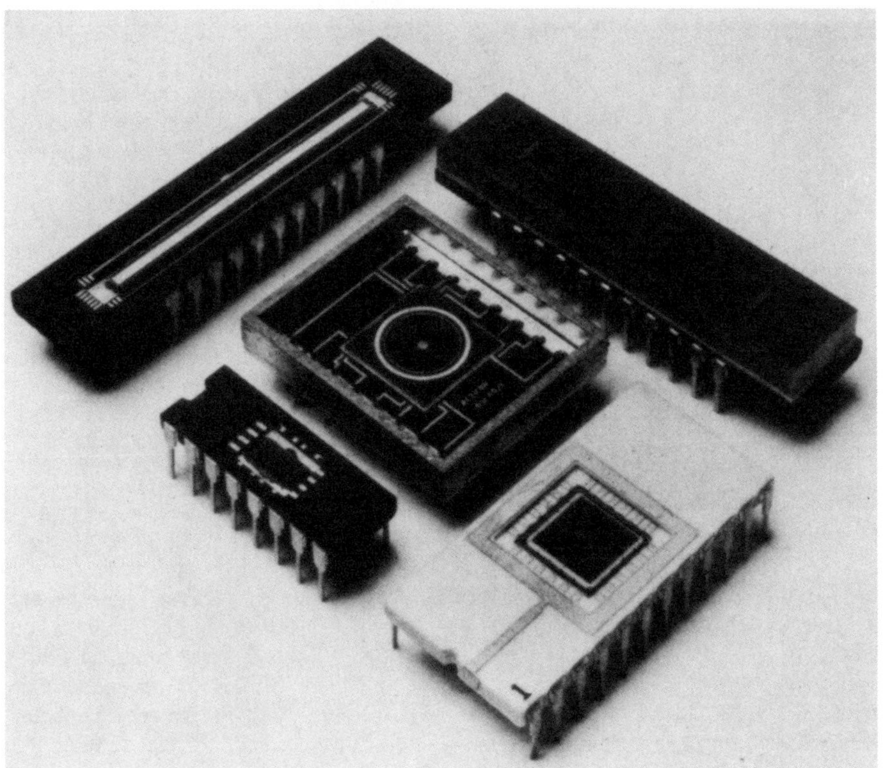

Fig. 5. Samples of silicon photodiode arrays: line, matrix (area), and circular. (Courtesy of EG&G Reticon)

also contain internal drive and amplifier circuits. The sensing elements are diffused pn junction photodiodes for high quantum efficiency and full silicon spectral response. Light incident on these diodes is converted to electric charge, which is integrated and stored on the diode capacitance until readout. The output signal is a discrete time-analog representation of the spatial distribution of light intensity across the array. Reticon devices operate over a wide range of scan rates, and speed can be traded directly for sensitivity because the lower the scan rate the longer the diodes can integrate charge.

Of the ten series of linear arrays listed in Reticon's product summary, two series are mentioned for non-contact measurement. The C series are available with 256, 512, or 1024 elements on one mil centers, and with 17 mil aperture width (width of sensing area). Sampling rates up to 10 MHz are possible. The EC series are identical in design, but only alternate diodes are sampled. Thus they are available with 128, 256 or 512 elements on two mil centers and 17 mil apertures. Sampling rates up to 5 MHz are possible.

Reticon also offers several types of area arrays and circular arrays. **Fig. 5** shows samples of these types.

Centronic, Inc., lists in their product summary photodiode arrays having 2, 4, 9, 12, 24, and 100 elements, with center-to-center spacing varying from 1.0 to 2.7 mm.

Hamamatsu Corp. lists an array of 32 elements with an element size of 0.585 × 1.3 mm.

Applied Solar Energy Corp. offers arrays of 2 to 33 elements, with element size in the latter of 0.125″ × 0.014″ and inter-element spacing of 0.004″. Custom arrays are offered up to 1000 elements, and they will make active areas of almost any shape and size from a few mils to strips three feet long, with inter-element spacing down to a few ten thousandths of an inch.

Measurement Applications

Photodiode arrays are used in applications that range from the prosaic to the far out. Dr. Robert Buckley of Hughes Industrial Products Div. says that their applications have included production-line type situations, such as handling machine parts, pins, and bolts. At the other end of the scale they have built a system for the quality control of hypodermic needles, using a HeNe laser to illuminate the tip of the needle. A 64-element photodiode array analyzes the two-dimensional diffraction pattern produced from the tip.

In any measurement application, the electro-optical system will include a light source, the detector array, some sort of controller to interpret the signals from the detector array, and some sort of handling system for acting on the decisions made by the controller. **Fig. 6** is a sketch of this kind of setup.

Among the production-line type of measurements are the diameter of fibers, wires, tubes, and rods; the edge position and centering of webs, belts, plates, ribbons, and strips; similarly the width and centering of the same types of items. These are mentioned by Reticon.

Mr. Robert Bible, Sr., of Automation Engineering, Inc., describes four systems that his company has developed. One measures top and bottom squareness, free length and spring rate of small springs, done at the rate of 1000 springs per hour. The optical setup, shown in **Fig. 7,** uses an object-telecentric system to provide a

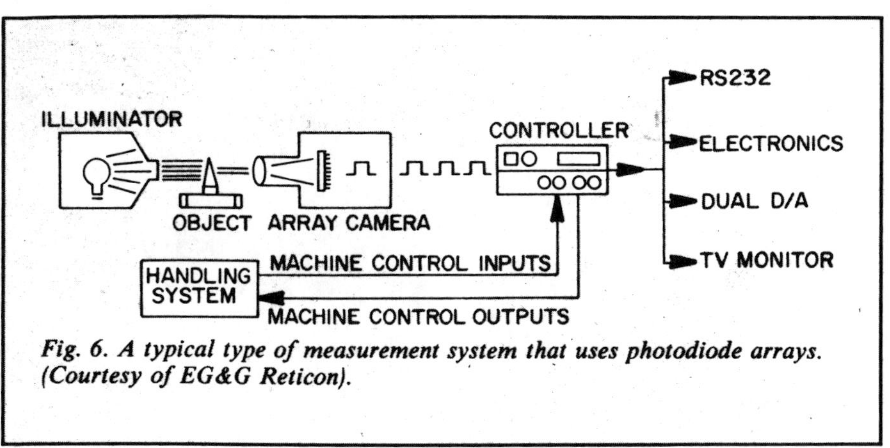

Fig. 6. A typical type of measurement system that uses photodiode arrays. (Courtesy of EG&G Reticon).

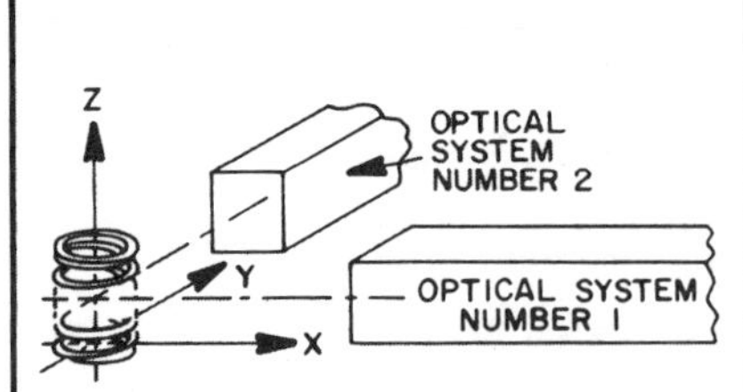

Fig. 7. Dual optical systems, using photodiode arrays, enable rapid measurement of the dimensions of small springs. (Courtesy of Robert E. Bible, Sr., Automation Engineering, Inc.)

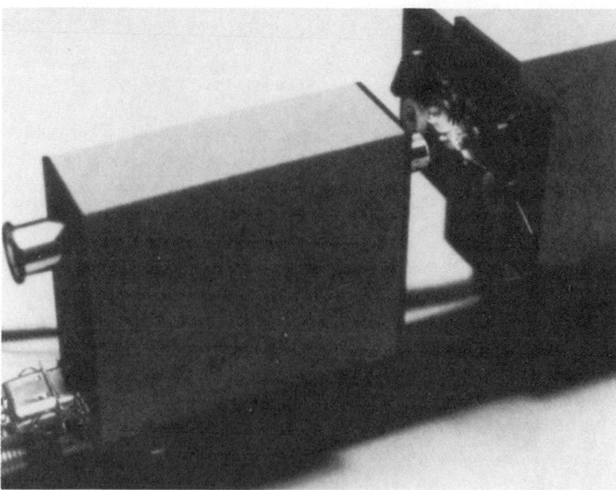

Fig. 8. This system measures wire diameters rapidly to a high degree of resolution. The line-scan camera is on the left; the flasher (light source) is on the right; the wire-feeding mechanism is between them. (Courtesy of Spectron Engineering, Inc.)

sharp two-dimensional image of the spring, which is more than 0.5″ deep. The photodiode detector array consists of 512 diodes spaced 0.001″ apart. The spring is backlit by a halogen incandescent lamp. The two optical assemblies make geometrically independent measurements.

The other three measurement systems using photodiode arrays are a small-parts measuring machine, a large-bore measuring machine, and an injector barrel machine. In each of these, 0.0001″ resolution is achieved.

Spectron Engineering, Inc., manufactures a high-resolution diameter-measuring system for wire or optical fiber. The electro-optical heart of this system is the CE410 electronic line-scan camera, shown as the box in the left portion of **Fig. 8.** On the right portion of the figure is the flasher unit, and between it and the camera is the wire-feeding mechanism. The camera is based on a 256-element self-scanned photodiode array, with telecentric optics. All elements of the array integrate simultaneously so that the image is acquired entirely during the duration of the flash.

Fig. 9 is a display of the signals produced by the array, showing the vertical line of diodes lit equally. The indentation toward the left outlines the diameter of the wire, where the diodes receive diminished signal or no signal.

The same CE410 line-scan camera is used in another instrument, the electronically-scanned processing reflection microdensitometer, in which the usual microdensitometer function is supplemented with an area analysis of densities over the test object.

Another instrument produced by Spectron is an automatic thread inspection system for threaded parts in a variety of sizes. The system tests pitch diameter, major diameter, and the length of four threads, all at a speed of 150 parts per minute.

Diffracto, Ltd., makes profile dimensional sensors for the measurement of high-tolerance piece parts, with resolution to 0.3 μm using driftless photodiode arrays. These sensors are available with both continuous and pulsed (stroboscopic) light source modules for high speed measurement. Up to several thousand readings per second can be taken, on the fly.

One of Diffracto's systems (a group of three single and three double packages) inspects up to 50 different automotive intake and exhaust valves for 12 dimensions including overall length, head thickness and keeper groove location, at a rate of

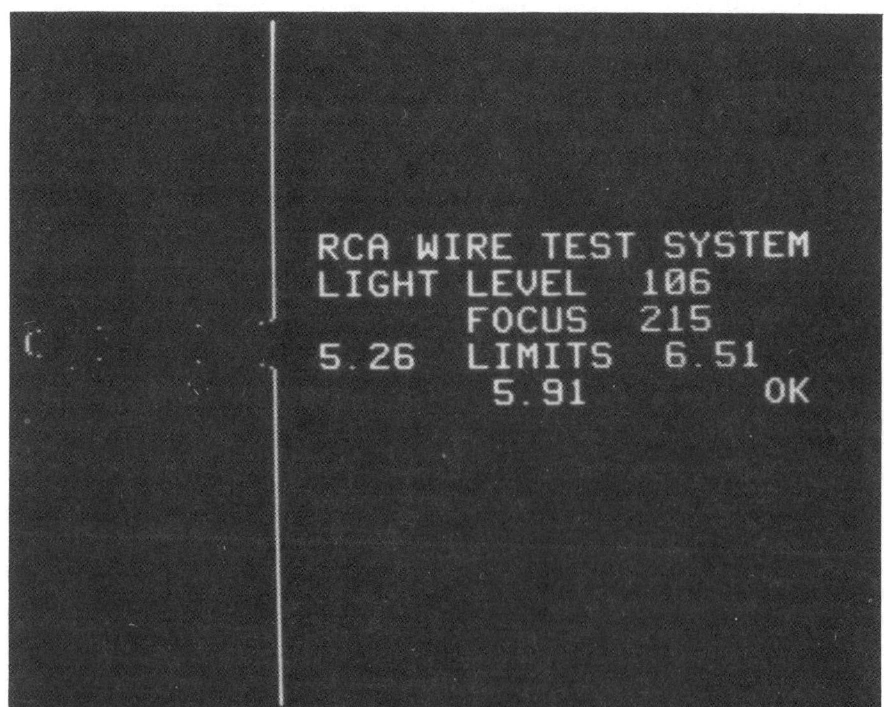

Fig. 9. The system in Fig. 8 produces this display of signals. The 256 photodiodes register in the vertical line, broken by the opaque wire whose diameter is delineated by the indentation to the left. (Courtesy of Spectron Engineering, Inc.)

4200 parts per hour. Resolution is 0.5 μm. **Fig. 10** shows this complete system. The parts enter from the left, are indexed through the sensing stations and proceed to the multiple reject gates.

Another of Diffracto's systems is the LaserProbe®, used to measure dimensions or displacements of objects to sub-micrometer resolution over a millimeter range (on some models). Based on the principle of optical triangulation, the units operate at speeds up to 1000 readings per second, using a HeNe laser light source, and achieving a resolution of 0.3 μm and a repeatability of 1.0 μm. In one application a fully automatic machine provides the contours of each of the four combustion chambers in an automotive engine head in sequence, and from the contours computes the chamber volume to an accuracy of 0.1 cm^3.

Fig. 10. A complete system designed to inspect up to 50 different automotive intake and exhaust valves for 12 dimensions, at a rate of 4200 parts per hour. (Courtesy of Diffracto Ltd.)

Conclusion

From this sketchy survey we can see that the photodiode array, in a variety of sizes and configurations, can be used with several different kinds of light sources to make accurate and high speed measurements in a multitude of applications. □

Reprinted from *Manufacturing Engineering*, Februrary 1986

Laser-based metrology and inspection systems—a look at their strengths, weaknesses, and basic principles

Measuring with Lasers

BY GREGORY T. FARNUM

Sparks. That's what usually comes to mind when the subject of lasers in industry is mentioned. Sparks shooting from a piece of metal that is being cut, drilled, or marked by a high-powered laser. But lasers are also making their mark on industry in a less dramatic way, as factory-ready systems using low-power lasers increase the speed and accuracy of noncontact measurement and inspection.

Of course, the laser can no longer be considered a new device. The first one was invented nearly 30 years ago. Older still are some of the optical procedures used in laser-based measurement. Increasingly, however, lasers are being linked with a variety of techniques and equipment, both old and new, in an effort to provide alternatives to traditional methods of gaging, part inspection and analysis, and machine tool calibration and alignment.

Close-up of an M550 scanning laser beam system (Anritsu America, Inc., Oakland, NJ) measuring the outside diameter of drawn glass fiber.

Fringes and frequencies

One of the earliest and most common metrological uses of low-power lasers has been in interferometry. The interferometer, a device that can determine distance and thickness by measuring wavelengths, has been around for a long time. The widely used Michelson interferometer, for instance, was invented in 1882, and a description of how it measures can help illustrate interferometry's basic principles.

First, a beam splitter divides a beam of light into a measurement beam and a reference beam. The measurement beam travels to a reflector mounted on the part whose distance is to be measured, while the reference beam is directed at a fixed reflector. Both are reflected back to the beam splitter where they are recombined into a single beam before traveling to their ultimate destination, a photodetector. This recombined beam has a high or low density, depending on whether the waves of the two returning beams are in phase (called constructive interference) or out of phase (termed destructive interference). In-phase waves produce a series of bright bands on the photodetector, while out-of-phase waves produce dark bands. These alternating bright and dark bands are called interference fringes. By making visible extremely minute differences in the lengths of the two beams' optical paths, the fringes allow distances to be measured in terms of wavelengths of light.

The achievement of interference fringes is largely dependent on the coherence of the light used. The availability of highly coherent laser light has made interferometry practical in far less restrictive environments than in the past, and the sometimes arduous task of extracting usable data from a close-packed series of interference fringes has been taken over by microprocessors. As a result, a process that had been largely confined to the optical industry and the metrology lab is now suitable for the factory, where its extremely precise distance measuring capabilities have been applied to the alignment and calibration of machine tools.

Interferometry, however, isn't simply a matter of bright and dark bands. "When people think of an interferometer," says Duane Christy, "they tend to think of the Michelson interferometer—the traditional fringe-counting device that they are familiar with from freshman physics class. The Hewlett-Packard laser interferometer doesn't count fringes at all." Christy is product manager of electronic products for Federal Products Corp. (Providence, RI), US distributors of the Hewlett-Packard 5528A, a widely used laser interferometer machine tool calibration system.

With the Hewlett-Packard system, the laser beam is Zeeman-split into two beams of different frequency and polarization. When the beams are recombined, any relative motion between the optics creates a Doppler shift in the frequency. This shift is then converted into a distance measurement. Hewlett-Packard refers to these systems as two-frequency interferometers.

The company's first two-frequency interferometer calibration system was introduced in 1970, largely in response to the problem of ambient contamination.

Christy explains: "Fringe counting interferometers can be affected by air turbulence, oil mist, and other conditions encountered in the workplace that affect the intensity of light. The success of the Hewlett-Packard systems," he adds, "stems largely from the fact that they are relatively insensitive to mist and turbulence."

Both sides of the fence

Gordon Bald, vice president of Diversified Manufacturing, Inc. (Lockport, NY) and a Hewlett-Packard laser interferometer user, has a rather unusual perspective on the use of lasers to align and calibrate machine tools.

Five years ago, Bald invited Hamar Laser Instruments, Inc. (Wilton, CT) into his plant to demonstrate the capabilities of its new Model 711 helium-neon laser alignment system. Helium-neon, a relatively safe, low-power laser that emits a visible red light at a wavelength of 0.6328 μm, is the laser most widely used in measurement.

"The alignment of machine tools by the use of levels, squares, tightwires, and that sort of thing—the way it's been done for the last 50-75 years—has been a difficult process," says Bald. "When I saw how effective Hamar's helium-neon laser was in aligning some of our machine tools, I recognized that this was the best thing for machine tool alignment I'd ever seen, and I've been in business for more than 35 years."

Enthusiastic over the possibilities of the new technology, Bald purchased a Model 711 and formed a second company, Laser Line, Inc., operating from his existing plant. He augmented his original purchase with a Hamar Model 800, designed primarily for spindle alignment, and a Hewlett-Packard laser interferometer.

"For our first couple of years," Bald explains, "getting people to understand that the older methods of alignment could be replaced with something as sophisticated as the laser was like pulling teeth. Now, people are beginning to see the advantages of laser alignment, both in accuracy and dollars."

According to Bald, the cost advantages of laser alignment result not only from the accuracy of the process, but also from its speed. "It's amazing how fast you can align equipment with a laser and know that the equipment is at the tolerances you want. With older methods—and I speak from my experience as a manufacturer—people have often been fooled into thinking that a piece of equipment was aligned to the tolerances it was designed for when, in actuality, it was never aligned to design specifications because alignment couldn't be verified without spending undue amounts of time and energy. And time and energy," he adds, "convert into dollars."

GLOSSARY

autocollimator. A special form of telescopic sight, including a light source and partially reflecting mirror, used in checking alignment and in measuring small angular motion.

coherence. Light waves that are of a single frequency and in phase with each other are said to be coherent.

divergence. The spreading out of a light beam. A light beam diameter that does not significantly change over a long distance is said to be stable, or of low divergence.

interferometer. An instrument that divides a single beam of light into two or more components, which then travel over different paths and reunite to produce interference. Interferometers are typically used for exact measurement of distance.

monochromatic. Light of the same wavelength or of a very small range of wavelengths. **ME**

The use of laser-based metrology systems has led Bald to another discovery: "Machine tools move out of alignment much faster than was once believed. Take, for instance, a three-quarter of a million dollar machine tool, aligned accurately and resting on a thick foundation. Heretofore, it was thought that the thing would stay aligned for a year or two, but we've found that even those large, well-aligned pieces of equipment require alignment more frequently." This situation, he feels, often leads to the misdiagnosing of production problems. "I hear people say, 'the problem isn't in the alignment, because we know the machine is lined up properly—it must be in the way we're using the machine, or in the tools we're using.' But often it's not. Often it gets down to the basic geometry of alignment."

LASERS VIDEOTAPE

Laser gaging is among the industrial laser applications featured in the Winter 1985 issue of SME's MANUFACTURING INSIGHTS videotape series. Entitled "Lasers in Manufacturing," the tape begins with a brief overview of the subject, then focuses on in-plant laser operations at Allen-Bradley, Cardiac Pacemakers, Inc., Spectrum Manufacturing, Inc., and various other firms. Interviews with industry experts offering their predictions for the future of laser technology conclude the tape. For more details, contact Connie Halderman, Video Communications Dept., (313) 271-1500, Ext. 429. **ME**

It's no secret, however, that laser-based systems are more expensive than the instruments they replace. Mark Haddox, vice president, corporate engineering for Jodon, Inc. (Ann Arbor, MI), makers of helium-neon lasers and peripheral equipment, feels that the cost of these systems must be viewed in relation to their capabilities.

"The expense is not simply because of the use of lasers," says Haddox. "It's because the systems are doing more than traditional methods. For one thing, we are seeing more and more microprocessor-based instrumentation in association with the laser. Overall, users are making measurements they couldn't make before, so it's hard to make a direct comparison with the old methods in terms of cost. Certainly, if we were to freeze the technology at today's level and simply concentrate on making its use more widespread, there's no doubt that the price of these systems would go down. However, requirements tend to increase as users learn more about the technology and its capabilities. As a result, it's very hard to predict where prices will go."

Scanning

The most widely used laser technique for inspection and in-process gaging is known as laser scanning. At its most basic level, the process consists of placing an object between the source of the laser beam and a receiver containing a photodiode. A microprocessor then computes the object's dimensions based on the shadow that the object casts. Other wrinkles exist, such as the use of a structured pattern of laser light for three-dimensional inspection, but the fundamental process remains a relatively simple one.

The noncontact nature of laser scanning makes it well suited for in-process measurement, including such difficult tasks as the inspection of hot, rolled, or extruded material, and its comparative simplicity has led to the development of highly portable systems. An example of the latter is the 1200 series laser bench gage manufactured by Zygo Corp. (Middlefield, CT). Designed primarily for batch lot measurements, the unit weighs only 26 lb (12 kg), with dimensions of 9 ½ × 24 ½ × 8 ⅝" (241 × 622 × 219 mm). The bench gage measures to resolutions of 0.0001 mm.

"We produce a sizable number of

bench gages," says Ken Wendt, laser product manager for Zygo, "and the market is growing very rapidly." He estimates that growth to be approximately 50% a year and believes it will continue for several years to come. "We're not even close to market saturation," he states. "We're just touching the tip of the iceberg."

And how does Wendt explain the dramatic growth in laser bench gage sales? "It's not because we're doing a better selling job; it's because it's very accurate and because people are now able to accept laser-based gaging as an alternative to contact gaging."

But just as traditional gaging is feeling the effect of sophisticated laser-based systems, these systems in turn are faced with the even more sophisticated technology of the emerging machine vision industry. The question arises: Is laser-based gaging and inspection merely a brief stop along a road leading from traditional gaging to vision systems?

A possible answer lies in the relative sophistication of the two types of systems and the steps needed to transfer an image into usable data. With laser scanning, processing the signals received from the photodetector is a relatively simple task. Video systems, on the other hand, must process received information by breaking down each picture, or frame, into its component pixels, digitizing those pixels, then analyzing the resultant data. If data-windowing or other limitation techniques are not used, the system's computer could be used to process literally millions of bits of information in order to analyze a single frame. As a consequence, video systems are generally slower than laser-based systems. In addition, poor lighting, distance from the object to be measured, and motion must all be compensated for. The effect of these variables on laser scanning systems is, in most cases, negligible. Video systems, of course, excel at complex measurements; however, the measurements involved in most gaging and inspection applications are not complex, and for these tasks, laser scanning systems would appear to be a logical, cost-effective alternative.

Extending the reach

As spectacular as laser technology often is, for industry, the laser is merely a tool—specifically, a source of heat or light for other systems. Lasers have been coupled with optical scanning devices and interferometers, but the list doesn't stop there. The laser-equipped measuring robot produced by Diffracto, Ltd. (Windsor, Ontario) is a case in point. Called RoboGage, the horizontal arm, Cartesian robot is designed primarily for three-dimensional measurement of sheet metal stampings such as automobile fenders and hoods. The measuring envelope is 6 × 4 × 2′ (2 × 1 × 0.6 m) for the RoboGage 642 and 6 × 3 × 1′ (2 × 0.9 × 0.3 m) for the 631 model. Overall accuracy for both models is ±0.004″ (0.10 mm).

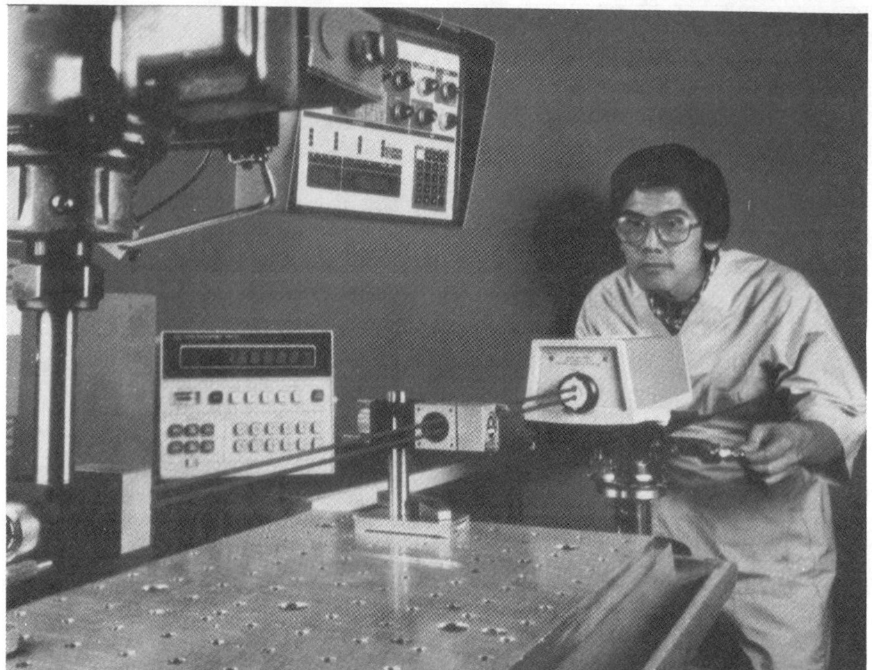

Measuring X-axis linear displacement with a Hewlett-Packard 5528A laser interferometer.

"The RoboGage isn't intended to be quite as accurate as a CMM," says Walt Pastorius, Diffracto's vice president of marketing. "Its major features are its programmability—the inspection of next year's model only requires a new database—and ruggedness. It's designed for difficult environments such as automotive stamping plants." The actual measurements are performed through optical triangulation by LaserProbes, Diffracto's miniaturized, solid-state, gallium arsenide laser sensors. The sensors, with a near infrared wavelength of 0.86 μm, weigh only 16-20 oz (454-567 g) and are designed to be mounted on existing measuring devices such as CMMs.

"Laser-equipped CMMs can get a lot more information about a part—and get it faster," says Leo Somerville, marketing and technical support manager for Renishaw, Inc. (Schaumburg, IL). "They also have the advantage of not touching the part." Renishaw, a manufacturer of metrology and inspection equipment, produces the OP2, a 4.2 oz (120 g) laser probe for CMMs. "Lasers enhance a CMM's ability to quickly scan three-dimensional contoured surfaces," he adds.

Are lasers, then, the answer for all CMM applications? Somerville thinks not. "The disadvantage of the laser systems—and of optical systems in general—is that they can only measure what they can see. Take gearboxes, for instance, or other parts with interior recesses that are difficult to illuminate; for these parts, a touch probe makes sense. Lasers also have problems with oil and corrosion on part surfaces. Consequently, for checking some parts, a laser probe would make sense, and for other parts, a touch probe would be preferred."

Holography

One of the most underutilized, and certainly the most exotic, laser-based inspection process is holography. Essentially a technique for taking three-dimensional photographs, holography was invented in the late 1940s but remained largely a scientific curiosity due to the absence of the highly monochromatic, coherent light source that the process demands. The invention of the laser stimulated renewed interest in holography, but because holographic equipment was both delicate and complicated, the process was rarely used in industry. Today, however, sturdy and simplified equipment has made holographic inspection a practical method of nondestructive testing.

The holographic technique most applicable to industry is known as holographic interferometry. Basically, it consists of superimposing a hologram of an object in an unstressed state on a second hologram of the same object in a stressed state, or on the object itself.* Heat or sound is commonly used to stress the object. If the object has not changed during the interval between the taking of the first hologram and the superimposition, the two wavefronts will coincide, but if change has occurred, the

* *An exception to this is time-averaged holographic interferometry, used chiefly in the study of vibrating surfaces. With this method, only one hologram is taken, that of the part in a stressed state.*

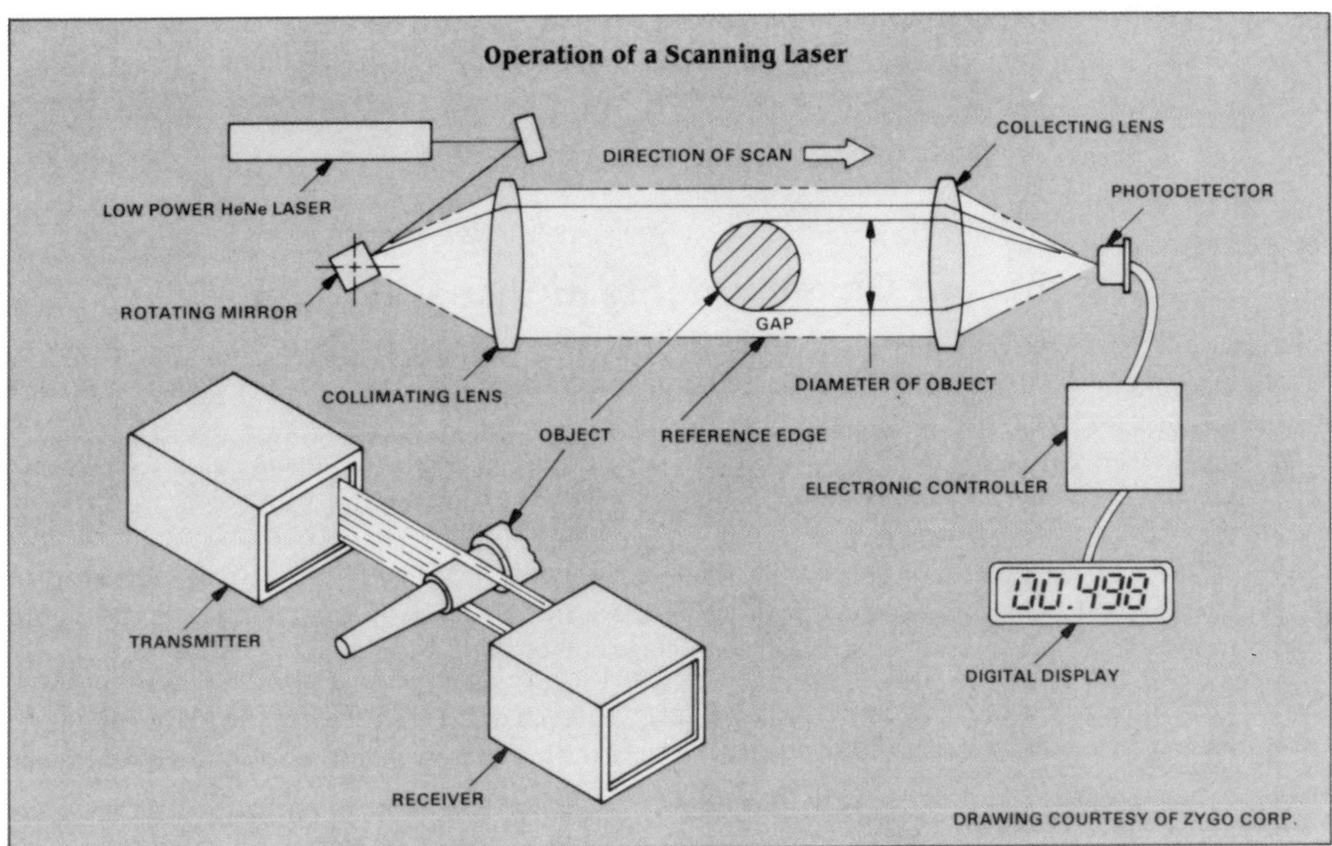

This simplified drawing illustrates how scanning laser gages operate.

two wavefronts will be at variance, and a pattern of interference fringes, resembling a topographical map of the object's surface, will result. Analysis of this fringe pattern reveals defects invisible to the naked eye and yields data on the object's strain and vibration characteristics.

Holography has been used to analyze a wide range of parts, including turbine blades, gun barrels, and computer components. Increasingly, the technique is being used in the auto industry. "Most of the major automobile manufacturers now routinely use holographic interferometry to analyze new parts that are likely to be major contributors to noise or vibration in the vehicle, or parts that are failing," says Jim Higgins, director of marketing for industrial laser products, Lumonics, Inc. (Tempe, AZ). He adds that fully assembled automobiles are sometimes examined in this way in order to study the vibration characteristics of large components such as door panels.

More than any other laser-based technique for measurement or inspection, holographic interferometry suffers from an "image problem." Many still consider it to be closer to science fiction than industrial reality. Higgins, however, believes this will change. "I think that we'll see a good many more applications in the future," he says. The reason? "It provides a source of data that isn't available any other way, particularly when looking at small or delicate structures."

Trends

Lasers offer distinct advantages in a number of metrology applications, but how widespread is their use? "It is by no means what you would call a well-established technology in terms of being widely used and understood," admits Jodon's Haddox, but he feels laser-aided metrology is following a pattern common to technological innovation in general. "As with any new technology, there will be a fairly small amount of it in application for awhile. The possible applications must first be developed, then demonstrated, in order to earn industry's trust. But once industry sees it and gains confidence in it, it eventually becomes an established technology."

Haddox believes laser-aided metrology's journey from innovation to established technology is occurring rapidly. "People have become much more aware of lasers as a metrological tool in the last few years. A lot has happened in the last two years with respect to lasers becoming more accepted, and I think it's going to go much further than it has." Gordon Bald agrees: "We're only scratching the surface. I can foresee the time when considerably more alignment tasks will be done with lasers, and lasers will be used extensively in the process control area as well."

For Duane Christy, the projected upswing in the use of laser-based systems is related to the changed manufacturing climate of the 1980s. "The recessionary period of the early '80s created a lot more interest in the elements of process control, such as the calibration and characterization of NC machining centers," he states. As examples, Christy points to the increased use of the laser interferometer as a calibration device by both machine tool builders and users. "Some builders have begun to offer, on installation, proof of accuracy using the laser interferometer, and a few major machine tool users are doing regular audits of their machine tools with it. This allows them to accurately schedule calibration and maintenance and to predict in advance the changing character of a machine.

"We see more and more small companies," Christy adds, "companies that own 5 or 10 NC machines, beginning to be aware of the added value they can get out of their machines if they can measure and characterize them accurately with the aid of lasers."

Whether or not these predictions are overly optimistic, only time will tell. It does seem plausible, however, to expect that the capabilities of laser-based measuring systems, coupled with their increasingly simplified operation and data analysis, will lead to their growing use in midsized—and perhaps even small—manufacturing operations. **ME**

Presented at the CASA/SME Sensors '85 Conference, November 1985

Recent Developments in Visual Gaging

by Pierre Trepagnier
Automatix, Inc.

This paper is about recent developments in visual gaging. Before beginning, however, it is appropriate to give the reader a sense of what is meant here by the phrase "visual gaging", which could be confusing. It is not the precision optical gaging done by old-line optics houses, such as Zeiss, nor is it interferometry. High precision techniques like these generally demand an environment more refined than the one we wish to deal with: the factory floor. The "visual gaging" discussed here uses the newly developed, and still rapidly evolving techniques of machine vision to monitor typical industrial processes. Images of parts to be gaged are acquired with a video camera, stored in a frame buffer, and analyzed using one of a number of relatively simple image analysis techniques. The output of this process is passed through some calibration function, which transforms it from the arbitrary camera space into the engineering units of interest to the user.

These visual gaging techniques are suitable for the harsh factory environment, and do not require clean rooms or temperature control. They are, however, generally not as precise as the previously mentioned measurement technologies. Precisions in the neighborhood of one part per thousand across the field of view can be achieved with relatively common solid-state array cameras (380 x 256 pixels); the resultant resolution depends on the size of the field of view chosen (or dictated). Typical gaging resolutions are in the range of 0.025mm (0.001"), at the relatively high precision end, all the way to 0.5mm (0.020") in relatively undemanding sheet metal applications.

II. OVERVIEW

Before proceeding to recent developments, it will be worthwhile to give the interested reader, who may not be active in the field, a brief review of its general state.

A. Fundamental Techniques

There are five techniques in general use to provide the information needed in visual gaging: *two-dimensional*, using a calibrated camera in a single plane; *structured light*, using an angled light beam to give triangulation in three-space; *stereo*, using two cameras to give triangulation data; *2-D plus structured light*, which combines those techniques to deal with a plane that can move; and *2-D plus autofocus*, which does the same thing by utilizing a moving, fixed-focus camera with either special hardware or software that detects when it's image is in focus.

Two-dimensional gaging is the simplest and was the first to be developed. It is shown schematically in Figure 1, and utilizes the fact that when a camera is perpendicular to the scene it is viewing, there is, in principle, a single scale factor, the magnification, which relates the scene to its camera image. This factor can be determined empirically, resulting in a calibrated camera. This method is well understood and highly effective where applicable; however, in many cases the objects of interest do not move only in a plane.

Structured light, such as the famous CONSIGHT system of General Motors[1] utilizes trigonometry to determine how far an object is away from a camera. As shown in Figure 2, an angled beam of light appears to move from side to side in the image as the object it's striking moves in and out. Structured light resolves the distance ambiguity, but at the cost of not illuminating the whole scene; only those parts of the object struck by the beam can be measured.

The topographical information given by structured light can also be used to analyze surface features that are otherwise hard to see. Figure 4, for example, shows a proposed system for measuring the position of a door header at two points. The structured light line is here analyzed to find where it drops off. (Care must be taken with large radius curves; small variations in reflectivity make it difficult to get repeatable results using this technique.

Stereo is geometrically similar identical to structured light, except that the light source of Figure 2 is replaced by another angled camera. If both cameras are looking at the same object, there is only one place where the one object responsible for both images could be. This can be calculated once the cameras have been calibrated. The chief drawback of stereo, aside from the expense of two cameras, is that there has to be an easily identifiable object that both cameras can unambiguously recognize (a "corresponding point").

Another approach to the three-dimensional problem is "2-D plus structured light". This method is useful when the features are confined to a plane, but the exact location of the plane is unknown. The situation is shown schematically in Figure 3. The location of the plane is established using structured light, and then "normal" 2-D illumination and image processing is applied once the location of the plane is determined. Either both structured light and area illumination can be present simultaneously, and separated by gray-level processing, or they can be toggled on and off and separate pictures taken. This method is made possible by full camera calibration, which allows the calculation of correct two dimensional calibration for any specified plane (including all perspective effects) on the fly. Its advantage over regular structured light is that, while the former restricts the object being gaged to whatever is illuminated by the structured light beam, the latter allows image processing over the entire field of view, with a concomitant increase in flexibility and accuracy.

Much the same idea lies behind the "2-D plus autofocus" technique. The camera hardware incorporates a commercial autofocus circuit - often used in advanced 35mm cameras - or special software to servo a fixed-focus camera which can move in its z (in/out) direction. Once the camera is in optimum focus, the position of the z stage is read out, while x-y processing proceeds normally. This technique is not subject to the depth-of-field problems associated with other 3-D methods; however, it requires complicated

---

(1) Holland, S.W. Rossol, L., Ward, M.R., "CONSIGHT-I: A Vision-Controlled Robot System for Transferring Parts from Belt Conveyors," in Computer Vision and Sensor-Based Robots, pp. 81-100, Plenum Press, New York, 1979

electrical and mechanical equipment, is relatively slow, and because of its cost is usually unsuited for multi-camera setups.

B. Illumination

Each of the techniques mentioned above is supported by a variety of illumination methods, each with its advantages and drawbacks. In general, all the "two-dimensional" techniques require a form of area lighting, while structured light requires illumination in a narrow sheet or beam.

Area lighting has been supplied by many sources, including incandescent, fluorescent, tungsten-halogen, infrared diode, and strobes; each has characteristic properties which can dictate its use in a particular situation. The choices for structured light are somewhat fewer; He-Ne lasers and IR diodes and laser diodes are often used, and strobe lamps occasionally. These sources are summarized in Table 1.

TABLE 1
Common Light Sources

Source	Advantages	Disadvantages
Incandescent	cheap, easily obtainable can be cycled off and on many configurations	low efficiency short lifetime heat
Tungsten-Halogen	high output	heat can't be cycled
Fluorescent	efficient, cool	few configurations can't be cycled flicker
He-Ne Lasers	highly collimated, infinite depth of field, simple optics	fragile glass, bulky expensive, Federal regulation, can't be modulated
IR Lasers	long lifetime, well matched to CCD sensitivity, compact, can be modulated	requires complex optics, highly divergent, Federal regulation, invisible beam
IR Diodes	same as laser diodes, no Federal regulation	low power, highly divergent, invisible beam
Gas Strobes	high power, high efficiency, freezes motion	bulky, requires high current HV, tricky to trigger, expensive, bad match to CCD sensitivity

C. Sensor Mounting Strategies

There are two basic methods to mount sensors in visual gaging: fixed or moving. Fixed sensors give an advantage in speed, and also accuracy unless the mechanical accuracy of the positioner is negligible compared to that of the optical sensor. They are generally the method of choice in on-line, 100% inspection applications where cycle time is the dominant consideration. On the other hand, they require stationary mechanical fixturing to hold the sensors in place; while the cost of the sensors can be expected to decline, that of the fixturing will not, and will represent the dominant cost of this form of gaging in the future. In addition, in cases where calibration to an overall coordinate system is necessary, a calibration or alignment fixture will have to be constructed and certified.

Moving sensors may offer several advantages - the complicated fixturing can be replaced by a robot program, simplifying part changeover, and of course the number of sensors is reduced. In high accuracy applications, with very small fields of view (e.g. 15mm) the number of fields of view required may make a fixed sensor strategy unrealistic. Moving sensors are intrinsically slower, however, and make the system dependent on either the absolute accuracy or repeatability of the mechanism, depending on the calibration strategy used (see below). Mechanical repeatabilities of less than 0.025 mm (0.001") are difficult to obtain and maintain in factory environments, as opposed to metrology labs. Moving sensor systems are predominately used in audit gaging applications, where cycle times on the order of 1 to 5 minutes are acceptable.

D. Calibration

Calibration is vital to the success of visual gaging, as without it there is no way to relate the measurements made in an arbitrary camera space by the image processing hardware and software to the world coordinate system of interest. In general, there are two calibrations, which should be kept at least conceptually distinct: the "local" calibration of the gaging module (camera[s] and light source), which produces a calibration function relating the camera image to a coordinate system which is referred to the module, and the "global" alignment of all these module-based systems into a uniform, overall system.

The local calibration of a gaging module depends only on the solid-state (usually CCD) camera and lens and the relative position of the structured light projector. When these are mounted on a common base plate to form the module, then the local calibration can be made with reference to the module alone, using its base plate (say) as reference zero, without knowing the eventual mounting position of the module. This practice is referred to as "off-line" calibration.

Figure 5 shows a schematic of an Automatix gaging module. Note how the module's base contains a dowel hole and slot which mate to the dowel pins in the mounting bracket. This feature allows any module to be mounted in a standard and repeatable way. Local calibration is performed by mounting the module in a special fixture which contains the standard mounting plate and presents calibration targets in a standard orientation with respect to the module. Thus, the local coordinate system is referenced to the dowel holes. Using least squares fitting techniques, the local calibration software then generates a "camera matrix" which transforms the sensor coordinates into this local coordinate system.

Global alignment is the process of taking all the module-based coordinate systems and combining them into the coordinate system of interest (typically, the entire part). This process can more or less tedious, depending on the needs of the application. For example, if all that is needed are relative measurements, or measurements in a "normal-to-metal" system, as is often the case in sheet metal gaging, local alignment will suffice. If, on the other hand, all measurements must be presented in a uniform, Cartesian system (e.g. the body-line system in automotive applications), proper alignment of all modules is more critical. One approach is to simply survey all modules into place using mechanical techniques, and possibly remove residual inaccuracies by comparing the results obtained on some specified part to careful measurements made of the same part on a hard gage or coordinate measuring machine (CMM). This technique is often called the "golden part" approach. Another possibility, to be described later, is optical alignment, using special alignment targets bolted onto a calibration frame. This latter method is particularly good at removing angular variations, which are otherwise difficult to deal with.

The "golden part" approach can also be used to advantage when a moving gaging module strategy is adopted. In a moving system, the local coordinate system of the module is translated about with a mechanical positioner [robot]. One method of global alignment would be to simply carefully align the local coordinate system of the module to the robot's world coordinate system, then add the robot's world position, as determined by its encoders, to the module's local coordinate result to determine the overall location in space. (This discussion has implicitly assumed a 3-axis Cartesian robot. It can be generalized in a natural way to the N-axis case; however mechanical wrists are rarely repeatable enough to allow high precision gaging.) The procedure outlined above is sensitive to the absolute error of the positioner. Small deviations can be removed to first order by comparing the results obtained on a golden part by these means with those obtained on the same part by mechanical means and subtracting off the deviations. This procedure yields answers that depend only on the robot's repeatability, rather than its accuracy.

III. SOFTWARE DEVELOPMENTS

In the past year, developments in gaging software have concentrated on increasing the functionality of systems. The two main thrusts have been to take over tasks that were once assigned to hardware, and to combine multiple tasks into a single station.

A. Constrained Rigid Body Rotation

An example of the first thrust, taking over hardware functions, is afforded by work on constrained rigid body rotation. One difficulty of 100% inspection (and hard automation in general) is that transfer mechanisms must be quite complicated in order to be dimensionally accurate. This fact, and the need to for high speed and maintainability, often make precision mechanical fixturing of a part impractical. The need to provide clearance on taper pins, for example, will often result in an additional 0.13 mm (0.005") fixturing error, while bearing run-out may add more. The end result is generally at least 0.2 mm (0.008") fixturing error, twice the typical gage resolution of 0.1 mm (0.004"). A mechanical solution to the problem would be to provide expanding mandrels for the master datum holes, or pusher cylinders to drive the part against hard stops. However, this strategy might be at best expensive to implement.

For some time, visual gaging, as well as vision-guided robotics, has used rigid body rotation to adjust for this uncertainty in part position. However, this has been usually performed as a simple calculation: three fixed points (e.g. net holes) are reconstructed on the object, they are compared to their locations when the part is in its nominal position, and a 6-dimensional offset calculated.

The situation need not be so simple, however. There may be more than 3 features, leading to an overdetermined system or, rather less felicitously, there may be no features at all. Consider Figure 6, which shows a door to be gaged in an automotive application. The square boxes arrayed around the edges depict gage points, which are measured by the 2-D plus structured light approach. The positioning mechanism has considerable play in it [0.4 mm (0.015")], so that the doors are not always placed in the same location, yet the edges being gaged are featureless sheet metal. The solution adopted was to consider a nominal body (door in this case) in its nominal position as a calibration device. When the actual door is measured, it is mathematically rotated and translated so as to match the nominal door in the nominal position "as closely as possible" in the sense of minimizing the sum of the squares of the residuals. The residuals chosen are simply the distances between the edges of the actual door in its current position and the nominal door in its nominal position. Thus, they do not depend on any fixed features. This rotated and translated position is then considered as the nominal position of the actual door. Finally, the measurements of the actual door in its nominal position are reported.

In researching the problem of fitting a rigid body motion to an overdetermined set of measurements, it was discovered that it had already been extensively studied in higher dimensional spaces by psychologists and statisticians interested in factor analysis[2] and named the "Procrustes Problem", after the mythical villain with the one-size-fits-all bed. It is conceptually similar to the problem of finding the coordinate system that diagonalizes an inertia tensor in classical mechanics, and results in an eigenvalue problem. However, relatively little work has been done in low-dimensional spaces.

In particular, the door problem is confined to motion in a plane, as a consequence of the method of fixturing. This feature is not uncommon, as many parts sit on a level plate, and can only move about on it. For instance, in the field of electronics, the mounting of surface mounted devices (SMDs) on their printed circuit boards provide a conceptually identical problem.

In the case of motion in a plane, the solution is particularly simple, and can be implemented in real time with much less computation than might be expected in an eigenvalue problem. At Automatix, we have found that positioning errors up to about 30x the gage resolution can be usefully be removed using this technique. That is, for a gage resolution of 0.1 mm (0.004"), positioning errors of roughly 3 mm (0.12") can be

---

(2) Schönemann, P. H., "A Generalized Solution of the Orthogonal Procrustes Problem", Psychometrika, v.31 no. 1, p.1 (March 1966). The author is indebted to W. W. Meyer of the G. M. Tech Center for suggesting this line of research.

accommodated. The reason for this limitation is twofold: first, after adding positioning error to the part tolerance, the gage point can drift outside the field of view, and second, certain local linearity assumptions will no longer be valid.

A disadvantage of any method that employs mathematical transformations is that the true positions and orientations of the gage points must be known in order to apply rotation and translation operators. Thus, even though only normal-to-metal outputs are requested in the case of the door, all the gage points must be defined in an overall coordinate system through a suitable master fixture.

B. Gaging Combined with Flaw Detection

The combination of multiple tasks increases the utility of visual gaging, and helps justify an investment in it. An example of this trend is the joining of gaging with go/no-go flaw detection. Flaw detection (or visual inspection) is one of the earliest applications of machine vision. At Automatix, we have had success in combining it with three-dimensional gaging using the 2-D plus structured light approach (Figure 3). This approach is particularly suitable because it permits the camera to be placed normal to the surface to be gaged, which is often the preferred position in flaw detection applications.

Consider an automotive application, in which suspension mounting holes are to be punched in a vehicle underbody. A visual gaging station just downstream could have two functions: process control monitoring on the X Y Z locations of the holes, as improper placement will prevent suspension alignment, and go/no-go inspection for misshapen holes, cracks, or burrs which could be caused by die wear or changes in the physical characteristics in the metal to be punched.

Backlighting of the holes eliminates the problem of interference between the structured light beam, which is at the side of the hole, and the area lighting used for 2-D gaging and flaw detection. Many flaw-detection techniques exist; one that we have found useful is to compute all radii from the center of the hole to its perimeter, than take the difference between the largest and smallest radius, resulting in a Total Indicated Runout on the hole radius.

It should be noted that because the purpose of flaw detection is the detection of small variations from the norm, the pixel averaging techniques used to increase resolution in gaging are inappropriate. For this reason, large fields of view may require multiple, overlapping cameras. Figure 7 is a graph if the the pixel image of one view of a large (45 mm) with a 1mm x 2mm crack in its flange. The crack is clearly visible at the bottom of the curve. Three cameras with overlapping 120° fields of view cover the entire hole.

IV. HARDWARE DEVELOPMENTS

A. Modular Packaging

One important advance has already been touched on in the earlier section on calibration: modular packaging. Several manufacturers have come out with modular structured light packages; two examples are shown in Figures 4 and 5. In addition to permitting the use of off-line calibration as mentioned above, the gaging module simplifies the fixturing problem by permitting the fixture to be constructed of tubing elements, jungle-gym style. Up-front engineering for camera setup is reduced, while alterations because of e.g. model changeover are simplified. Figure 5 shows a 6 degree-of-freedom mounting bracket in some detail. In the future, it can be expected that more use will be made of modular fixturing, and additional modules, with extended capabilities, will be added to the already existing families.

B. Optical Alignment

The flexibility offered by modular packaging is to some degree lost if the process of aligning the local coordinate systems to a global one is extremely tedious. One way to avoid this difficulty is optical alignment. An optical alignment target, which consists simply of an array of features, is shown in the inset in Figure 5. (Here, the features are holes, as it is easy to determine their centroids using visual techniques.)

Optical alignment is a byproduct of the process of off-line calibration. This, it will be recalled, consists of the calculation of the "camera matrix" for a module by placing a standard calibration target in a standard position. It has been shown that the camera matrix can be decomposed mathematically into pieces that describe the purely optical properties of the camera/lens system, and other pieces which simply represent the orientation of the camera with respect to its calibration coordinate system.[3] When the alignment target is viewed by the module in its final mounted position, the off-line matrix elements which characterize the optical system are held fixed, while those that represent the orientation of the module are varied so that the matrix represents the new scene. In other words, once a module has seen its calibration target in its standard position, if it sees it later in any other position, it can calculate how it must have moved. Modules do not have to be placed in any fixed orientation with respect to their gage points; they can be placed anywhere convenient (given the depth-of-field limitations) and can determine their own locations.

Of course, the positions of the alignment targets themselves must now be known in the overall coordinate system; either a special fixture to hold them must be constructed, or they can merely be attached to a suitably-reinforced "golden part", and measured with a CMM. The technique of reinforcing a production part to use as the "golden part" is particularly cost-effective, as it bypasses the need to design and fabricate a special calibration master. In many cases the procedure of using calibration targets is more convenient than directly surveying in the module, especially when possible later modifications to the fixture are considered. It also creates a checking fixture, should the accuracy of the gage come into question later for any reason, e.g. because of an accident.

V. CONCLUSION

We have presented several recent developments in the field of visual gaging. These developments have been in both the fields of hardware and software; however, they have been

---

(3) Gnapathy, S; "Decomposition of Transformation Matrices for Robot Vision" in Proceedings of the International Conference on Robotics, p. 130, IEEE, Piscataway, N.J., 1984.

linked by the common thread of increasing the cost-effectiveness, ease-of-use, and utility of this technology, making it ever more the method of choice for solving the factory measurement problem and increasing the quality of manufactured products.

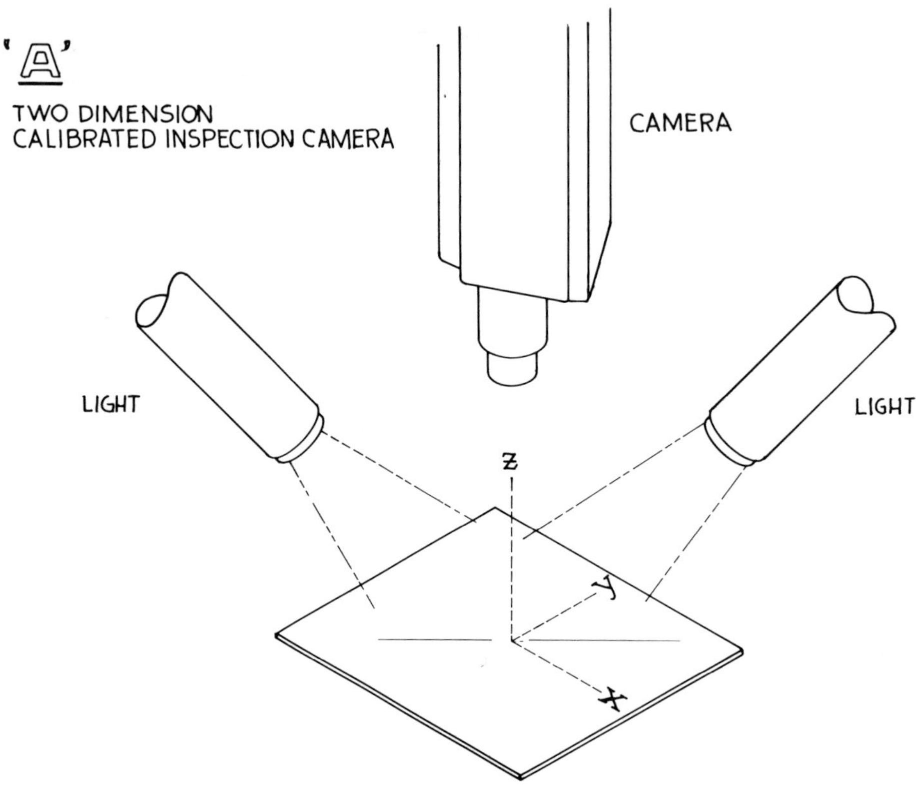

FIGURE 1: **Two-dimensional inspection camera**. Any feature known to be in the calibrated plane may be accurately measured.

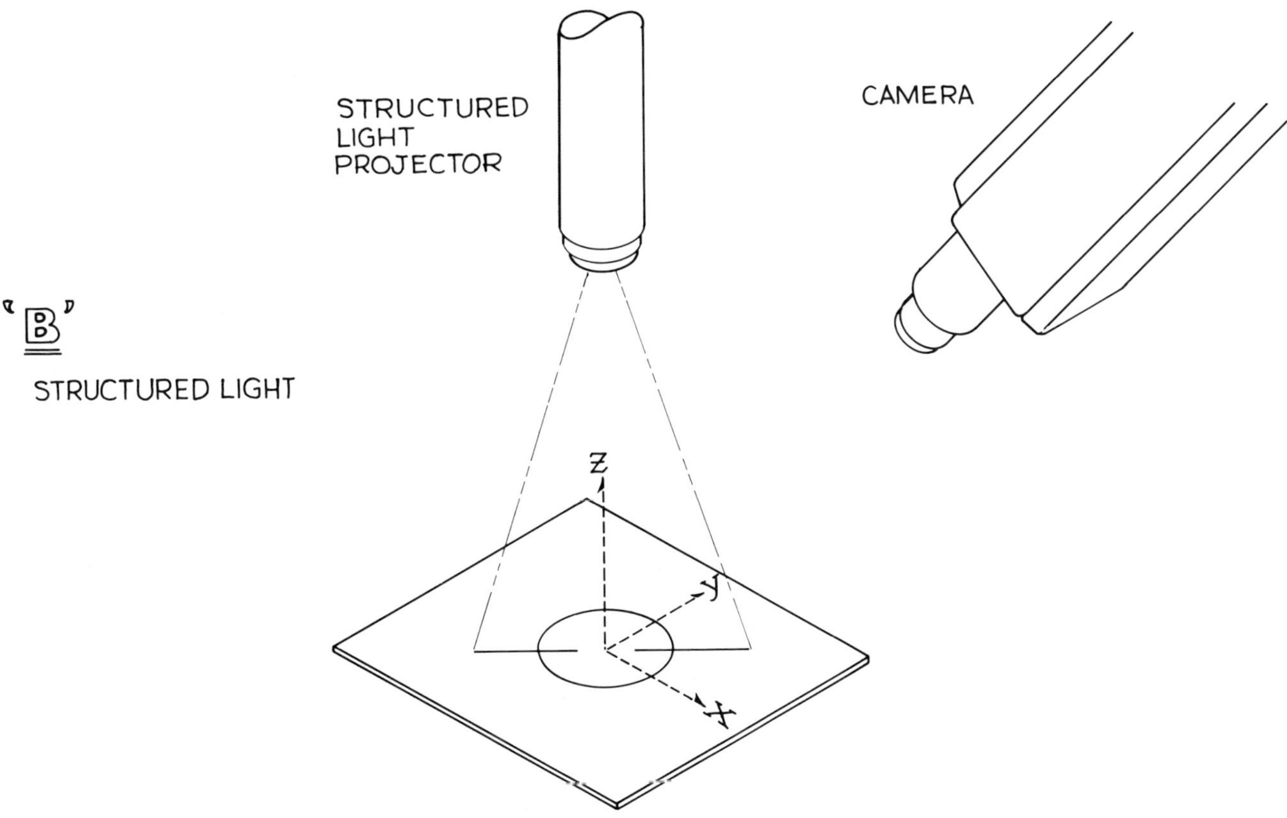

FIGURE 2: **Camera with structured light.** The structured light plane may be used to give the three-dimensional location of any feature it strikes.

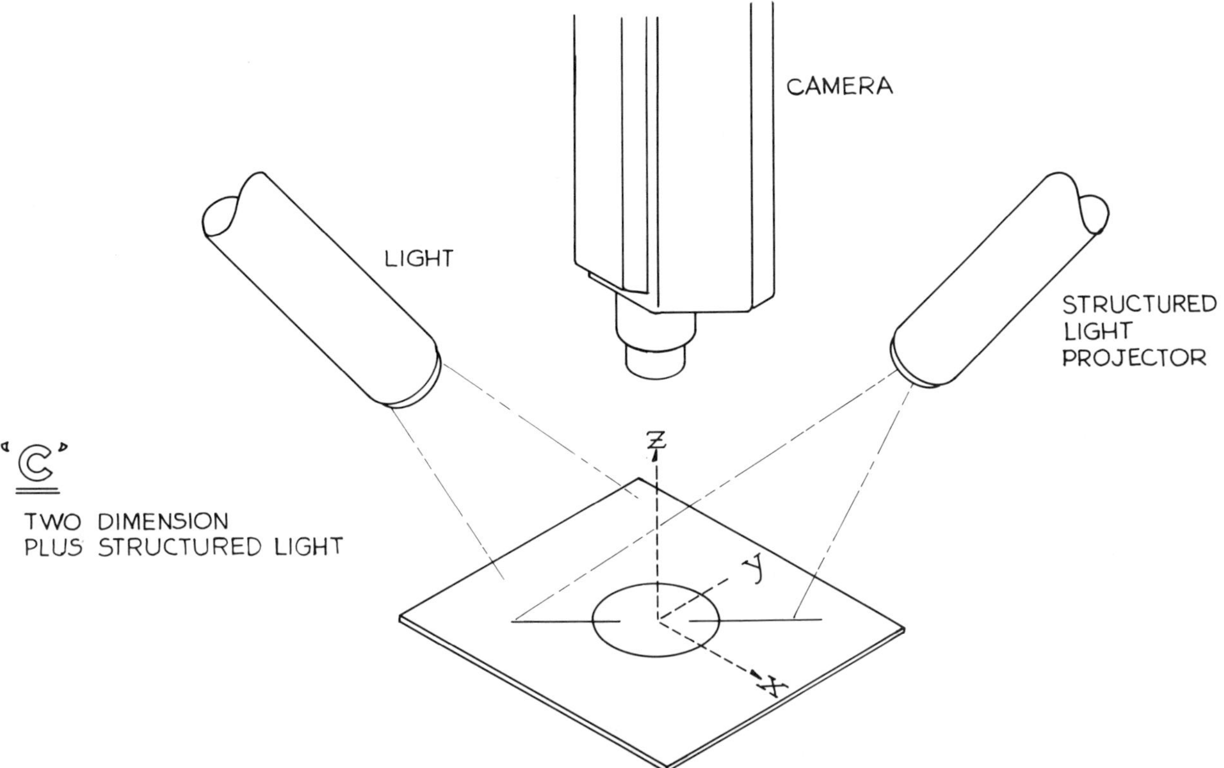

FIGURE 3: **Two-dimensional camera plus structured light.** A calibrated two-dimensional gaging camera and structured light projector are combined in the same field of view.

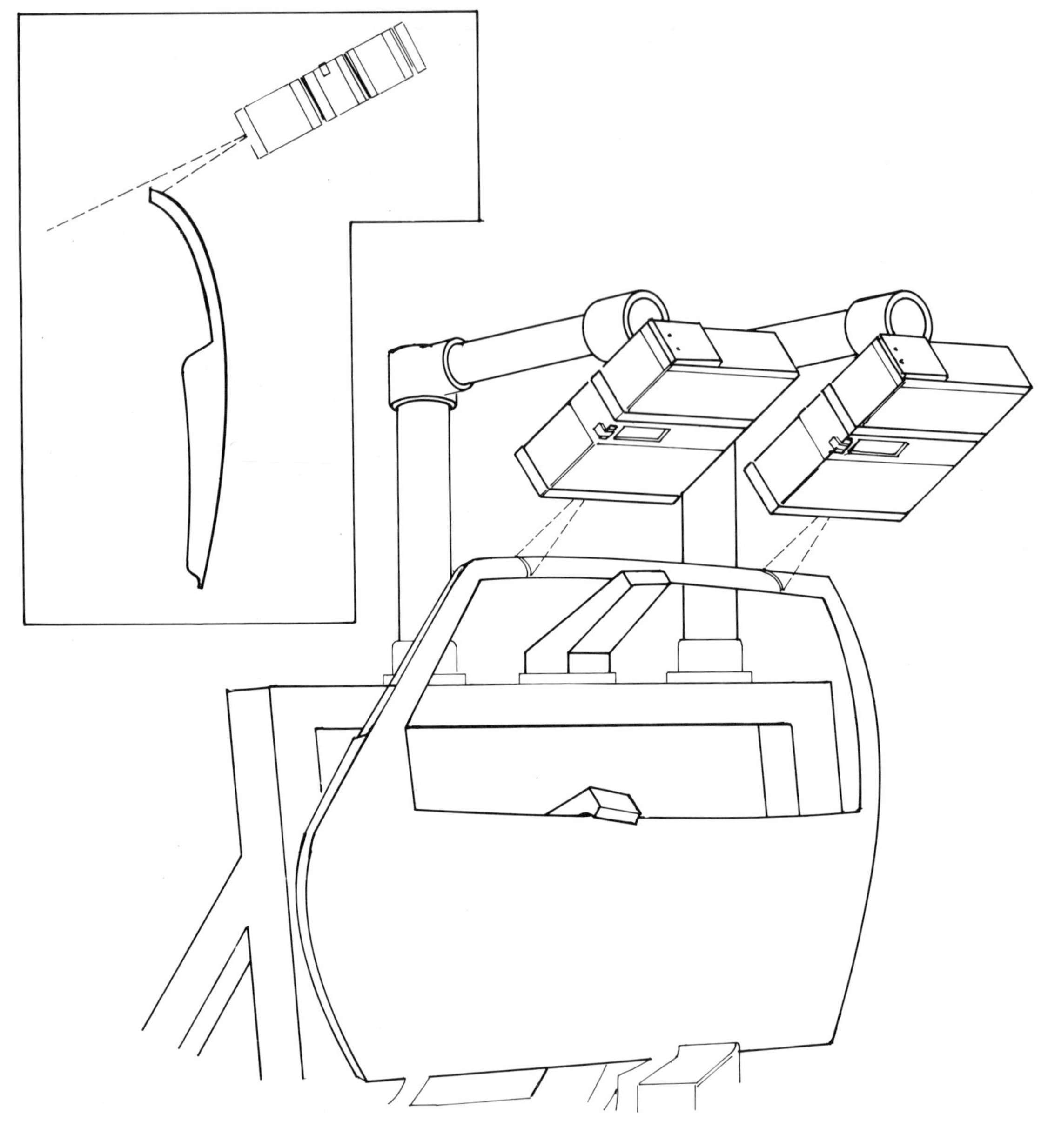

FIGURE 4: **Typical structured light gaging system.** This proposed system will use two structured light plus camera gaging modules to measure two points on a door header.

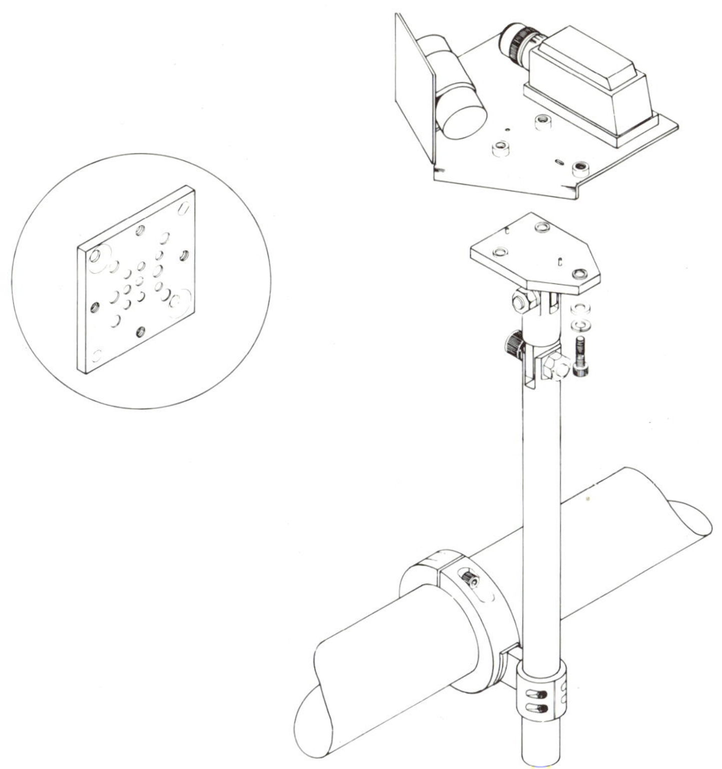

FIGURE 5: **Schematic of a structured light gaging module with mounting bracket.** This Automatix module is shown without its protective cover for clarity. Inset: An optical alignment target.

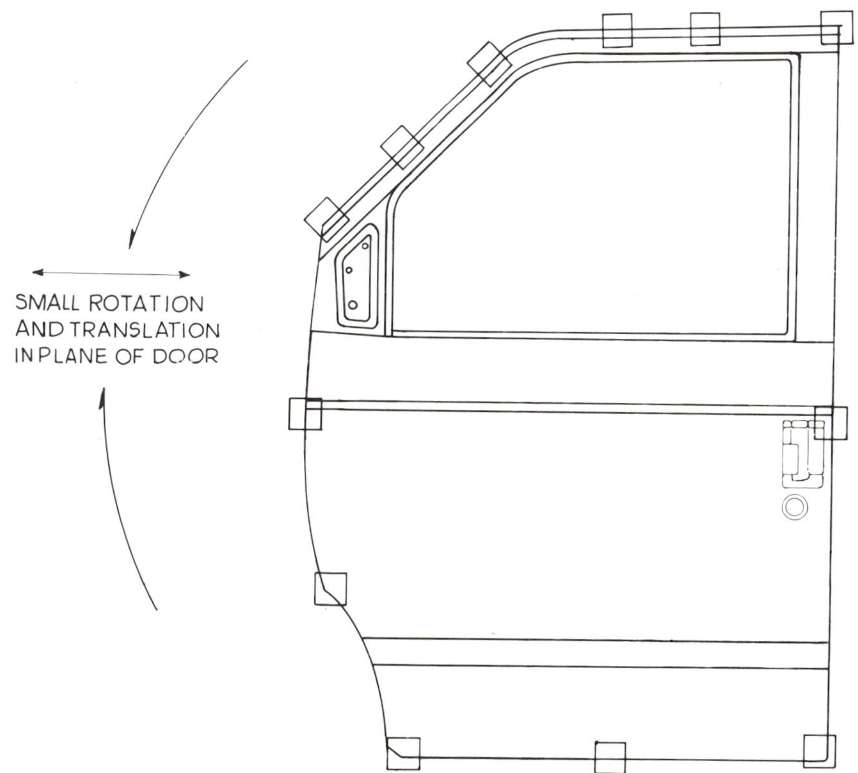

FIGURE 6: **Schematic of a door to be gaged.** Each overlaid square around the edge represents a 2-D plus structured light gage point.

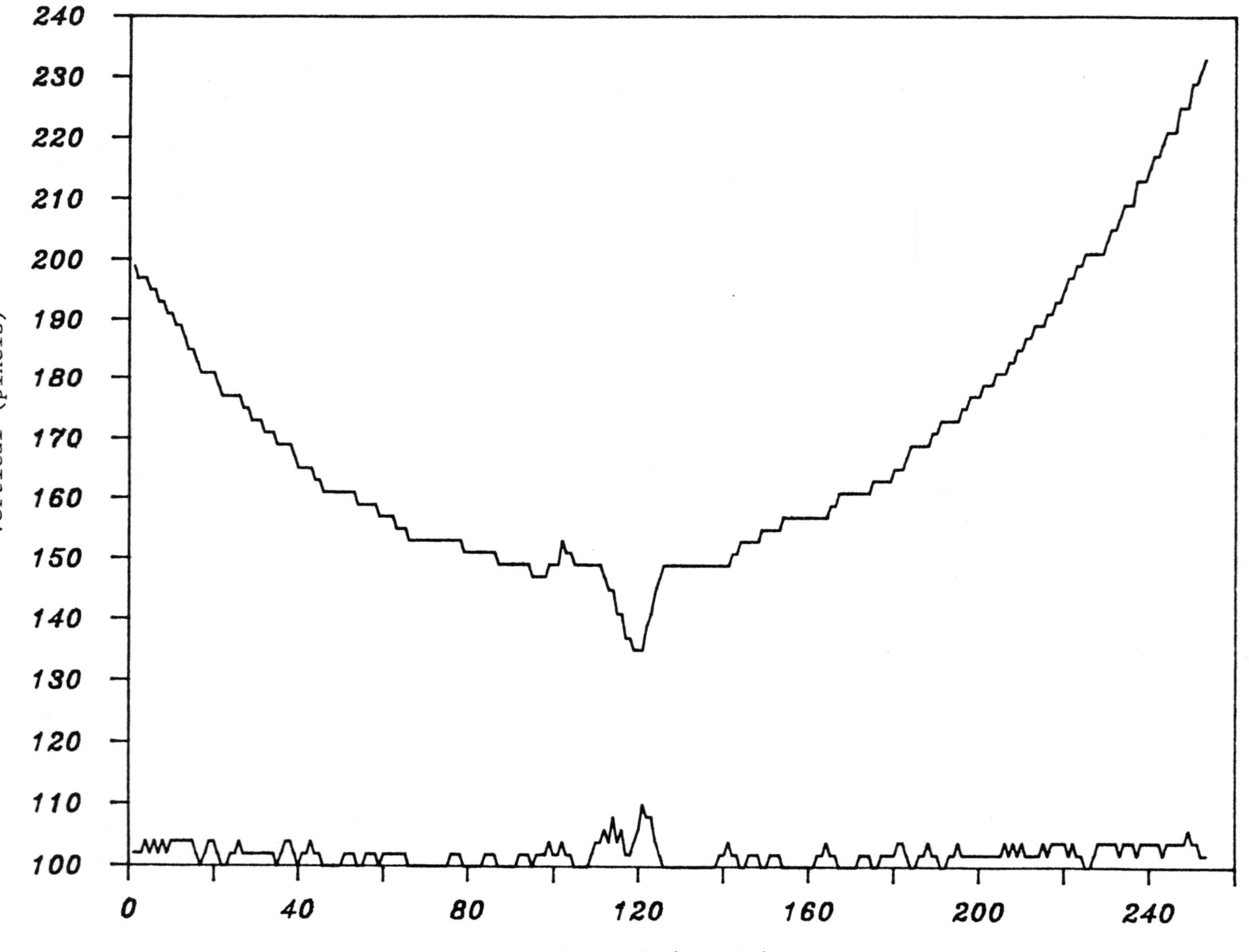

FIGURE 7: **Flaw detection.** The graph shows the perimeter of a 45mm flanged hole with a 2mm x 1mm crack in the flange.

Presented at the MVA/SME Vision '85 Conference, March 1985

The Role of Vision in a Dimensional Control Strategy

by Wayne S. Wilson
General Motors Corporation

Introduction

In the last several years the manufacturing environment of the assembly plants of General Motors has undergone numerous changes, with even more anticipated in the future. These changes toward increased automation have shown the need for a total review of current systems used in dimensional control. A total system strategy, rather than a fragmented approach, needed to be considered in the development of an effective overall dimensional control program.

Several factors must be considered by the dimensional control engineer in the development of a strategy. As we move the assembly plants from manual to automatic, the need for systems to monitor the build for tooling changes has increased. The need for improved quality and designs that dictate closer tolerances demand more sophisticated inspection and statistical process control techniques. Strong cost competition demands that we become more creative, flexible, and cost conscious in our decisions. In addition, because we must be able to bring new products to the market quicker, we require tooling (inspection) systems that are more adaptable.

To date, the checking systems used in the assembly plants have been predominantly manual, treated as separate identities, limited in the amount of data collected, and largely dependent on the human element for interpreting and sorting data. Tooling costs for design and construction of these conventional check fixtures have risen sharply in recent years, and due to the inflexibility of these tools, represent a recurring cost in each product update.

Vision in the last few years has made significant strides in development to the point where it can now play a major role in a dimensional control program. The range of choices to the engineer vary greatly not only with numerous manufacturers but also with the technology employed, from the simplest of vision (binary-part presence) systems to the most complex algorithms. All of these have their place and should be considered in putting together a total approach.

Groundwork laid in previous years with other vision systems in General Motors, particularly the assembly plants, provided much of the foundation on which the systems to be discussed today were developed. These included ROBI (Robot Operated Body Inspection) which was installed in eight assembly plants in 1981. ROBI consists of a pair of electric robots on precision tracks with dual laser, range finding type of sensors. Another system was the silicon bronze mig welding robots installed in the Janesville plant in 1982. This system used vision to locate the gap to be welded in space, determine its width and adjust the robot path and the mig weld schedule accordingly. From these systems and others installed in General Motors it was determined that vision had developed to the state of being a usable tool.

What will be discussed in this paper is a review of several vision systems installed in the new automated body shops in the Buick-Oldsmobile-Cadillac Division of General Motors and how they relate to an overall dimensional control program.

Several underlying themes are stressed throughout this paper. Systems should be flexible, adaptable, reconfigurable, and user friendly. Systems should be geared to the operator/engineer who will use the system and the data it generates.

Regardless of how complex a system is internally, it must, on the surface, appear to be as easy as 'child's play'.

In the assembly plant there are four general areas, which dimensional inspection can be broken down into, that must be considered as part of a total system approach.

Off-Line Tools

In-Process Parts Monitoring

On-Line Assembly System Monitoring

Secondary Sources of Data as a Result of Vision Controlled Processes

Off-Line Tools

The foundation of any dimensional control program is an adequate off-line area for detail analysis of individual parts, subassemblies, or the final assembly. The role of these systems can not be readily replaced by current vision systems but are crucial to the success of the program. The off-line tools might consist of conventional fixtures (such as the Combination Inspection Metal Matchcheck fixture) used in conjunction with portage layout equipment or more sophisticated Coordinate Measuring Machines, which are highly accurate and math data controlled.

The off-line tools should be assigned the task of doing the detail analysis, answering the question Why one is building; the vision systems task can best be defined as determining How one is building. Where vision systems can monitor a limited number of fixed points on 100% of production, the portage equipment or CMM's can check an unlimited number of points on a few assemblies determining curvatures and other features.

In addition, the off-line tools can serve as the master for the vision systems. While the use of 'tooled masters' can be effectively argued as a calibration method, a statistical mastering method is preferred. This means taking the average of several parts, as determined by the vision system, comparing it to the average generated by the off-line master, and adjusting for the differential. The advantage of this is that it does not disrupt the manufacturing process, is less costly, and adds a degree of flexibility to the vision system. It should be recognized, however, as a statistical method, not an absolute method subject to sample size and the repeatability of the master. From a vendors view it also requires a vision system that once set up is fairly stable.

In-Process Parts Monitoring

In addition to the off-line tools for qualifying assemblies it sometimes becomes necessary to have 100% in-process parts monitoring. Due to the critical nature of some components, it is essential at time to 'sort' discrepant parts as received to prevent them from entering the manufacturing process. To accomplish this with the traditional method of separate fixtures and operators is costly from a fixture, floor space utilization, and operator point of view. Vision can play a significant role in this area.

The most cost effective method of parts monitoring is to incorporate checking features in the actual assembly fixtures at the earliest point possible in the system; preferably at the first station the part is used. The easiest way is to incorporate simple go-no go physical features that prevent discrepant (off location or over sized) parts from being loaded to the tools. This, however, provides only a limited one dimensional check.

If this method is inadequate, then more sophisticated vision systems can be used. These can consist of simple, low cost, systems that simply check the presence or absence of a part. There are two examples of this installed at the Detroit assembly plant.

Dash Panel (Figure 1)

At the end of the automated dash line, there is a station to check the presence or absence of 8 small brackets installed on this line. A robot picks up the part, rotates to the vision check station, the part is checked, and if good the robot continues to the accumulator. If the part is discrepant, it is unloaded to a repair area. The vision system consists of a separate mini-controller for each camera and light source which makes it easy, in this instance, to readily add and delete cameras as designs change. The system is a basic binary one, providing only go-no go output.

Sometimes, however, more information is required from a system than simple part presence. In these instances more advanced vision systems can be used. It is often necessary to have dimensional data from a processing perspective as well as to have documentation for the supplier or for determination of possible shipping damage. Two examples of these types of systems are installed at the Orion assembly plant.

Cadillac Front Body Hinge Pillar Inspection (Figure 2 & 3)

The front body hinge pillar of the Cadillac 'C' car is subassembled at a fabricating plant with a part termed the hat section. This part's width, curvature, and placement is important to the appearance of the final vehicle as the door matches one surface and the windshield glass and molding match another surface.

In order to verify the dimensional accuracy of the parts before they entered the manufacturing system further, cameras were installed in the first fixtures in which the parts were used. Dimensional data is obtained on the location of the part for in/out, fore/aft, and width at two check sections through the part.

The operator loads the part and then closes the clamps. While the operator is getting the weld gun, the part is inspected. If the part is correct, a green light indicates to the operator to continue the operation. If the part is discrepant, the part is unloaded. In addition, dimensional data is available for review to determine if shifts have occurred in the fabricating process that can be accommodated in the assembly process.

In order to control cost in this type of application, it was necessary to optimize as much as feasible the utilization of the vision controller. The system is comprised of eight separate tools split between two vision controllers. The four tools per controller are comprised of two sets which are separated approximately 80 feet physically with 180 feet of cabling.

Two problems that had to be overcome in the installation, that should be considered in planning other installations, were the communication problems of transmitting vision information that distance, as well as protecting the cameras from the weld splatter. The latter problem was resolved with covers and shields.

One limitation of this system, though in this application is not a concern, is that since the vision system controller was not a multitasking operation system, retrieval of dimensional data could not be done while the fixtures were used, if inspection of the parts was to continue. A second concern that should be considered in future applications is the physical distance the tools are separated from the controller from human factors perspective. Though the technical problems were overcome, the controller is a considerable distance from the remote site for diagnostic work.

The system generated a secondary benefit which should be mentioned that further justified its installation. In setting up the system we were able to assess the repeatability of the tools themselves. It was determined that two of the tools had been setup differently and were significantly less repeatable than the others. This was corrected and the vision system was used to verify the correction and to subsequently monitor the tools themselves.

Door Header Inspection (Figure 4)

Another system that demonstrates the use of vision to monitor for off-dimension parts is the inspection of the door headers on the Cadillac 'C' doors. This system inspects for the in/out and height location of the header in two places. As in the previous application the cameras/sensors are installed in the first fixtures in the door final assembly process.

Off-line the door hinges are loose assembled to the door. The door is then loaded to a two-stage fixture. The door headers are checked to see if they are correct as received. The hinges are then tightened, and the headers are rechecked to verify no distortions occurred in that step of the process. If a discrepancy is noted, when the door indexes to the next station, the hinge piercing operation is not performed, and the door is unloaded for repair. In this case the vision system performs the dual function of checking incoming material and the assembly process.

As with the front body hinge pillar, costs were controlled by sharing the vision controller with several tools. In this application there are three pairs of tools separated by 150 feet with the controller located in a central point.

In all of these applications the future flexibility of the equipment needs to be mentioned. Whether it be the simple part presence or the other monitoring systems, the vision components are reusable. If a bracket being inspected was eliminated, the vision system for that part could be readily relocated to another application. If it was determined after a period of time that processing was in control and it was no longer necessary to monitor the front body hinge pillars or the door headers in each tool, then those components could be moved to a more critical area that may have been identified.

On-Line Assembly System Monitoring

The on-line assembly systems are meant to monitor the build of the key assemblies on the automated lines and to provide a warning when tools break, wear, or shifts occur due to changes in incoming metal. When operations were manual, the operator physically closing a clamp, served as a built in inspection system for tool breakage or major changes, functions that now must be accomplished automatically.

The underlying philosophy of the on-line systems installed is the strategic placement of monitoring systems that can 100% monitor the build of the body-in-white assemblies as well as provide the necessary Statistical Process Control data required to control and improve the dimensional toleranes of the body. The systems indicate which areas of the body-in-white build needs to be concentrated on, in order to reduce the overal build variations. Their locations in the system are a function of where it was felt they could do the most good weighed against costs and distance from potential sources of problems.

A valid point can be made that SPC does not advocate 100% inspection. ROBI (the Robot Operated Body Inspection system) was a statistical process control system. The system could be programmed to go through a sampling profile of the vehicles creating the profile of a composite vehicle. It could be set to check 'x' number of windshields, 'n' number of decklid openings, 'z' number of door openings, etc., until all openings were sampled and then repeat the process. However, changes in vision technology altered the cost picture resulting in the fixed camera system being lower cost than the robot system. This, combined with fewer jobs required to master and the ability to use more than one vision technique, led to this direction.

By going to 100% inspection several additional abilities were picked up at no extra cost. These include knowing immediately if a tool breaks versus waiting for the next sample, the advantage of immediate access to a sufficient sample size to resolve a problem, rather than having to wait for the data to be collected, and the advantage of potentially using the data for other purposes, where data from every job would be required, such as robotic or tool control.

There are two examples of this type of system that will be reviewed.

LDSF (Laydown Sideframe) (Figure 5)

There is a vision-based inspection station in the laydown sideframe systems at the Wentzville and Detroit assembly plants. The Wentzville systems consists of 15 camera/sensors in each of the four automated lines (two pair). Coupes and sedans are built in each line with 10 points inspected on the sedan and 8 on the coupe per side, with 30 seconds of available cycle time.

The systems are located in the last station on the automatic lines. The principal function of these systems is to monitor their respective assembly lines to prevent discrepant assemblies form reaching Robogate, the automated framing station. Complete SPC capabilities are available to provide early warning to trend shifts in the assemblies.

Individual specification limits for each check point, along with a warning limit criteria, is entered by the user. If an assembly has a point that exceeds the specification, the assembly line is stopped; an alarm light is initiated; and the system, sensor, and discrepancy is automatically transmitted to a printer.

The system utilizes only a limited number of clamps (fore/aft and height only) taking advantage of the vision system for the other dimension. This significantly reduced the amount of tooling in this station, resulting in less cost and congestion.

Due to the fact these systems were more directly tied to the assembly process, separate controllers were provided for each system. However, a common printer is shared. The systems were being interlinked in order to permit an engineer to access the data from any of the systems while located at anyone of them, speeding up the problem analysis phase.

The hardware/software features of the system are common to the Automatic Dimensional Check station to be reviewed next.

ADC Automatic Dimensional Check Station (Figure 6 & 7)

The Automatic Dimensional Check station, or ADC system as it is called, is located at the end of the completed body-in-white automated welding lines to check all critical opening dimensions on the vehicle. In addition, all the net holes and studs for the door and fender processing are monitored to verify those processes. Three of these systems are currently installed at Wentzville, Orion, and Detroit.

The Wentzville system consists of 100 sensors monitoring approximately 140 key data points on the body-in-white, plus numerous relational measurements such as the width of the windshield opening; on four different make/models. These checks are made in less than the 30 seconds of available cycle time. All checks are dimensions controlled directly by the plant in one of the previous tools.

The readings are made normal to metal (90 degrees to the surface) as this is the way the matching components see the job. However, the ability exists to have X, Y, Z data if preferred.

The bodies are on an automated line with 40 different pallets which are mastered to +/- 0.1mm. The overall system accuracy considering vision, pallet variations, and pallet positioning is +/- 0.25mm (3 sigma). One of the major problems with our previous systems was the occasional pallet with the locating pins significantly off from the design location due to damage. To correct that situation, the two primary gauge pins are monitored by sensors as to their location. If a pallet pin is found to exceed a specified tolerance, the system indicates that it has located an incorrect pallet, and the data for that vehicle is not entered into the data base, retaining the statistical validity of the other data.

Several vision techniques are utilized in order to optimize the performance and cost of the system. They were selected on a hierarchy of the simplest to the more complex. The four principal types of sensors which were used are:

Range Sensors - These sensors were used where only one dimensional direction was required, such as in/out. They were used where it was possible to mount the sensors normal to the body metal. They were used only in locations where there was sufficient surface area so that normal part variation does not result in the laser beam hitting an internal or external corner. This was a problem experienced with our previous systems.

Feature Sensors - These sensors are used when two dimensional data is required to check the pierced holes or door attaching studs.

Contour Sensors - These sensors consist of a plane of laser light and are used in three different applications. The first is when only one dimensional data is required but the sensor could not be mounted normal to metal. The second instance is where two-dimensional data is required and the sensor could be mounted to view both surfaces directly. In the third case, when clearances to the body were insufficient to allow the sensor to be mounted so it could view both surfaces, a plane and edge measurement are made. It should be mentioned, one of the main design criterias was no indexing sensors.

Surface Sensors - These consist of two planes of perpendicular laser light and are used when it was necessary to obtain three-dimensional data, such as the holes for the lock striker attachment.

Several features of this system highlight the theme of reducing design and construction costs through easily modifiable, adaptable systems that permit them to be expanded or reduced and accommodate changes in build allocation without incurring expensive redesign costs. The system was made easy to add, delete, or relocate check points as experience or need indicated through the use of what has been termed 'Jungle gym' construction techniques. If data at one point proved to be stable across time, that sensor can be readily relocated to a more critical location or even removed and used in some other location in the plant.

User configurable graphics permits the plant floor person to obtain a graphic representation that he can relate to.

Specification and trend warning limits are set by the plant floor person to permit the gradual narrowing of the specification band as the process comes more under control.

Often a relational measurement means more to the user than the actual dimensions, such as the parallelism of the backlite opening. The system permits the user to define these types of relationships and track them with alarm limits as if they were an actual data point.

With this much data it became important to present the data in the most comprehensive, meaningful method possible keeping the amount of paper to a minimum. In addition, it was necessary to sort the data to a manageable level of points that one could handle. This was accomplished by the frequency distribution report which shows the distribution of the data for each point in an opening but contains the needed statistical information as well. To sort the data, the exception report only reports those points that had data outside the warning limits. Additional reports, including mean and rage control charts, are available for more detail analysis. The data is kept on 3600 jobs and can be reported by time-to-time, job sequence number to job sequence number, the last 'x' quantity of jobs, or for a particular job. The data is available while the system is on-line inspecting bodies.

In order to increase the overall effectiveness of these tools, two types of intergauge networking is being developed. The first is a direct communications link to the laydown sideframe systems to permit access of the data from any system and the comparing of dimensions of one system to the other. The second network permits acquisition of the data from any locale in the plant or Central Office were a terminal is available. This will speed up the problem analysis phase plus provide needed data for future designs.

Secondary Sources of Data as a Result of Vision Controlled Processes

A valuable source of data that should not be overlooked by the dimensional control engineer is the utilization of the data obtained from vision controlled robotic processes. One of the inherent side benefits of vision controlled robots is the ability to collect, store, and analyze dimensional data. This option is available from most manufacturers for a nominal extra charge.

An example of this is the vision controlled robot system installed at the Wentzville plant to silicon-bronze mig weld the door facing joints. (Figures 8 & 9) The system consists of two stations with two robots located in each station. The cameras and laser light sources are located in the first station.

When the job enters the station; the width, depth, and location of all four joints (two on a coupe) is determined and the vision offsets are passed to the two robots in station. In addition, the appropriate weld schedule and wire feed rate information for that gap condition is relayed to the mig weld controllers. The data for the other pair of robots is transmitted while the job is shuttling. Since the positioning accuracy of the two stations is within +/- 0.1mm of each other, downloading the respective vision information performs satisfactorily.

From a dimensional control engineer's perspective the system stores the data relative to the width, gap, and location by make/model with complete SPC reporting capabilities. This information can then be compared to the Automatic Dimension Check station to access whether the gap variation is part variation or panel location variation.

Conclusions

The dimensional control engineer has a wide selection of systems to choose from in putting together a total dimensional control system. Vision can play an effective role in such a program.

The engineer needs to base the selection of what is needed on the job needing to be done. He must keep in mind that the simpler it is, the easier it is to use and the lower the cost will be.

The engineer has several opportunities to control costs by utilization of existing tools and stations; the sharing of controllers' low cost/reuseable fixturing; and flexible, user reconfigurable software.

In addition, the engineer should not overlook sources of data that may already exist that are being or could be collected by vision controlled robotic processes.

Intersystem communications can be used to tie the subsystems into an overall system to increase the effectiveness of the whole.

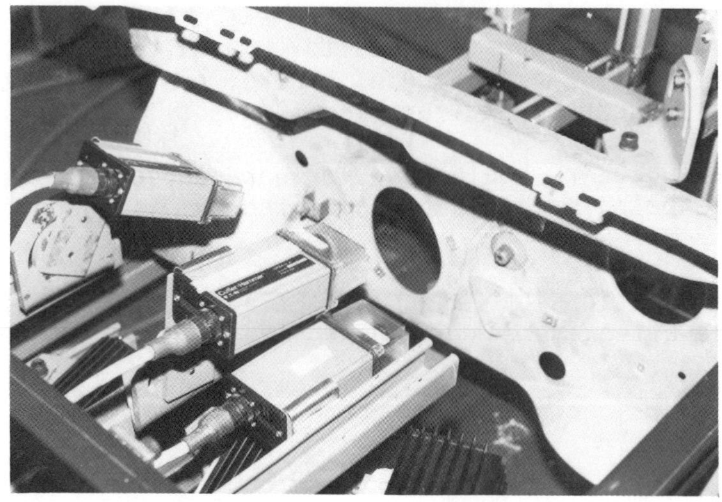

Figure 1: Dash Panel - Parts Presence

Figure 2: Front Body Hinge Pillar - Part Monitoring for Dime

Figure 3: Front Body Hinge Pillar

Figure 4: Door Header Inspection

Figure 5: Laydown Sideframe On-Line Inspection

Figure 6: Automatic Dimensional Check Station

Figure 7: Automatic Dimensional Check Station

Figure 8: Silicon Bronze Mig Welding - Door Facing Joints

Figure 9: Silicon Bronze Mig Welding

Presented at the MVA/SME Vision '86 Conference, June 1986

Three Dimensional Measurement with Machine Vision

by Kenneth R. Pelowski
Perceptron

INTRODUCTION

Dimensional measurement is becoming an essential element in the machine vision industry with applications in inspection, gauging, robot assist, and many other fields. Most of these applications require three dimensional measurement. This paper discusses a three dimensional measurement technique for machine vision. The technique is capable of measuring the cartesian axis X, Y, and Z to an accuracy of plus or minus one hundred microns over a repeatability of three sigma. Some of the current objects being measured by this technique are hinge holes, punched holes, slots, and tapped holes. (Figure 1 illustrates the implementation of this technique on an automotive door hinge).

Fig. 1

Mathematical Assumptions

Virtually all algorithms have certain assumptions which are required to reduce the mathematical computations. The two major assumptions in this approach are:

1. There are no significant deviations due to the thickness of the lense.

2. The perspective error, due to the tilt in the array at a constant distance Z from the sensor, is minimal. In other words, at a constant Z from the sensor, the X and Y movement is linear with respect to the rows and columns.

IMAGE EXTRACTION

This three dimensional technique uses a "Surface Sensor" to extract the images. This sensor extracts two different images. Both images are captured using external lights to enhance specific image features. The laser light source highlights the object's surface, while the strobe defines the object's centroid.

The sensor is enclosed in a machined aluminum casting which is shaped for heat dissipation and mounting accuracy. The internal supports fixture the imaging array, optics, and laser to the required focal length and field of view. The sensor's micro-controller manages the image exposure and lighting sequence. The micro-controller also manipulates the calibrated data that is stored in memory. This data is created during an automatic calibration process insuring measurement accuracy. Calibrating the sensor during assembly eliminates the need of any plant floor calibration process. This also allows the interchange of sensors during plant maintenance.

Fig. 2a

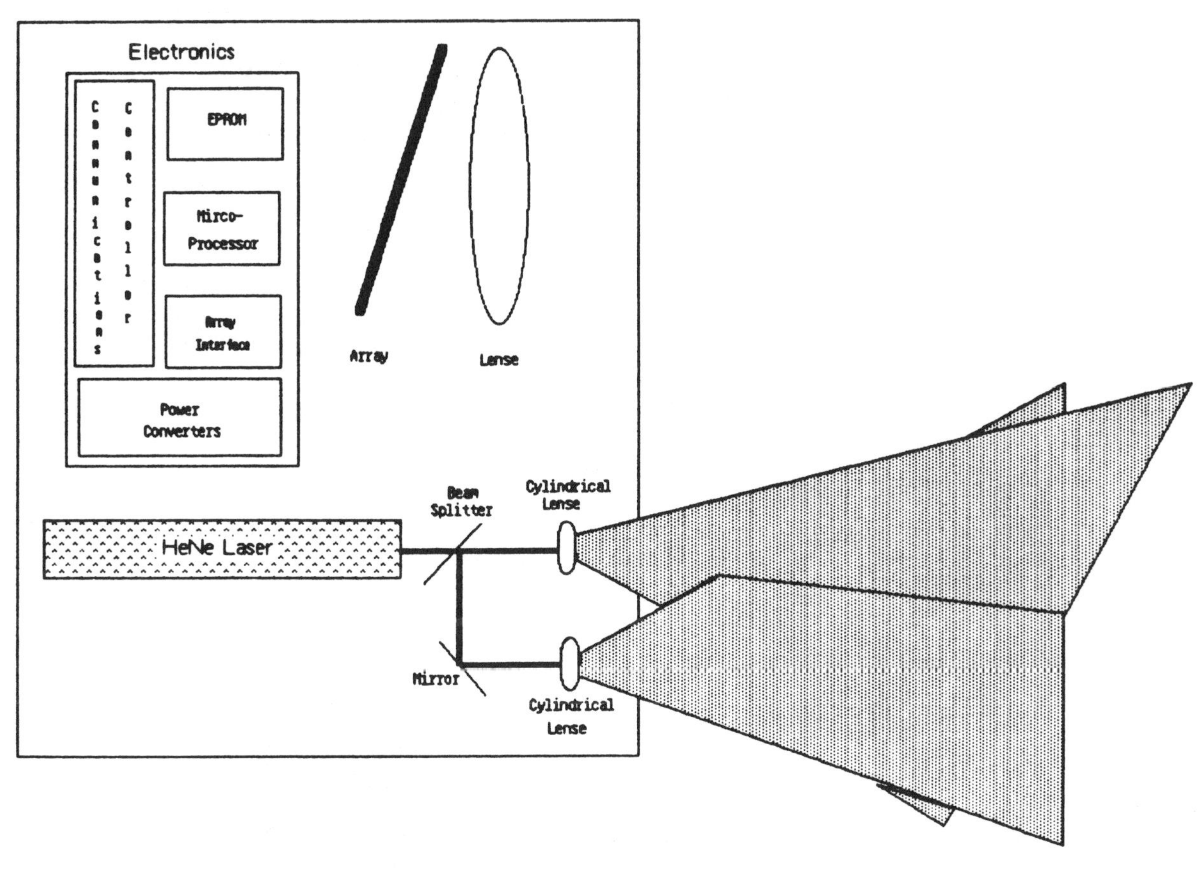

Fig. 2b

The Z-Axis measurement is based on triangulation. The surface sensor utilizes a photo-diode imaging array and a laser beam that is structured into two intersecting planes. These planes form a cross-hair as they reflect off a surface. There is also an external light source, a strobe, which enhances specific features. The image acquired with the strobe contains information that when combined with the laser image provides the three dimensional position of the object's centroid. (Figure 2 contains drawings of the sensor and examples of the images extracted).

IMAGE ANALYSIS

This algorithm uses both images to determine the object's centroid in three dimensions. The laser image defines the plane of the object based on the sensor's calibrated data.

$$\text{PLANE: } A\,x + B\,y + C\,z + D = 0$$

The strobe image calculates a line (in three space) through the object's centroid and the sensors's optical origin.

$$\text{LINE: } T = (x - x\_origin) / x\_slope = (y - y\_origin) / y\_slope = (z - z\_origin) / z\_slope$$

The intersection of this line and plane is a point (X, Y, Z) on the plane going through the centroid. (See figure 3).

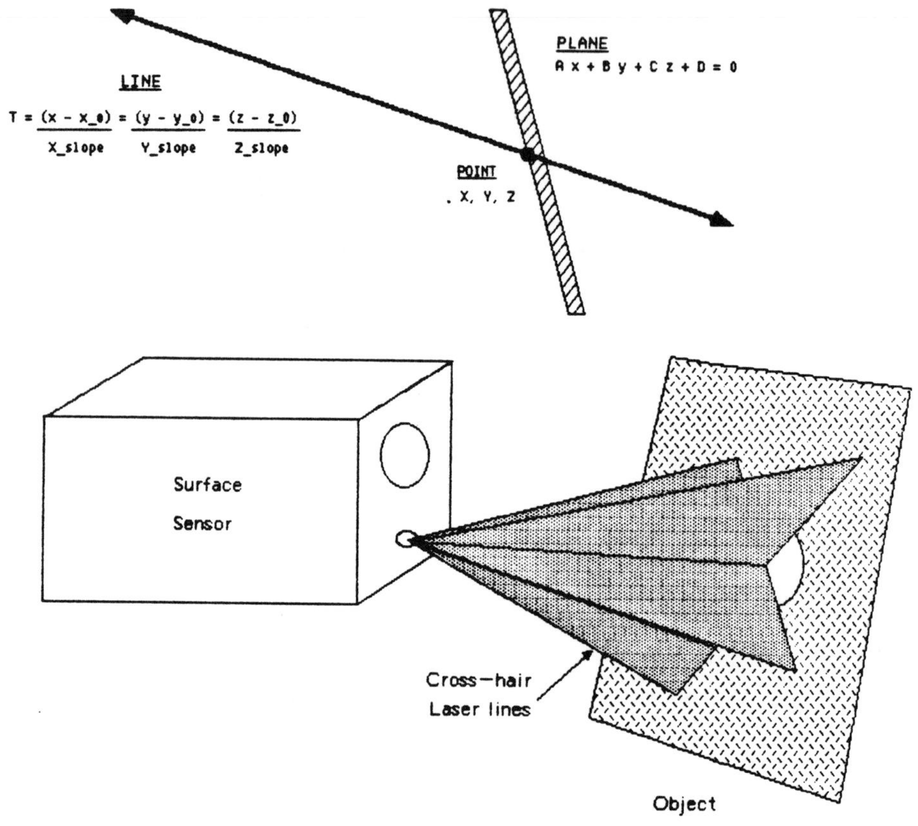

Fig. 3

PLANE

The plane of the object is determined by the surface sensor. First, the sensor is instructed to grab a laser image with the appropriate exposure. Once the image is extracted, it may be necessary to filter the image. Passing the image through filters eliminates most of the noise generated on the image. The filters also enhance the Gaussian distribution of the laser line. After the image is filtered, the next step is to determine the area to be analyzed for the plane (See figure 4a). Assuming the diameter of a good object is known, two ellipses can be generated around the object's centoid. The ellipses produce a windowed area which defines the area to be analyzed for the plane. Once the window is determined, the laser edge points can be determined by many techniques (Sobel, Convolution, etc). It is very important that the edge points be in pairs for each laser transition in the image.

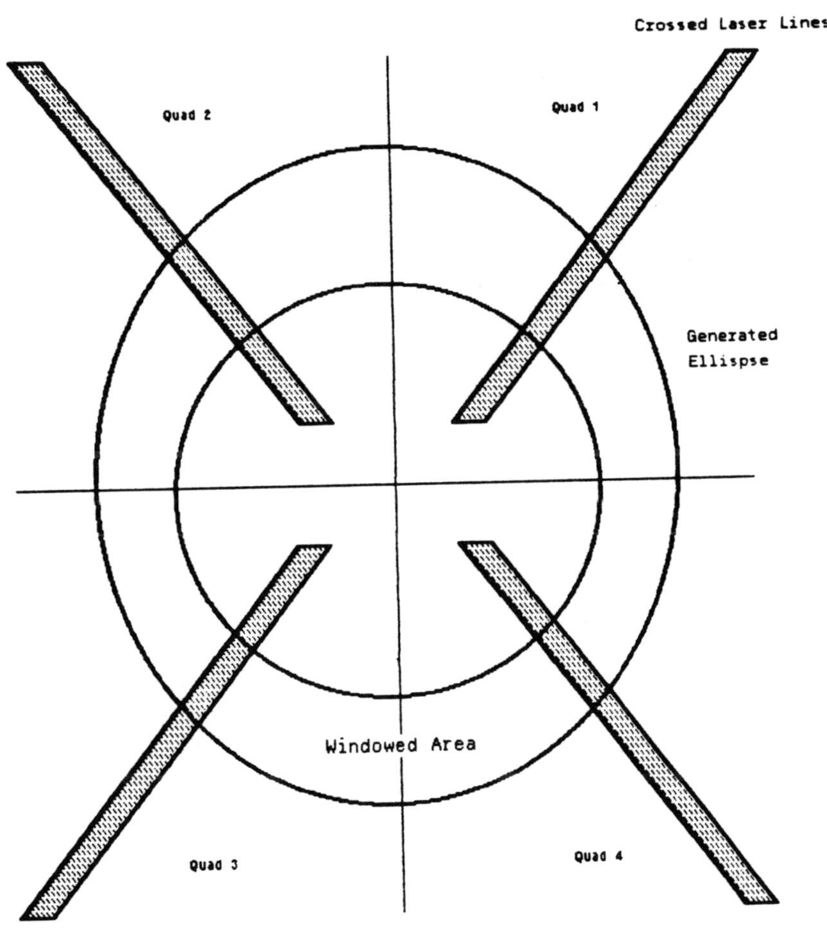

Fig 4a

The image is now divided into four quadrants. Once the transition points are paired for each laser line, these pairs can then be used to determine a sub-pixel point for each laser transaction (see figure 4b). These sub-pixel points can then be rectified from the calibrated data stored in the sensor. This converts the sub-pixel point (row, column) into an X, Y, Z location. Note that this coordinate system is based on the sensor and its calibrated data. Next, take the same amount of rectified points in each quadrant. This ensures equal emphasis for each quadrant. Using these rectified points, a linear regression plane fit routine can be used to calculate the plane of the object in three space. When the plane of the object has been determined, the next step is to find the line passing through the optical origin and the object's centroid.

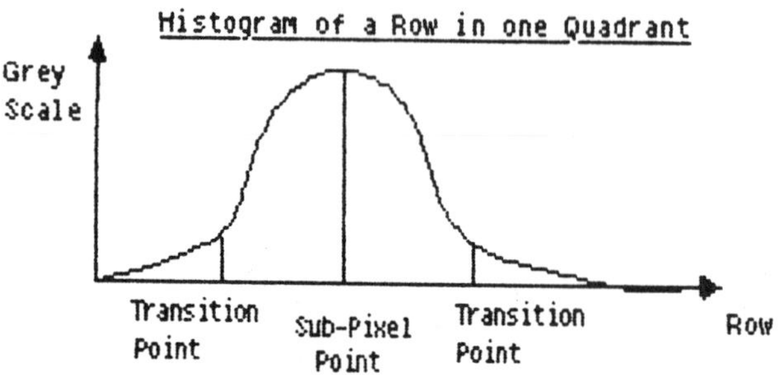

LINE

The sensor has a unique coordinate system to calibrate the relationship between the array's laser image and the X, Y, Z calibrated data. This coordinate system is perpendicular to the sensor enclosure. The line is found using another optical coordinate system. Therefore, the next step is to calculate a mathematical mapping from the sensor coordinate system to the optical coordinate system.

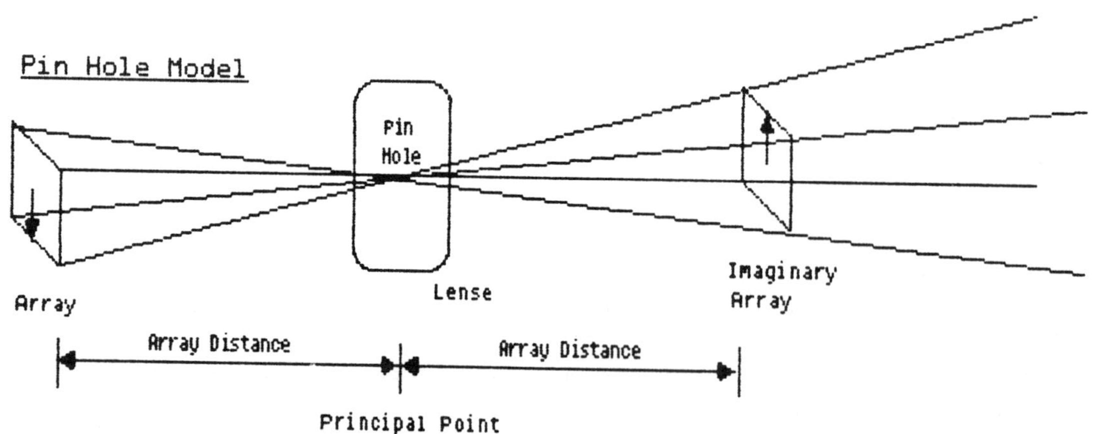

Fig 5a

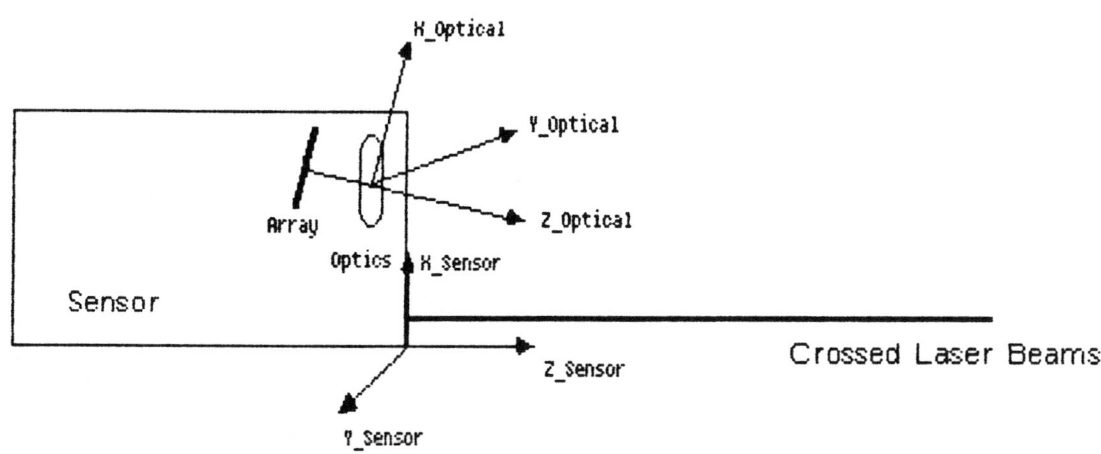

Fig 5b

The optical coordinate system is based on the theoretical pin hole model (See figure 5a). The pin hole coordinate system has an origin at the principal point of the lense. The Z-Axis is extend through the center of the array. The pin hole's Y-Axis is defined as movement along the rows, while the X-Axis is defined as movement along the columns (See figure 5b).

If the distance between the array and the principal point are known, a line can be defined in three space. This line is the movement from the center of the array, change in rows and columns, with respect to the optical origin (See figure 6). Therefore, the line is calculated with the following information: principal point in sensor coordinates, array distance, the optical XY, XZ, and YZ planes with respect to the sensor coordinate system.

Array Distance

As figure 7a shows, the distance from the sensor to a point in the field of view is Z. This distance is the calibrated data point stored in the sensor. Therefore, any image point, row and column, can be rectified from the calibrated data which will be in the sensor's field of view. That distance is equal to the total conjugate length times the angle of the array.

COS(Theta) = Total Conjugate Length / Distance Z.

Figure 6

Object
Centroid
Delta Col
Delta Row
Principal Point
Array Distance
ROWS
COLUMNS
Imaginary Array
OBJECT
CENTROID
Expanding
DELTA COL
LINE
DELTA ROW
OPTICAL ORIGIN
(X_0, Y_0, Z_0)
ARRAY DISTANCE
Imaginary Array

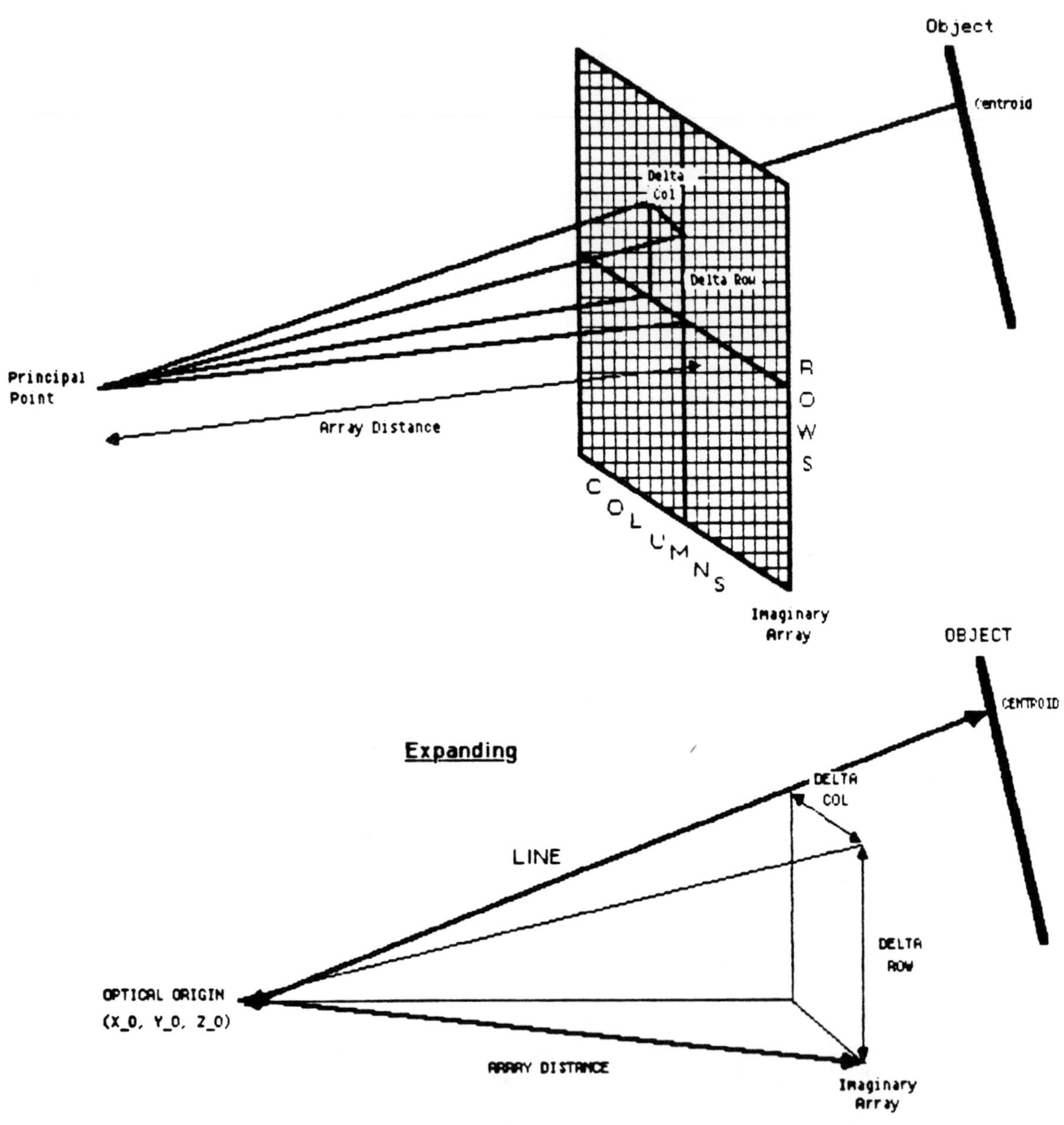

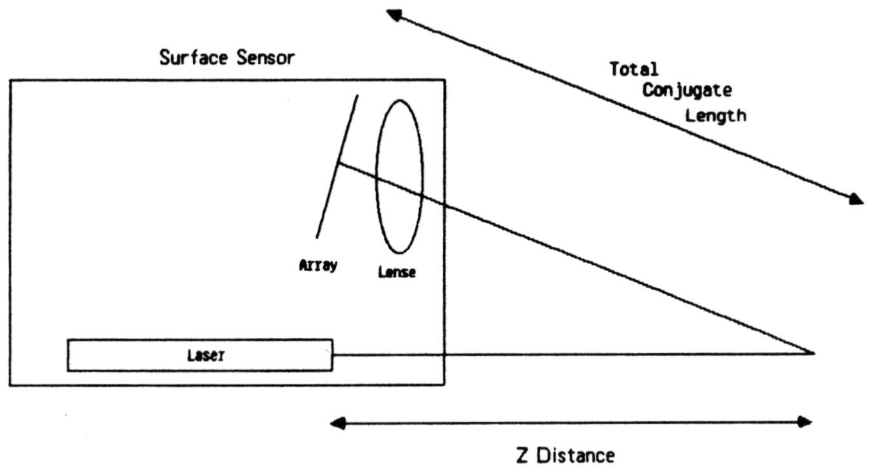

Fig 7a

With any optical system, the change in Y-Axis movement divided by the total conjugate length (TCL) minus the array distance will equal the change in column movement divided by the array distance as shown in figure 7b.

$$\frac{\text{Delta Y}}{\text{TCL - Array Distance}} = \frac{\text{Delta Columns}}{\text{Array Distance}}$$

Therefore, with a little mathematical computation the Array Distance equals,

$$\text{Array Distance} = \frac{\text{Delta Columns * Total Conjugate Length}}{\text{(Delta Y + Delta Columns)}}$$

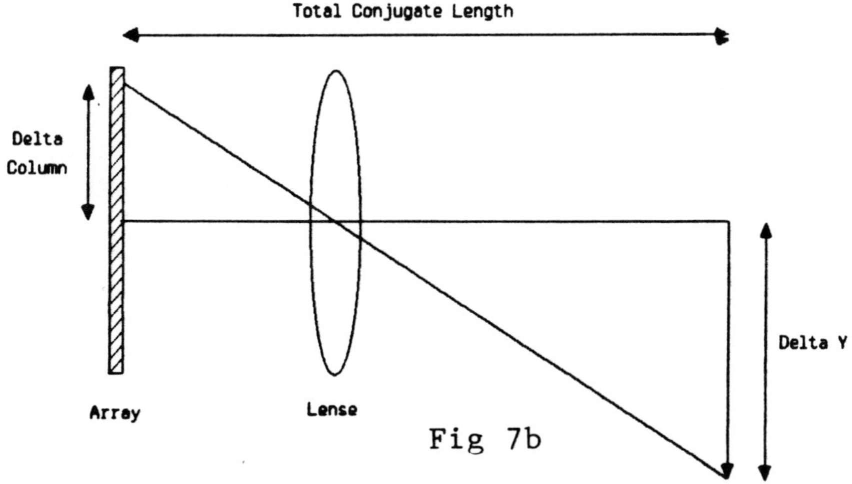

Fig 7b

Optical Planes

The optical planes define the angular differences between the optical coordinate system and the sensor coordinate system. The optical Y-Axis is stated as the movement along the rows of the array, while the X-Axis is the movement along the columns of the array (See figure 8).

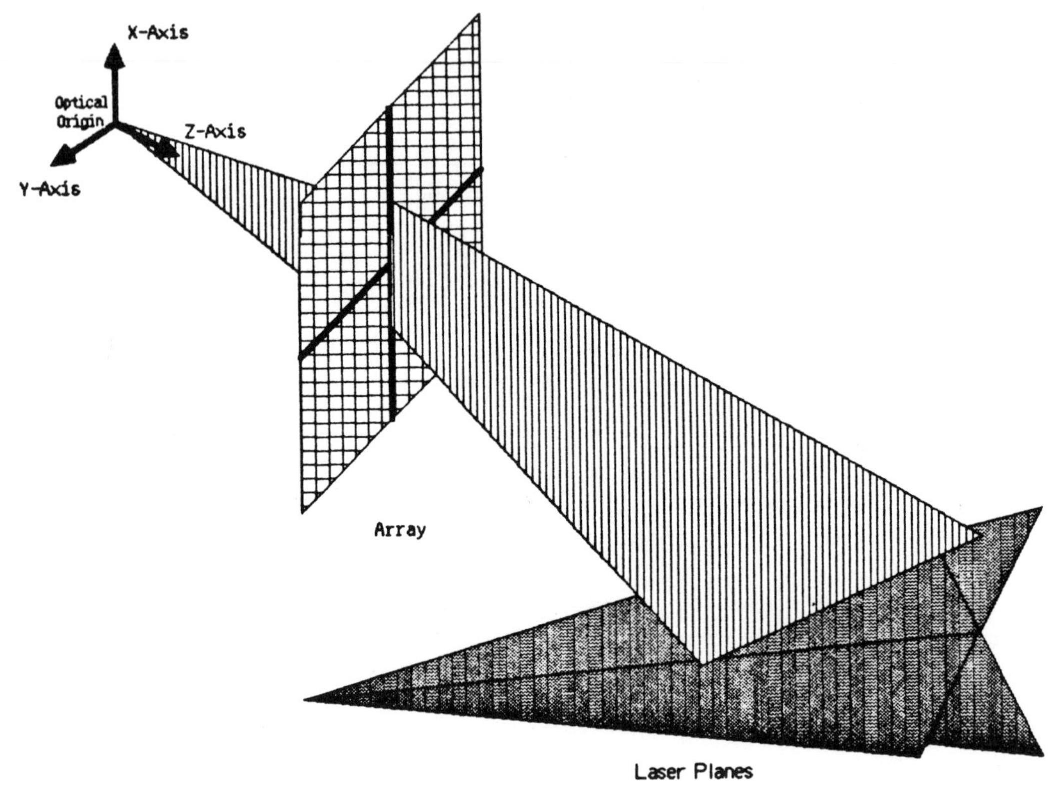

Fig. 8

If all the image points are rectified along a single column, the result is a set of object points. These points are all on the optical coodinate's Y-Axis. Therefore, the XZ plane can be calculated with a linear regression plane fit routine. The same process can determine the XZ plane. Once these planes are calculated, the XY plane can be determined with the cross product.

XY Plane = YZ Plane X XZ Plane

Principal Point (Optial Origin)

The optical origin is calculated with trigonometry. Several points are rectified at different parts of the array. Next, the rows and columns of the points are converted into real dimension. (Figure 9) shows the trigometric relationship that exists after rectification. Note:

D1 = Hypotenuse of Array Distance and Delta Row
= SQRT (SQR(Delta Row * 0.027) + SQR(Array Distance))

D2 = Delta Columns of rectified point Pt A

D3 = Delta Columns of rectified point Pt B

D6 = Distance between two points in space (PtA and PtB)
= SQRT(SQR(X1-X2) + SQR(Y1-Y2) + SQR(Z1-Z2))

D4 = SQRT(SQR(D2) + SQR(D1))

D5 = SQRT(SQR(D5) + SQR(D1))

Based upon simple trigonometry the above information is used to determine the corresponding angles. D7 and D8 can then be calculated using the Law of Cosines. Using several areas of the array, the optical origin can be calculated.

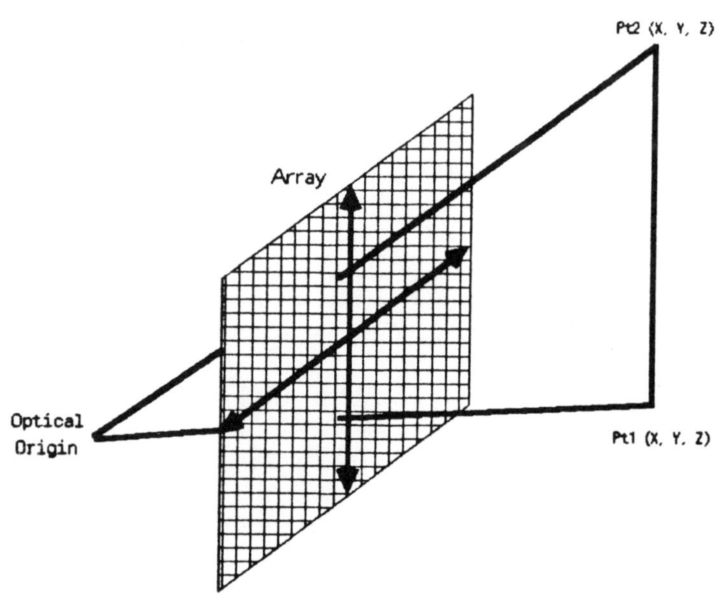

Fig 9a

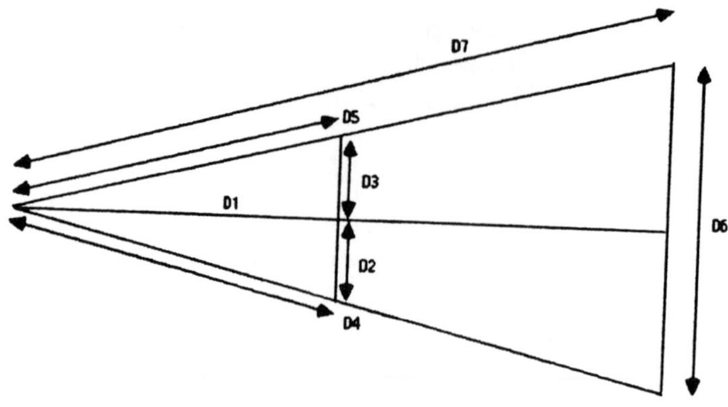

Fig 9b

At this point, the optical coordinate system is known, enabling the line through the centroid to be defined. One procedure to determine this line involves first instructing the sensor to extract an image using the strobe to highlight the object's image. Once the image is extracted, it may be necessary to filter the image. Many types of filters are known to elimate noise (such as high and low pass filters).

After the image is filtered, the centroid of the object is found. One technique to determine the centroid was discovered at the Standford Research Institue and is commonly called the SRI algorithm. This technique determines the centroid using the area of the image with respect to its position on the array. The centroid's position can be used to determine the line:

$$\text{Tline} = \frac{(\text{X\_optical} - (\text{Delta Row} * 0.027))}{((\text{Delta Row} * 0.027) - \text{Array Distance})}$$

$$\text{Tline} = \frac{(\text{Y\_optical} - (\text{Delta Column} * 0.027))}{((\text{Delta Column} * 0.027) - \text{Array Distance})}$$

$$\text{Tline} = \frac{(\text{Z\_optical} - \text{Array Distance})}{\text{Array Distance}}$$

Note that the line and plane are in different coordinate systems. Therefore, before the intersection can be calculated, the line must be mapped into the sensor coordinate system. This is accomplished by multiplying the line by the inverse matrix of,

$$
\begin{bmatrix}
xzPlane\_A & xzPlane\_B & xzPlane\_C & OpticalOrigin\_X \\
yzPlane\_A & yzPlane\_B & yzPlane\_C & OpticalOrigin\_Y \\
xyPlane\_A & xyPlane\_B & xyPlane\_C & OpticalOrigin\_Z \\
1 & 1 & 1 & 0
\end{bmatrix}
$$

This transformation into sensor coordinates can be intersected with the linear regression plane found earlier. This intersection is a point X, Y, Z in real space.

Conclusion

Because of the time required for the calculations, the above algorithm may seem impractical for real time applications. However, most of the calculations are performed during the set-up. The only computations during the running between parts are the linear regression plane fit, SRI algorithm, mathematical transformation, and the intersection calculation. The algorithm runs in one second using a Motorola 68000 micro-processor at 8 MHz to perform the calculations. If a higher density array is used with a better calibrating process, it is conceivable that the specification of this algorithm may be reduced to 50 microns at three sigma.

This technique is being used at several automotive plants today and is planned to be installed at many other sites. Accurate three dimensional measurements will assist the machine vision industry to achieve its projected status as one of the best tools for measurement and quality control.

CHAPTER 5

OFF-LINE METROLOGY

Presented at the Society of Photo-Optical Instrumentation Engineers'
Third International Conference on Robot Vision and Sensory Control,
November 1983

Vision-based Parts Measurement System

by Frank L. Skaggs
Corporate Manufacturing Technology Center

Abstract

A computerized parts measurement system (PMS) controls cost and reduces value-added waste in the production of multilayer printed circuits, printed circuit layer art work, and general mechanical parts at Texas Instruments by automatically inspecting these parts for dimensional correctness before assembly.

Employees can learn to operate all features of the menu-based, user-programmable system in about three days. The PMS uses advanced computer-vision technology to increase accuracy, repeatability, and throughput while eliminating the adverse effects of inspector-to-inspector variances. It is also compatible with automated factory and fabrication cell concepts.

The PMS noncontact optical metrology concept has demonstrated productivity-enhancing quality control inspection automation for many types of parts in sizes 45x36x12 inches (x,y, and z) with accuracies to 0.0002 in. while increasing part throughput many orders of magnitude.

The problem, approach, and results

As industry modernizes, the use of automated assembly methods continues to increase at an ever-increasing rate.[1,2] The cost of producing assemblies using automated methods is greater than the cost of manual assembly methods if the individual parts do not fit because of poor dimensional processing and inspection before assembly. If a part does not fit, means must be found to eject the part from the process hopefully before the machine jams or breaks, stopping the assembly process.

When a dimensioning defect occurs early in the production cycle but is not detected until much later, time for process, energy, materials, and human labor have been wasted. The more process steps between defect occurrence and detection, the greater the accumulation of value added waste.

Also, if the assembly process is preceded by an automated fabrication facility, the need for prompt (real-time) dimensional data is critical to overall factory control as early in the fabrication process as possible. Step-by-step dimensional inspection of all critical part features would seem to be the obvious solution, but monitoring by human inspectors is expensive. When an assembly involves hundreds of parts, it usually is impractical. In processes involving several steps per second, it is impossible.

The dimensional inspection of printed circuit boards, for example, is a unique problem. By its very nature, a two-dimensional part requires the use of some form of noncontact measurement. Traditional methods of measurement such as plug gages, calipers, and even the more recently available computer controlled coordinate measuring machines (CCCMM)[3] can at best determine the location of a hole and, if the hole is large enough or time is no problem, the hole's diameter. Such methods are slow. Several of these systems have been developed to inspect printed circuits for cosmetic and functional defects.[4,5,6] All of these systems were designed primarily for detecting etch pattern anomalies and defects, with no emphasis on the dimensional quality of the layers or assembled boards.

Other factors further complicate the dimensional inspection problem. Most parts are not, in fact, two-dimensional but possess topological properties such as warp, etched pattern height variances, and in many cases intentionally produced reliefs. It is necessary in such cases to be able to measure, on request, some of a part's z-axis (or depth) properties at high accuracy. The PMS concept is a method of combining rigidly accurate coordinate measuring techniques with digital image processing technology to allow a computer to emulate a human inspector.

The PMS approach to the problem

The Texas Instruments PMS concept addresses traditional questions of absolute accuracy and repeatability while increasing throughput, reducing the effects of variable human acuity, and adding compatibility with new automated factory and fabrication cell concepts.

The PMS is a new, noncontact dimensional inspection system capable of automatically inspecting a wide range of parts from printed circuit board layers to molded, stamped, or die-cut metal and plastic parts. The system can "see," locate, and measure such features as: holes, slots, printed lines, conductors, etch patterns, edges, corners, geometric patterns, etc.

The PMS applies binary computer vision to images from its television camera to convert what it "sees" to digital form. Its computer routines extract information about features in the images to compute distances, hole sizes, and relative positions of the features. Measurement data can be stored, displayed, printed, or compared against nominal values, tolerances, or both. This information can be used to signal variations from a standard, to collect statistics about such variations, or to measure and record part dimensions.

Results with PMS

The highly automated nature of the PMS in concert with the extremely fast vision processing technology has produced an inspection system capable of rapid throughput without the need for additional inspectors as product volume increases. Table 1 lists a few examples of typical measurements and the average time to execute the measurement and display the results if a reject is encountered.

Table 1. Typical Measurement Cycles

Measurement Performed	Time(s)
Length (~0.1 inch)(x,y plane)	0.8
Hole diameter (~0.25 inch)	1.8
z height (0.01 inch)(z-axis)	3.6

The PMS increases throughput by about one-half to three orders of magnitude depending on the complexity of the inspection and the needed amount of data reporting and statistical analysis. Direct computer analysis of measurement data greatly reduced overhead costs. One PMS application currently is demonstrating its ability to increase the reliability of the inspection process, increase part throughput by four orders of magnitude, and save about $120,000 annually in quality control overhead costs.

In another application at Texas Instruments, an operator can program a typical 8 x 8 in. printed circuit board layer for 50 measurements in less than 30 minutes while viewing on the monitor either an analog or digital picture of the scene scanned by the television camera. The program, stored on a high-speed data cassette, will measure each part automatically in about three minutes. The resultant productivity increase has been evaluated as 10 to 1 over the previous semimanual optical comparator methods.

The increased measurement speed and reduced data processing time accelerate system payback, thus offsetting equipment capitalization much more effectively than nonautomated inspection methods. This system feature becomes even more important when the PMS is integrated into existing or planned computerized production control systems.

System description and discussion

The PMS is a noncontact inspection system using a distributed architecture for computer image processing and general system control. The most important functions of the system are image processing performed by a Texas Instruments 16-bit (word) minicomputer and the main system control, including application program machine control language generation, editing, and execution; accumulation and processing of dimensional data; and the proper timing control of the system's peripherals such as the CRT displays, printers, and programming console TMS9980 microprocessors. All of the control functions are performed by the TM990/102 16-bit microcomputer, supported by 64K-byte dynamic RAM and 128K-byte nonvolatile magnetic bubble memory. Figure 1 is a block diagram of the main system blocks, without regard to their actual physical location. The physical distribution of the components is shown in Figure 2. Figure 3 shows a typical system (left to right), a 960B vision computer, programming console, and an inspection table system with printed circuit board/layer holding fixture. Figure 4 shows the largest system currently on line, with 48 x 36 x 12 inch measuring ranges.

Programming console

The primary means of PMS human interface is the programming console panel. The part programmer enters an inspection program through a combination of strategically located front panel buttons and logically ordered and displayed menus shown on one of two TV monitors. The console also provided key interfaces such as the vision monitor, which displays either an analog or digital (binary) picture of the scene viewed by the TV camera; the cassette tape unit used for the entry of operating system software to the system's RAM and magnetic bubble memory and for recording or reading part programs; the hard copy printer; and the TM990/102 system control computer and its peripheral interface circuits.

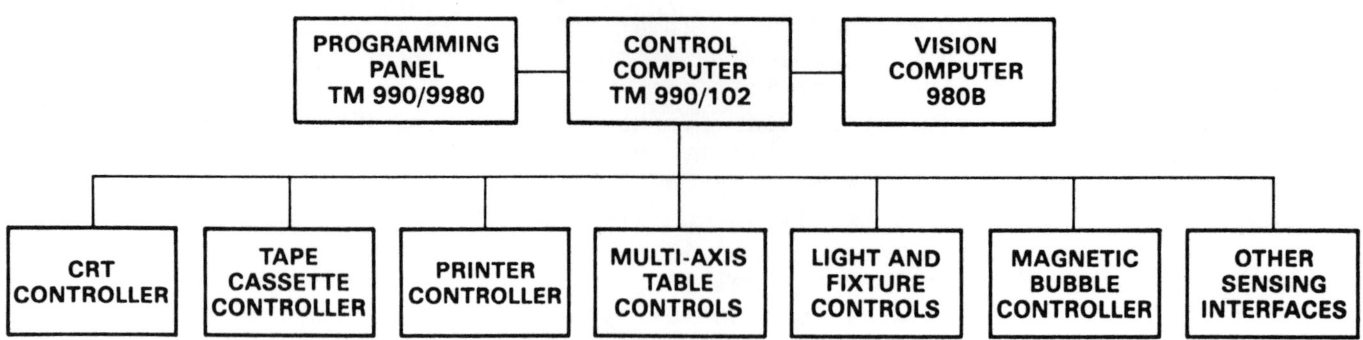

Figure 1. PMS major components

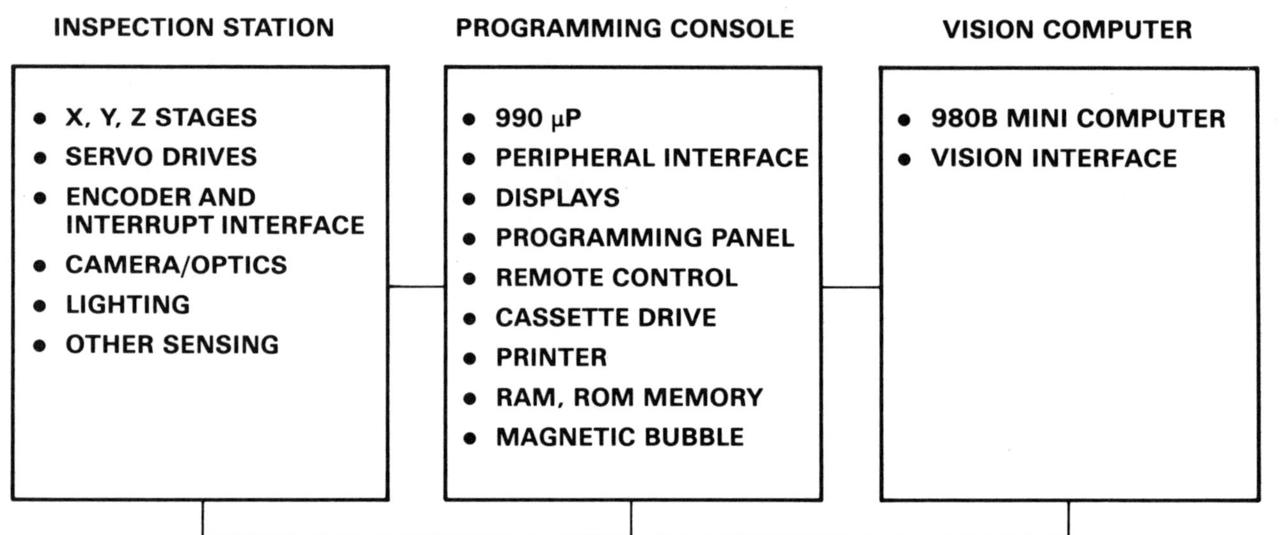

Figure 2. Physical distribution of components

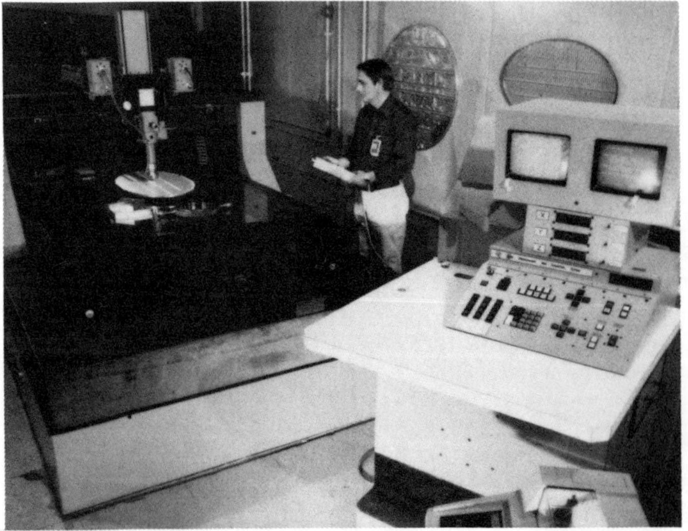

Figure 3. Typical PMS system layout

Inspection station

The inspection station is the subsystem on which all parts are placed before generation of part program or the actual automatic inspection of a part. This subsystem contains the PMS optics/camera assembly; precision x,y, and z translation stages; special lighting; the x, y, z dc servomotor drive amplifier and the clock interface board for decoding the x,y, z position encoder signals; and the interfacing of the TM990/102 control register unit's (CRU) commands issued or received by the inspection station. The designs of the x,y, z electronics located in the programming console and clock interface electronics in the inspection station are general enough that 3-axis inspection stations with measurement ranges up to 48 x 36 x 12 inches have been built and used as shown in Figure 4.

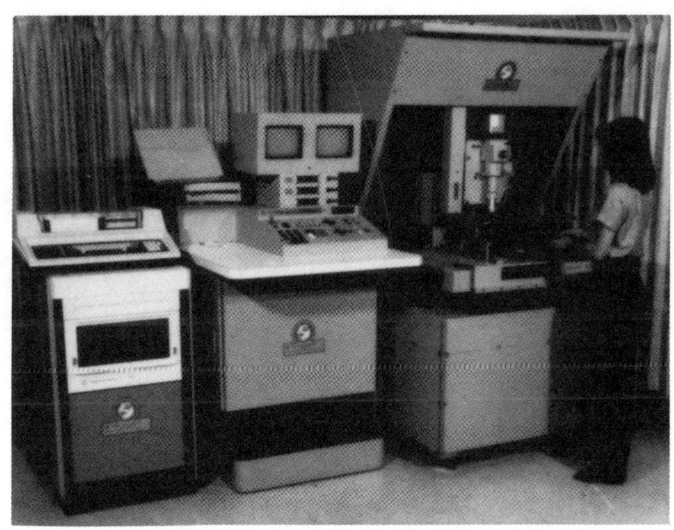

Figure 4. 48- × 36- × 12-inch PMS

Vision computer

The heart of the noncontact vision inspection capability is the vision computer. Television images generated by the camera/optics assembly at the inspection station are received by the computer vision interface where they are preprocessed before binary picture processing.[7,8,9] The decision concerning when to process a television image or what process is to be performed by the computer is the programmer's choice, and all such decisions are stored as part of the TM990/102 Machine Control Language (MCL) on the data cassette at part programming time.

The image processing procedure results in a determination of parameters such as:

Where is an edge in the television image?

What is its orientation?

Is its contrast and signal-to-noise ratio of sufficient quality to continue the MCL sequence as written or should some other process be followed, such as resorting to semimanual inspection?

The parameters determined by the vision computer are then returned to the TM990/102 microprocessor where other critical data such as table position or changes in table position are combined to produce such final dimensional calculations as:

Part length
Hole diameters and position
Radius of arc
Conductor width
Part z-height relative to reference plane.

Currently, PMS engineers are working to expand the system's geometric tolerancing capabilities to include cyclindrical and noncylindrical true position, perpendicularity, and concentricity measurements.[10,11]

Actual monitor displays for a printed circuit board layer conductor are shown in Figure 5. The analog or full gray-scale picture is in Figure 5A, and the binary thresholded equivalent is shown in Figure 5B. The system's cursor boxes and crosshairs are shown superimposed.

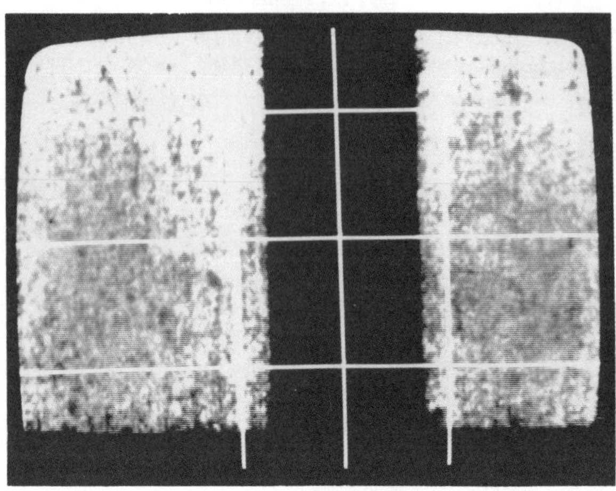

(a) ANALOG VIEW

(b) DIGITIZED VIEW

Figure 5. Actual monitor displays for a conductor

Programming

The PMS is programmed thorough a high-level menu-based language. The menus prompt the operator through any selected type of part measurement that inputs the data required to carry out that part measurement. The key steps in setting up the system are given in Table 2 for the case in which the system has previously been powered up and successfully calibrated.

Table 2. Part Programming Steps

I. Palace the part on the x, y, table
II. Define the alignment procedure
III. Define part datum (0.0)
IV. Program the system by teaching or entering drawing coordinates via the table step function
V. Validate the program by calling auto/run
VI. Edit measurements or basic binary picture parameters as needed
VII. Store the program on cassette tape for future use

Programming may be performed on line through menus or off line through a CAD/CAM system. During on-line programming, for example, the main menu (Figure 6) allows the operator to choose which portion of the system to use. To create or edit a part measurement program, for example, the operator selects "Program A New Part" from the main menu and the monitor will respond by displaying the alignment and origin menus shown in Figure 7.

Each time the operator makes a choice, the system branches to a new screen where the next logical decision is made. The typical decision flow for a feature-to-feature distance measurement is shown in Figure 8. During the programming sequence, the vision monitor displays the feature that will be seen when the program is run, the window size that will be selected to process the picture, and the digital threshold for the processing. The BCD displays will show the actual table position.

MAIN MENU

0 SINGLE STEP PROGRAM
1 CASSETTE SUBSYSTEM
2 PROGRAM A NEW PART
3 LIST PROGRAM STEPS
4 STATISTICAL PACKAGE
5 CHANGE PROGRAM NAME
6 X-Y TABLE UTILITY
7 TRACE FUNCTION UTILITY
8 MANUAL MEASUREMENT
9 PRINT SETUP FOR AUTO-RUN

SELECT ITEM, PRESS -ENTER-

Figure 6. Main system menu

PROGRAM A NEW PART 2.

PART ALIGNMENT

MOVE TO POSITION # 1
SELECT FEATURE TYPE
0 EDGE
1 CIRCLE CENTROID
2 CIRCLE 4 POINT
3 FORCE 0 DEG. ALIGNMENT
4 MAX-MIN POINT SEARCH
5 CENTER OF ARC
6 MORE FEATURES
7 DONE

SELECT ITEM, PRESS -ENTER-

PROGRAM A NEW PART 2.

PART ORIGIN

MOVE TO PART ORIGIN
SELECT FEATURE TYPE
0 1ST ALIGNMENT POINT
1 CIRCLE CENTROID
2 CIRCLE 4 POINT
3 INTERSECTION OF TWO LINES
4 MANUAL ORIGIN
5 MAX-MIN POINT SEARCH
6 CENTER OF ARC
7 MORE FEATURES

SELECT ITEM, PRESS -ENTER-

Figure 7. Part alignment and part origin menus

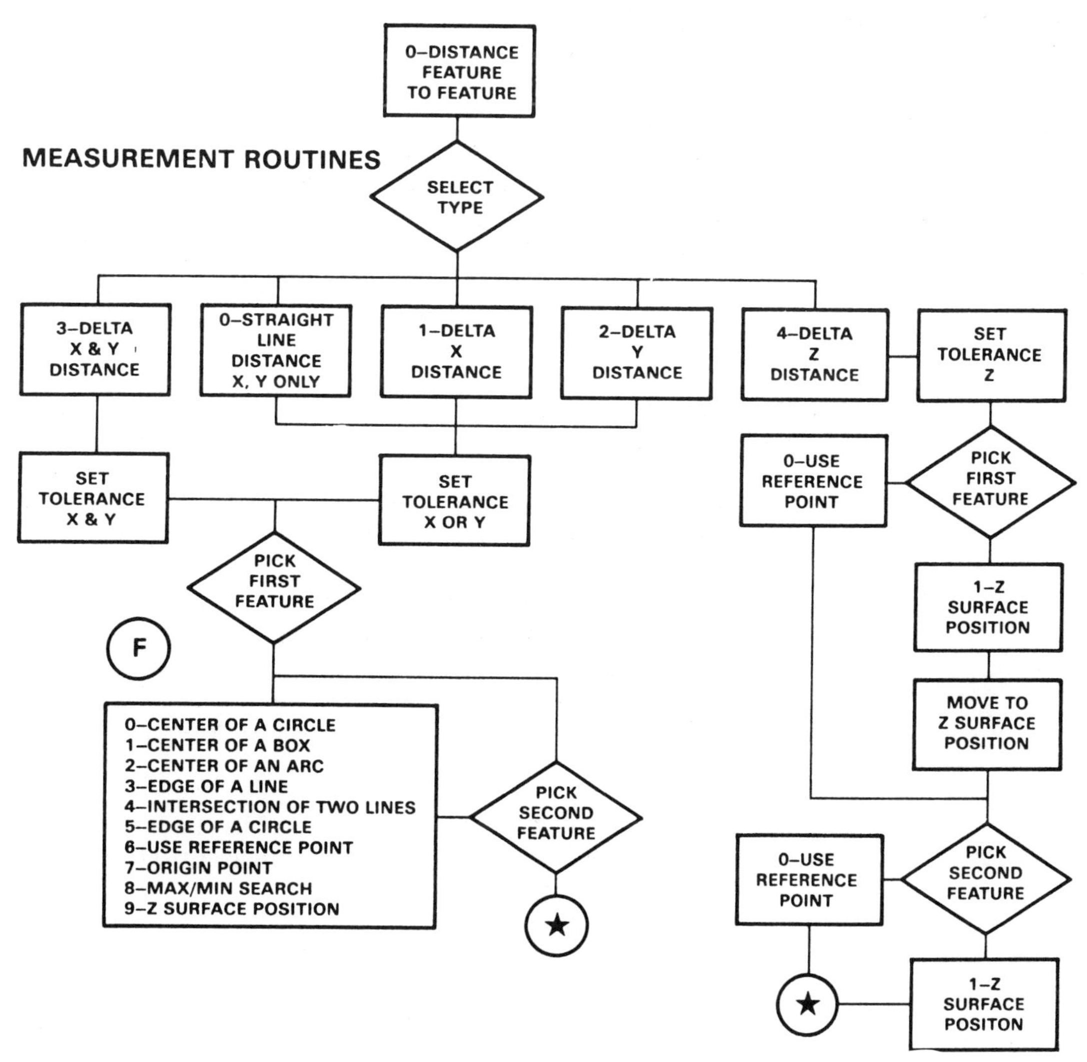

Figure 8. Typical part program step decision tree

The PMS also may be programmed off line using an augmented version of the automatically programmed tooling (APT) NC programming languages. APT code may be generated manually or interactively using CAD/CAM equipment. The APT code is then compiled on a 990/DX10 system and downloaded to the PMS. This off-line programming feature allows the PMS to operate in a fully automated group technology or flexible machining system environment. It effectively closed the loop between part design, production, and inspection by providing the capability to generate both production and inspection programs from the same design database.

Parts programs that have been either off line or on line may be edited at any time and subsequently saved. It is possible to make a dummy run through the program to tweak the thresholds, table position, and window sizes if necessary. It is also possible to run a parts program on one PMS even though it may have been created on another PMS (if the systems' hardware architectures are compatible).

The program editor allows the operator to change any portion of a part program should he detect a mistake or a need to respond to a system error message. This feature also allows operators to modify measurement routines to accommodate engineering changes to parts already programmed. Furthermore, the system's programmability allows a single PMS to inspect any of a number of different types of parts, allowing for maximum inspection capability with minimum capital outlay.

Measurement speed and throughput

Comparisons performed against conventional methods would show that inspection times may be reduced by 4 to 20 times, depending on the dimension to be measured. If the time necessary to perform tabulation of the data and computation of part reject statistics is included, measurement efficiency is increased by 40 to 60 times.

Accuracy

Two primary parameters by which any dimensional inspection system is judged are its measurement accuracy and repeatability. Accuracy is defined as the degree of agreement of the measuring instrument to an accepted standard. Repeatability is the ability of the instrument when used in its normal manner to repeat previous measurements and typically is expressed as one, two, or three standard deviations (sigma) from the mean measured value.

With gages, calipers, and optical systems using manually positioned cursors, both accuracy and repeatability are greatly affected by the individual visual acuity and the spread of manual skills among the operator group. Indeed, operator-to-operator variances increase repeatability standard deviations by at least four times the basic accuracy of the instrument used.

The combination of highly accurate x, y, z translational stages, precise position encoders, and system programmability removes most of the human causes of reduced accuracy and repeatability.

There are two major contributors to inaccuracy and reduced repeatability in PMS. The first is table distance moved to encoder readout agreement, and the second is camera/optics assembly resolution. For the printed curcuit board PMS,[12] the table-to-encoder disagreement does not exceed +/-0.0002 inch, and the camera/optics assembly can easily resolve 200 lines pairs per millimeter in the object plane. A comparison of the absolute accuracy of the x, y stage assembly to an NBS-traceable laser interferometer distance measuring device is shown in Figure 9. As can be seen, the net error envelope for a 10-inch PMS never exceeds +/-0.0002 inch.

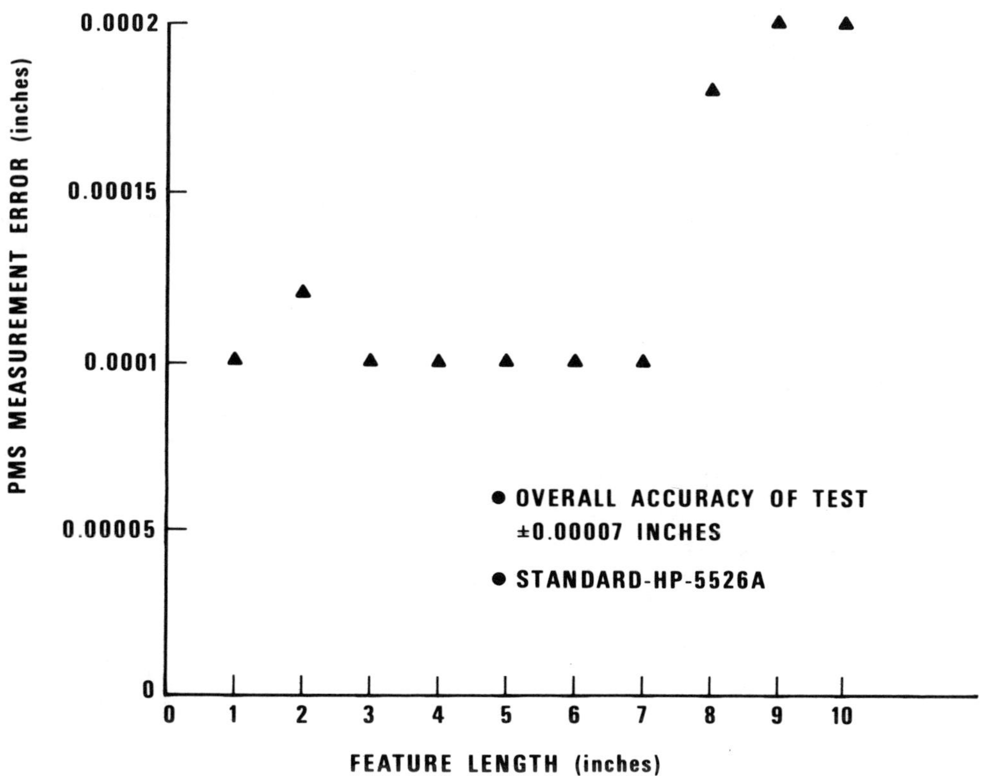

Figure 9. PMS accuracy (compared to NBS-traceable laser interferometer)

Some of the other overall accuracies of PMS when measuring a complete length or feature are shown in Table 3. In no case did the three-sigma repeatability exceed 0.0003 inch.

Table 3. Typical Measurement Accuracies

Feature	Accuracy
Length	
(up to 2.0 inches)	± 0.0001 inch
(up to 10.0 inches)	± 0.0002 inch
Hole diameter	
(up to 2.0 inches)	± 0.0001 inch
(up to 10.0 inches)	± 0.0002 inch
Hole location	± 0.0002 inch
Radius of arc	
(up to 2.0 inches)	± 0.0001 inch
(up to 10.0 inches)	± 0.0002 inch
Width of a conductor	± 0.0001 inch
Angle between two lines	± 0.10 degree

Optics, camera, and lighting

The image forming components of the system (camera, optics, and lighting) are designed to optimize images of desired part features. Different parts require different techniques, so the system includes a good deal of flexibility.

Magnification is selected to provide the necessary resolution with low light levels. Resolution requirements of from 50 to 200 line pairs per millimeter have been provided to inspect parts ranging from glass masters for PC artwork to molded three-dimensional parts. The optical system included a provision for doubling (or halving) the primary magnification to allow changing either field-of-view size or resolution per pixel.

Three types of illumination are provided. Back side illumination aids inspection of objects like artwork and parts with through holes. Front side on-axis illumination is used for z-axis autofocus on machined parts and x/y measurements of surface features with high foreground/background contrast. Front side off-axis illumination permits inspection of surface features such as slots and blind holes. The illumination system also includes neutral density filters and, where needed, spectral filtering to optimize image contrast. The television camera uses a Plumbicon imager adjusted to provide the necessary linearity for critical measurements while retaining its characteristically long imager life and resistance to target burn.

Autofocus and z-height measurements

The key to the PMS z-height measurement capability is its accurate autofocus capability. Once a part surface has been located automatically, the height of that surface is remembered by the TM990/102 microprocessor. The distance between two such surfaces is the height. The method of locating a surface is by the detection of the surface texture and an optimization of the transition or binary picture element counts between the light and dark regions of the part surface. Figure 10 shows a drawing of a typical magnified cross section of a part surface brightness as a function of distance along the surface. If the binary picture threshold (black-to-white decision line) is set as shown by the dotted line, a transition can be defined and counted with the maximum number of transitions occurring at optimum focus.

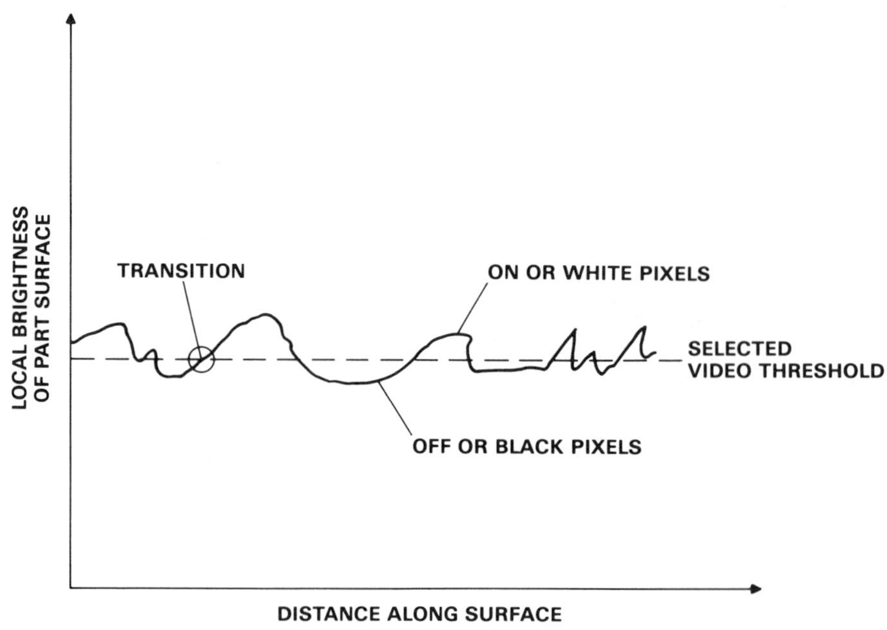

Figure 10. Part surface brightness as a function of surface position

By constructing a histogram of those transition counts or the binary picture elements remembered as a function of the optical column z-height position, the maximum count location (histogram peak) may be determined through analysis of the histogram shape. Figure 11 shows a typical histogram of white above-threshold picture element counts versus z-height. Typically, a surface z-height may be determined to an accuracy of +/- 0.00015 inch within three seconds of the time that the search begins near the nominal height position.

PMS in the automated factory

The computerized PMS is highly compatible with automated factory and fabrication cell technology. When a CAD operator designs a part, he creates a database that often is used by methods and tooling personnel to create NC or CNC machine instructions used to actually build the part. As mentioned in the programming section, PMS designers have carried this a step further by developing a method to allow quality operators with access to the design database to create PMS measurement programs off line. This not only reduces the human effort involved in the task but also ensures that both "build" and "inspect" programming proceed from the same data. The PMS inspects a part based on its design dimensions, not its manufactured dimensions.

The accuracy and repeatability of the PMS is ideal in an environment where large production lots of the same part are to be measured. With a PMS, the number of samples in a lot can be increased, in some cases up to 100 percent, while still maintaining throughput. It also offers superior data collection and statistical analysis capabilities over conventional inspection systems. The ability of the PMS to do real-time measurements on line makes it an ideal addition to the automated factory environment.

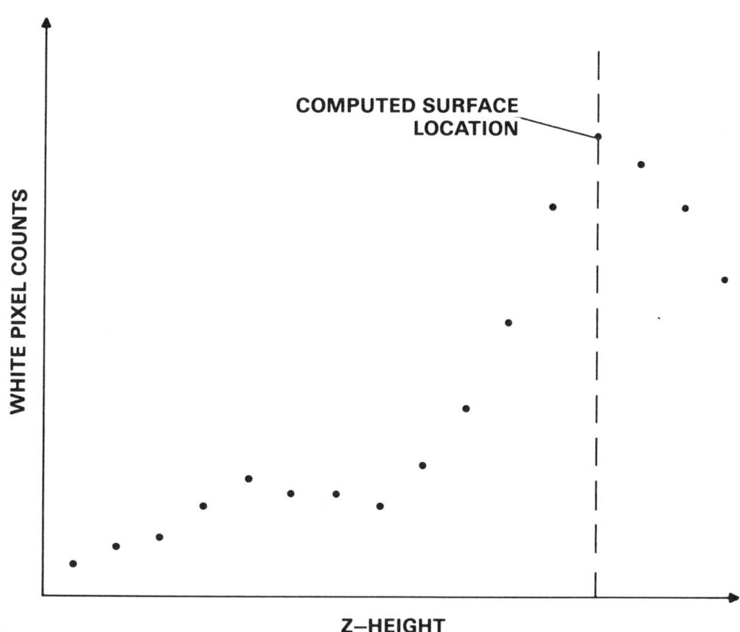

Figure 11. White pixel count versus *z*-height histogram

In the automated factory environment, the PMS can be configured as a standalone measurement station or it can be programmed to communicate with a master control computer. The PMS can communicate with other equipment through an optically isolated interface such as the Texas Instruments 5MT series input and output modules or an RS-232C communications link at up to 9600 baud.

Conclusion

The PMS is an innovative approach to printed circuit board dimensional inspection that not only performs the measurement of board holes and router features but through the use of carefully selected optics and special computer vision processing can extract conductor widths, radii, heights, and gaps with accuracies better than +/-0.0003 inch. Such concepts have been extended to large mechanical parts of up to 48 x 36 x 12 inches.

This approach locates manufacturing defects at each stage of the multilayer circuit board production process and reduces manufacturing overhead by controlling costs and value-added waste. It also offers superior data collection and statistical analysis efficiency and will integrate directly into the totally automated factory and fabrication cell concept.

References cited

1. Treer, Kenneth R., "Automated Assembly," Society of Manufacturing Engineers (1979).
2. Proceedings of the 1st International Conference on Assembly Automation, UK; IFS (Conferences) Ltd. (March 25-27, 1980).
3. Roth, Edward S., "Gaging: Practical Design and Applications," 163-171. Dearborn, Michigan: Society of Manufacturing Engineers (1981).
4. Hara, Yasuhiko: Akiyara, Nobuyuki: and Karasaki, Koichi, "Automatic Inspection System for Printed Circuit Boards," Industrial Applications of Machine Vision, IEEE Computer Society, 62-70 (1982).
5. Sterling, Warren M., "Nonreference Optical Inspection of Complex and Repetitive Patters," Proceedings, Vol. 281, Society of Photo-Optical Instrumentation Engineers (SPIE), 182-190 (1981).
6. Restrick, Robert C., III, "An Automatic Optical Printed Circuit Inspection System," Proceedings, Vol. 116, SPIE, 76-81 (1977).
7. Geisselmann, H. Tropf H; and Foith J.P., "Some Applications of the Fast Industrial Vision System S.A.M." Industrial Applications of Machine Vision, IEEE Computer Society, 73-79 (1982).
8. Keferstein, Claus, "Practice-Oriented Use of the TV Camera in Automated Linear Metrology," Proceedings: 5th International Conference on Automated Inspection and Product Control, Institut fur Produktionstechnik and automatisierung (IPA), Stuttgart, Germany, 67-78 (1980).
9. Hewkin, P.F. and Phil, M.A.D., "An Approach to the General Purpose Robot Vision System," Proceedings of the First International Conference on Robot Vision and Sensory Controls, 313-322.
10. ANSI Standard Y14.5 (1973).
11. Foster, Lowell W., "Geo-Metrics II," Addison-Wesley, Reading, MA (1979).
12. Skaggs, Frank, "Computerized Parts Measurement System for Printed Circuit Boards," Proceedings of the 2nd Annual International Electronics Packaging Conference, 361-375 (1982).

Presented at the RI/SME Robotics Systems in Aerospace Conference
March 1985

3-D Vision and its Applications in Aerospace Manufacturing

by Gerard Betz
Robotic Vision Systems, Inc.

INTRODUCTION

The widespread use of robotics in industry and in particular the aerospace industry will to a large extent depend on the integration of robots with various types of sensors (i.e. 2-D vision, 3-D vision, force, torque etc.) and intelligent sensory processors. The need for sensory controls in the aerospace industry is driven by the small lot batch manufacturing environment of that industry.

Although the field of sensor control of robots is a new field some systems have been developed and are in use. In the area of 3-D vision, RVSI has produced a series of sensor systems, the first of which was installed in 1981 and is still in operation. Since then, the 3-D vision sensor has evolved and it is now a modular product which can be configured to accomplish a variety of tasks.

The core of the Company's 3-D Robo Sensor® vision systems is the electro-optical Vision Module sensor (Model 210 Series). This sensor, introduced in the spring of 1984, is designed to be easily reconfigurable so that it may be optimized for each task without an extensive design effort. The Robo Sensor and some of its aerospace applications will be described in more detail in the following sections.

VISION TECHNOLOGY

There are several types of vision systems which are in use today. The most commonly used type is 2-D vision using either binary or gray scale processing. A large and complex body of software has been developed over the years to perform the functions of part location, and measurement based on 2-D data input. Some systems being introduced now offer a tool-kit of part location and measurement algorithms. As robust as these algorithms are, they can often be confused or lack necessary data. As can be seen by observing Figure 1, a 2-D sensor cannot tell the difference between a small object close up, a large object further away or a flat picture of the object. A 2-D sensor can also be confused due to unexpected contrast variations as shown in Figure 2. There are other cases such as seam tracking for sealant application or welding where 3-D information is required. For these cases and to complement 2-D vision systems, RVSI has developed a family of 3-D vision sensors which have been integrated into turnkey factory cells.

FUNDAMENTALS OF 3-D VISION

The sensor systems produced by RVSI operate on the principle of optical triangulation. As shown in Figure 3 when a single spot of light is shown on a surface the distance to the surface can be determined by observing the image of the spot on the surface with an imaging device at known angle and distance to the projector. If the projector is not normal to the surface there is a cross coupling between the X and Y axes. The data obtained from this type of sensor is truly 3-D but it is from only a single point on the surface.

In order to improve the data rate, a plane or sheet of light, is projected onto the surface, as shown in Figure 4, and viewed by an array imaging device. Now 3-D data is obtained from a slice through the part being measured. In order to

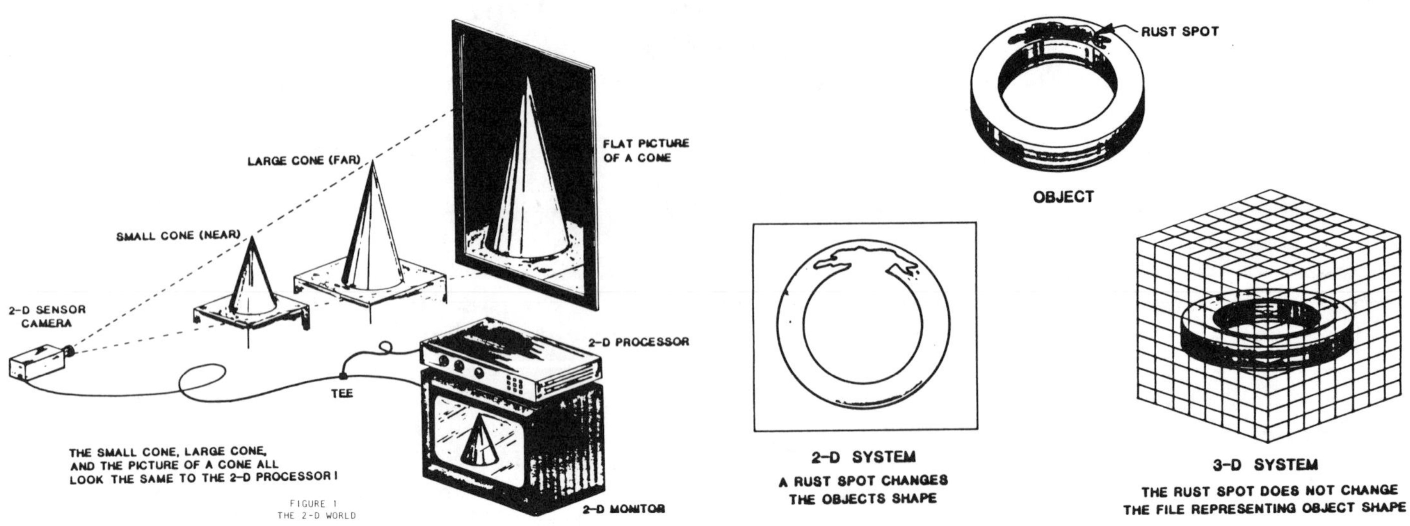

FIGURE 1
THE 2-D WORLD

FIGURE 2
A 3-D SYSTEM

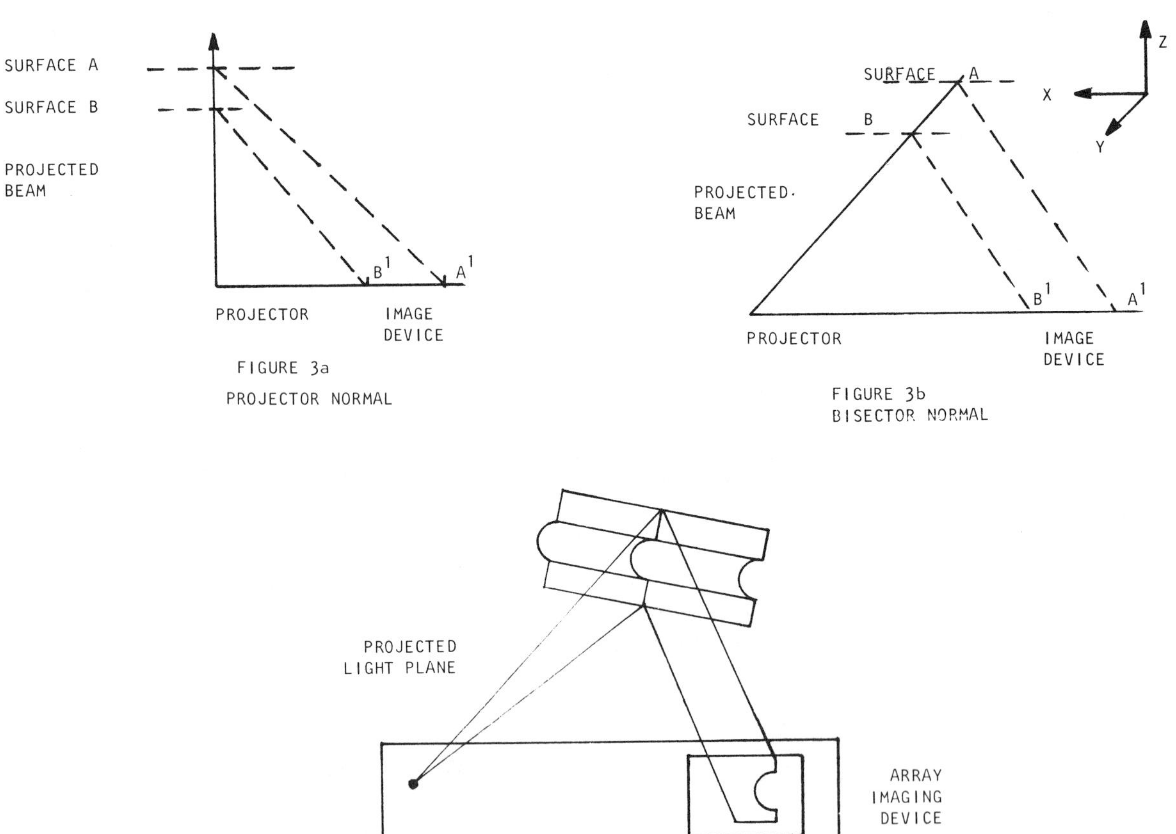

FIGURE 3a
PROJECTOR NORMAL

FIGURE 3b
BISECTOR NORMAL

FIGURE 4
PROJECTOR/CAMERA MODULE

measure the whole surface there must be relative motion in the X axis between the sensor and the part. Since this type of sensor measures 3-D data, but only from a single plane, it is often called a "2 1/2 D" or "line" sensor.

True 3-D volumetric measurements can be made, without relative motion, by projecting coded sequences of light planes onto the surface. This type of sensor has been developed and was used commercially in 1977; however, it required that the image be recorded on photographic film which after development was read into a computer for processing. A real time verson of the sensor was developed and demonstrated in the summer of 1984. A further description of the volumetric sensor will follow the application section. All the applications which are presented herein use a line sensor.

CONTROL APPLICATIONS

Sealant Application

Sealant is widely used in the construction of an aircraft to prevent both moisture and fuel leaks. RVSI has conducted tests to verify the ability of 3-D vision to control a robot to be used for sealant application. As part of this technology demonstration, software was developed to detect the joints to be sealed. Typical cross sections are shown in Figure 5. The test was conducted using a Cincinnati Milacron 746 robot and an RVSI Robo Sensor. Although a sealant dispensing system was not used for the test, the ability of the vision control system to position the robot to within 0.050 inches was demonstrated by the use of pointers and dial indicators. During the test, the part was translated and rotated in all axes with displacements up to 3/4 inches. The system operated in a two pass mode whereby a vision pass was used to measure the spacial deviation between the taught path and the actual part. This data was then used to calculate a new path which was downloaded to the robot for execution.

Cross-sectional 3-D data from a typical fastener is shown in Figure 6. Sufficient data is available to locate the position of the fastener as well as the orientation of the surrounding surface.

Workpiece Position and Orientation Compensation

Due to the small manufacturing lots typical in the aircraft industry, it is essential that set-up times be minimized and further it would be highly desirable if generic tooling could be used. These objectives can be met by using 3-D vision to locate a workpiece within a robot cell. The actual function of the cell is not critical for this discussion, it could be routing, drilling, sealant application or any other suitable processes. Figure 7 shows a robotic sealant cell which was developed for the auto industry. In this cell 4 vision sensors scan the car as soon as it enters the sealant cell and use the vision data to generate position and orientation corrections to the robot paths. Four sensors were used for this application to rapidly acquire the spatial location of the car body due to the short time in station on the auto line. In the case of the aircraft industry, with its slower production rates, a single sensor mounted to the robot could be used to locate key features on the part or fixture and use the same software to generation the correction data for the robot. An illustration of such a cell is shown in Figure 8.

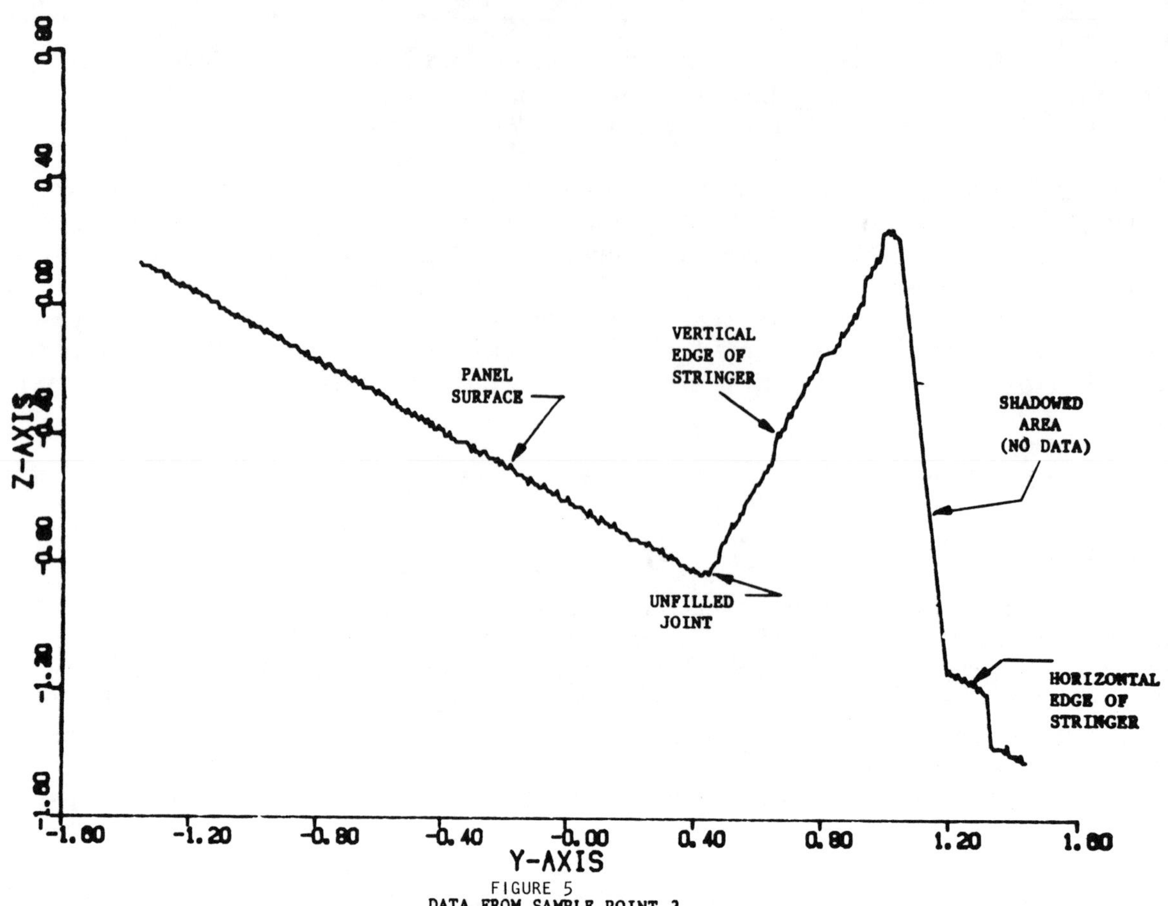

FIGURE 5
DATA FROM SAMPLE POINT 3

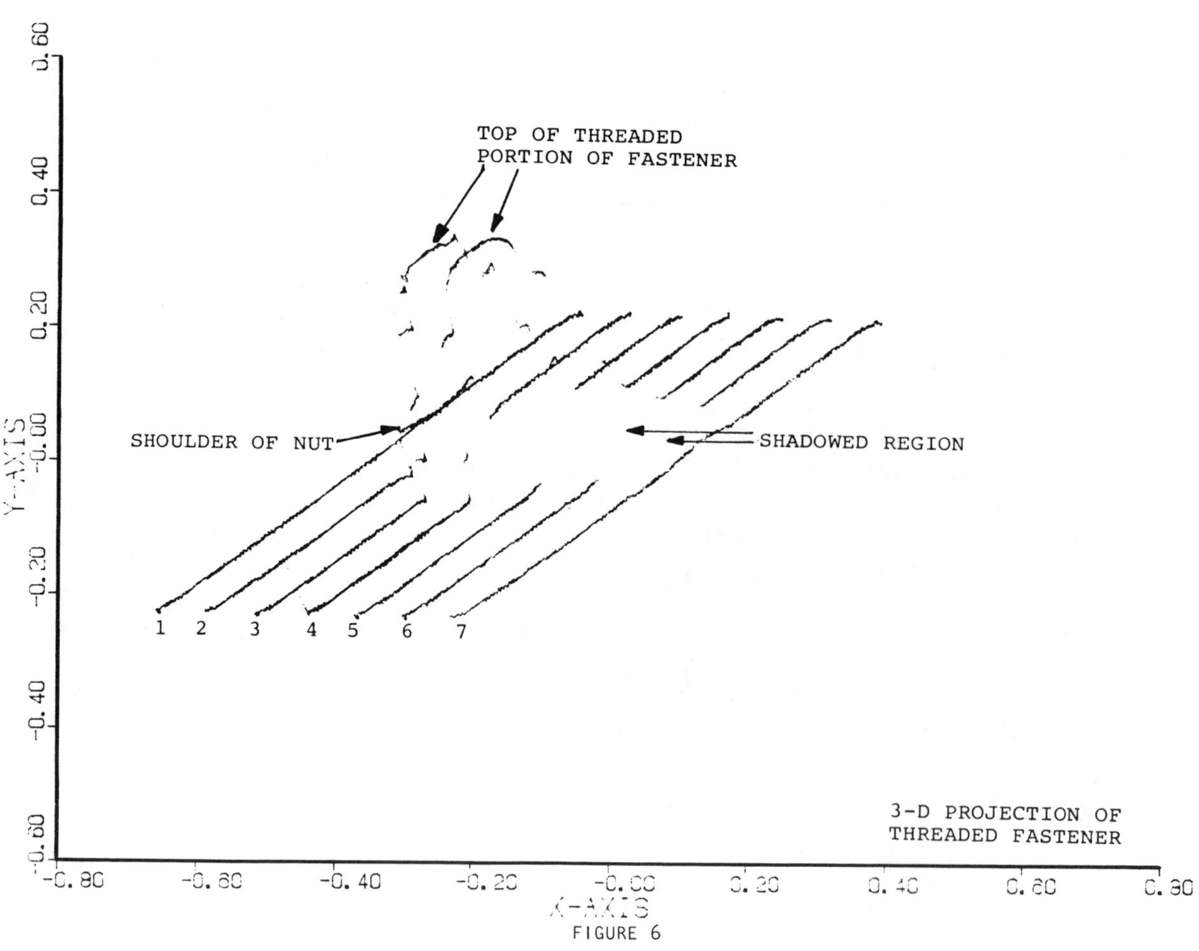

FIGURE 6

FIGURE 7
ROBOTIC SEALANT CELL

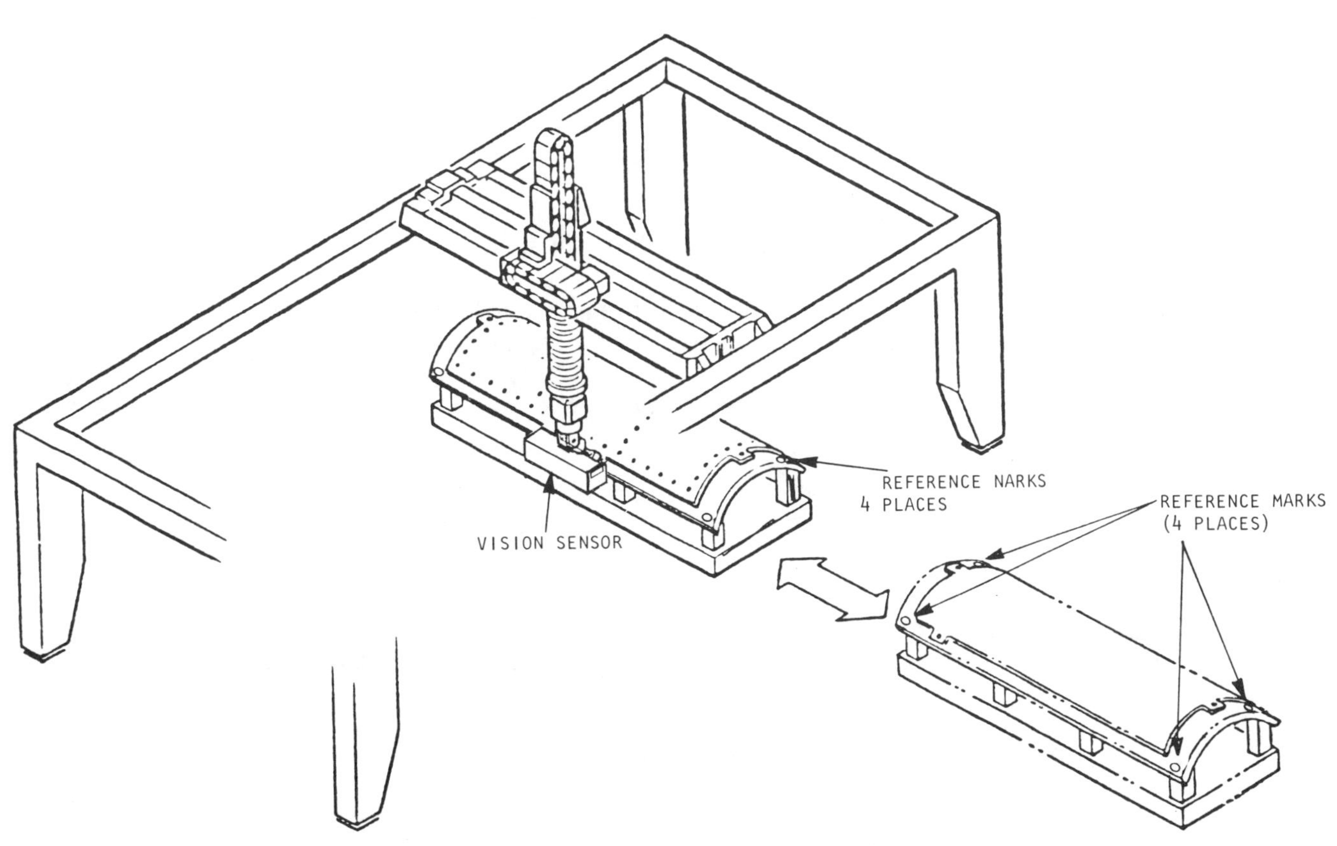

FIGURE 8
Routing Cell

Real Time Robot Control

In certain operations it is not possible to first scan a part and generate a program to later be executed by the robot. One reason for such a requirement would be if the operation being performed on the part causes the part to change shape or location. For such cases it is possible to track feature marks on the part and in real time to generate update data to the robot. RVSI is presently under contract from the US Navy to develop a sophisticated single pass welding tracker, the technology derived from the program will be applicable to real time control of other process such as those used in the aircraft industry.

Robot Path Generation

Aircraft manufacturing is typified by small batches of many different parts. This is a significant drawback to the use of robots since paths for each part must be taught and verified. Even if the part design is on CAD, robot accuracy limitations often necessate the trimming of the robot path generated on CAD simulators. 3-D vision can greatly shorten the time associated with robot path generation and verification. For example, if master parts or manual templates exist, a robot with a vision system mounted to the tool plate can be rapidly taught several key features on the part to be used to generate a vision scan sequence. The data obtained during the vision scan is then used to generate the final robot sequences. In like manner, path generated off line or on another robot can be executed by a robot with a 3-D vision sensor mounted to the tool plate which will use the sensor data to measure this displacement between the expected and actual locations of the part. This data can then be used to "customize" the path for each individual robot.

Aircraft Assembly

RVSI is a member of a team headed by Grumman Aerospace Corp. which is developing a robotic cell for the assembly of aircraft bulkheads. A pictorial of the cell is shown in Figure 9. The cell uses 2-D vision to locate the parts in kit trays. If the 2-D cannot, with a high confidence, identify and locate a part, the cell controller will request that 3-D data be acquired in a particular region of the tray in order to resolve any ambiguity. The 3-D sensors to be used in this system are true volumetric sensors.

3-D vision will also be used to measure the position and orientation of each part in the robot gripper and to locate features on the bulkhead for part placement. Since the system uses a true volumetric sensor, the required position and orientation measurements can be made without the need for scanning the sensor over the part.

INSPECTION APPLICATIONS

Low Tolerance Inspection

Many parts fabricated for the aircraft industry are manufactured to comparatively low tolerance (approximately 0.030 inches). Inspection of such parts is readily achieved by using a robot based vision inspection system. Figure 10 shows a Robo Inspector which uses a Cybotech V15 robot to translate a 3-D vision sensor. The photograph shows the system inspecting an engine frame. The system measures 40 features on the part and compares them to allowable tolerance and

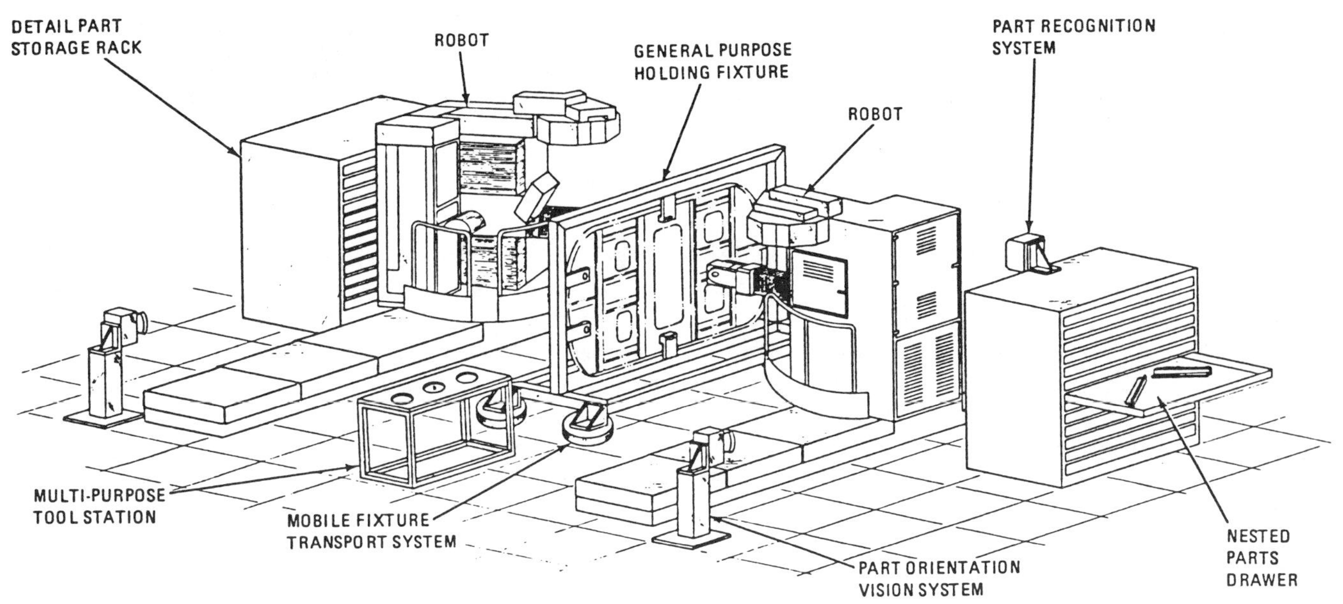

FIGURE 9
AUTOMATED SUBASSEMBLY OPERATIONS

FIGURE 10
ROBO INSPECTOR USING CYBOTECH V15 ROBOT
TO TRANSLATE A 3-D VISION SENSOR

reports out of tolerance conditions. The system is accurate to 0.010 inches and can measure the 40 feature in 2 minutes. Measurements can be made relative to either the holding fixture or other features on the part or both. Since the measurements are made relative to other features, set-up time is vertually eliminated. The inspection volume can easily be changed by selecting a robot to match the volume and accuracy requirements of the part being inspected. A system of the type would be applicable to the inspection of sheet metal and tubing parts in the aircraft industry.

Composite Tape Layup Inspection

More and more composites are being used in the fabrication of aircraft. 3-D vision is a useful tool for certain inspections of composite parts. Figure 11 and 12 show the unprocessed vision data obtained by scanning an uncured composite sample fabricated of 0.006 inch composite tape. Figure 11 shows a gap between two plys of tape. The data could be further processed to either inspect partially completed parts or in real time to measure the gap or lap for control of the tape laying machines. Figure 12 shows the results of scan across several plys of tape. Each 0.006 inche step is clearly visable. The data could be used to verify proper part construction.

Precision Inspection

It is possible to use 3-D vision sensors to perform accurate inspection. The Robo Sensor has an accuracy of 0.001 inch for a 1 inch filed of view (field of view is defined as the height of the inspection plane of light, that is the Z axis shown in Figure 4). Since robots do not have repeatability in this range, the vision sensor must be mounted to a more precise translation device. Figure 13 shows 4 sensors mounted to a custom designed translation unit. The system was manufactured for the Cummins Engine Co. and is used to inspect engine castings. The system makes 1250 measurements on the block in 40 minutes. These measurements are compared against the blueprint tolerance and exceptions are reported. The system also analyzes the data to determine if a shifted datum will bring all points into tolerance and reports out the new datum coordinates if one exists. It is also possible to mount the sensor to precision machine tools or large coordinate measurement machines to perform precision measurement tasks.

Sealant Inspection

As part of the robotic sealant application evaluation, RVSI also conducted limited tests to determine the ability of 3-D vision to inspect the applied sealant. The tests were performed on several parts. Data representing a cross section through the part in the unsealed state was shown in Figure 5. Figure 14 shows a cross section of a properly sealed region and Figure 15 shows a region with a void in the sealant. The test results from these and other samples indicated that 3-D vision could be used to measure sealant defects. Before a final determination of the practicability of using such a system can be made, the defects in sealant must be quantitatively defined and more rigorus tests performed, however, the initial results are promising.

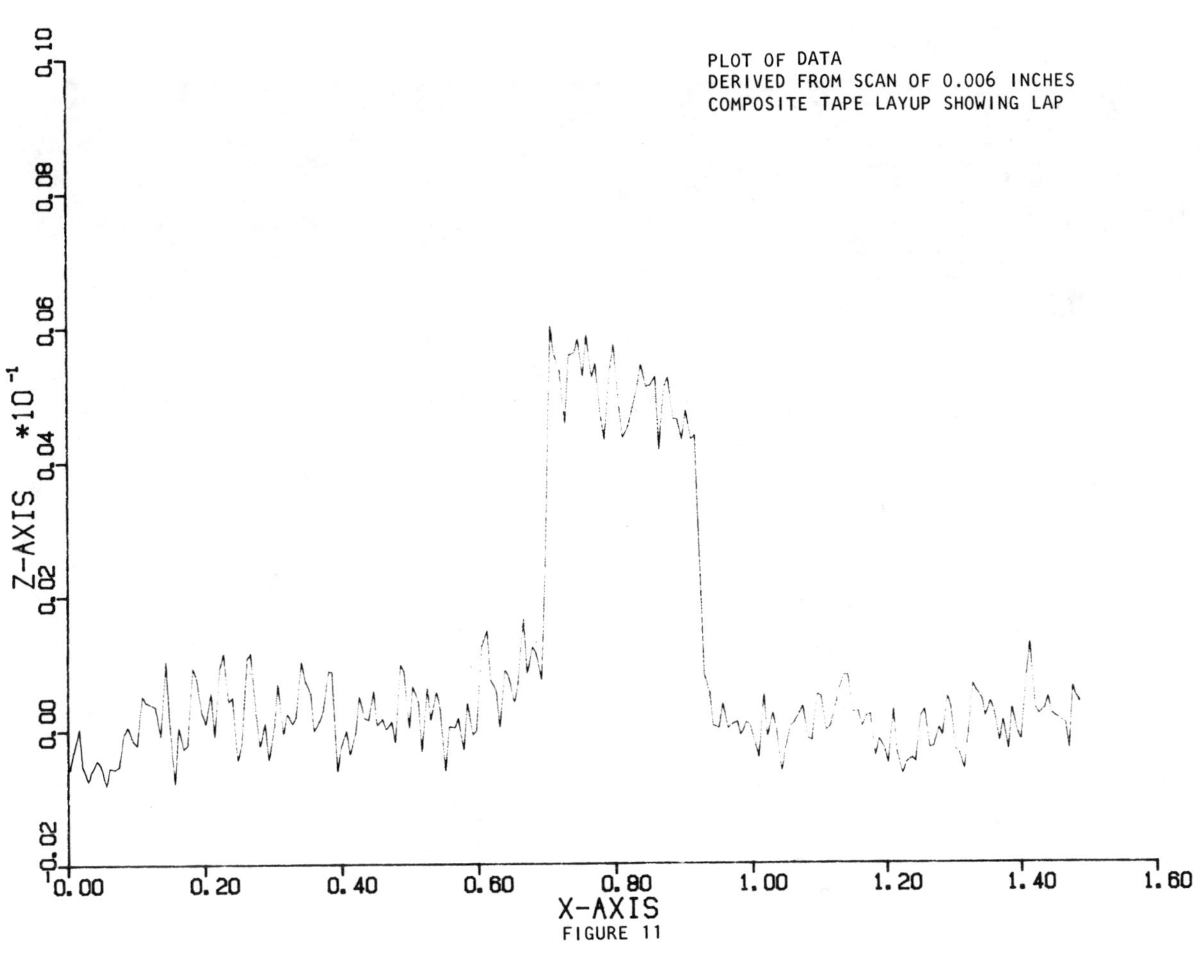

FIGURE 11

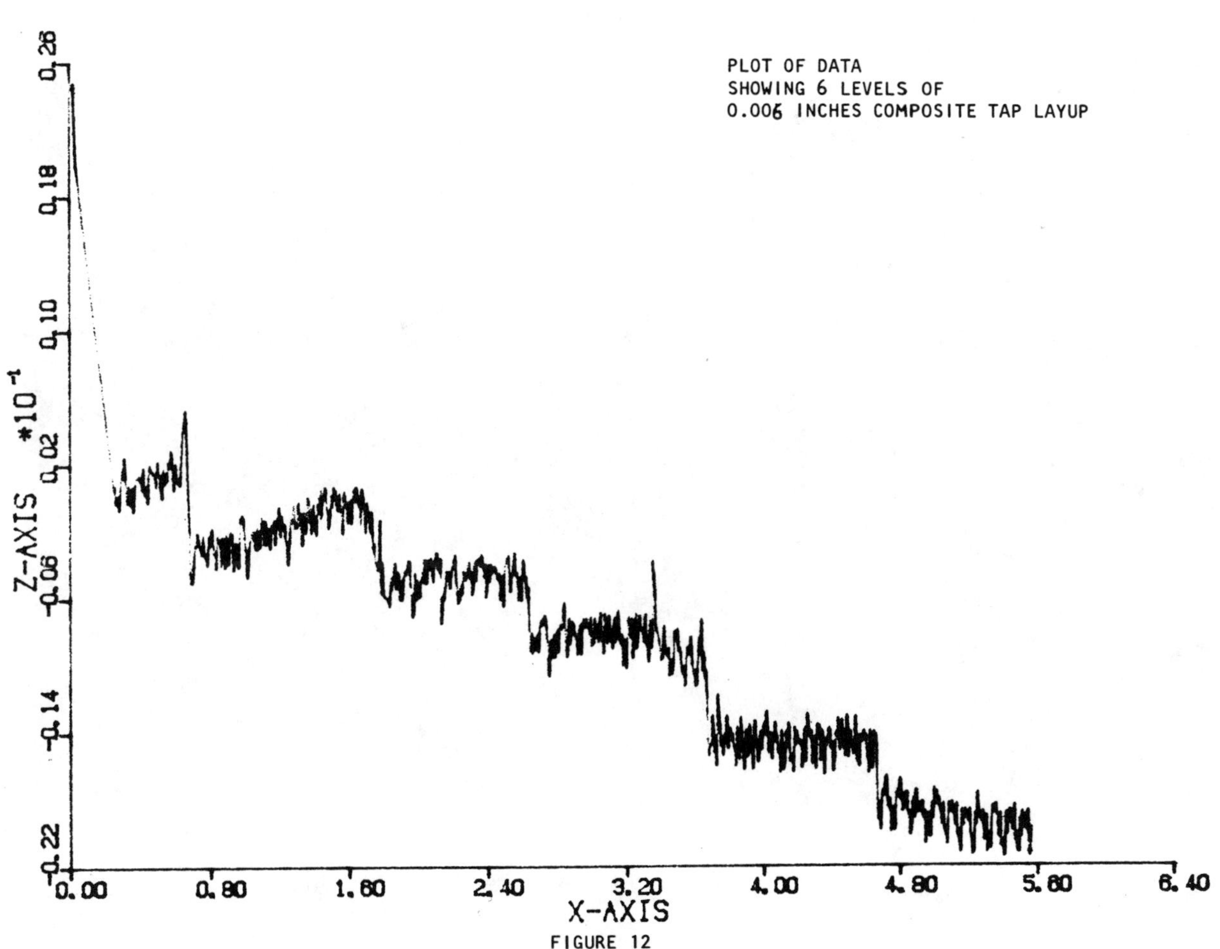

FIGURE 12

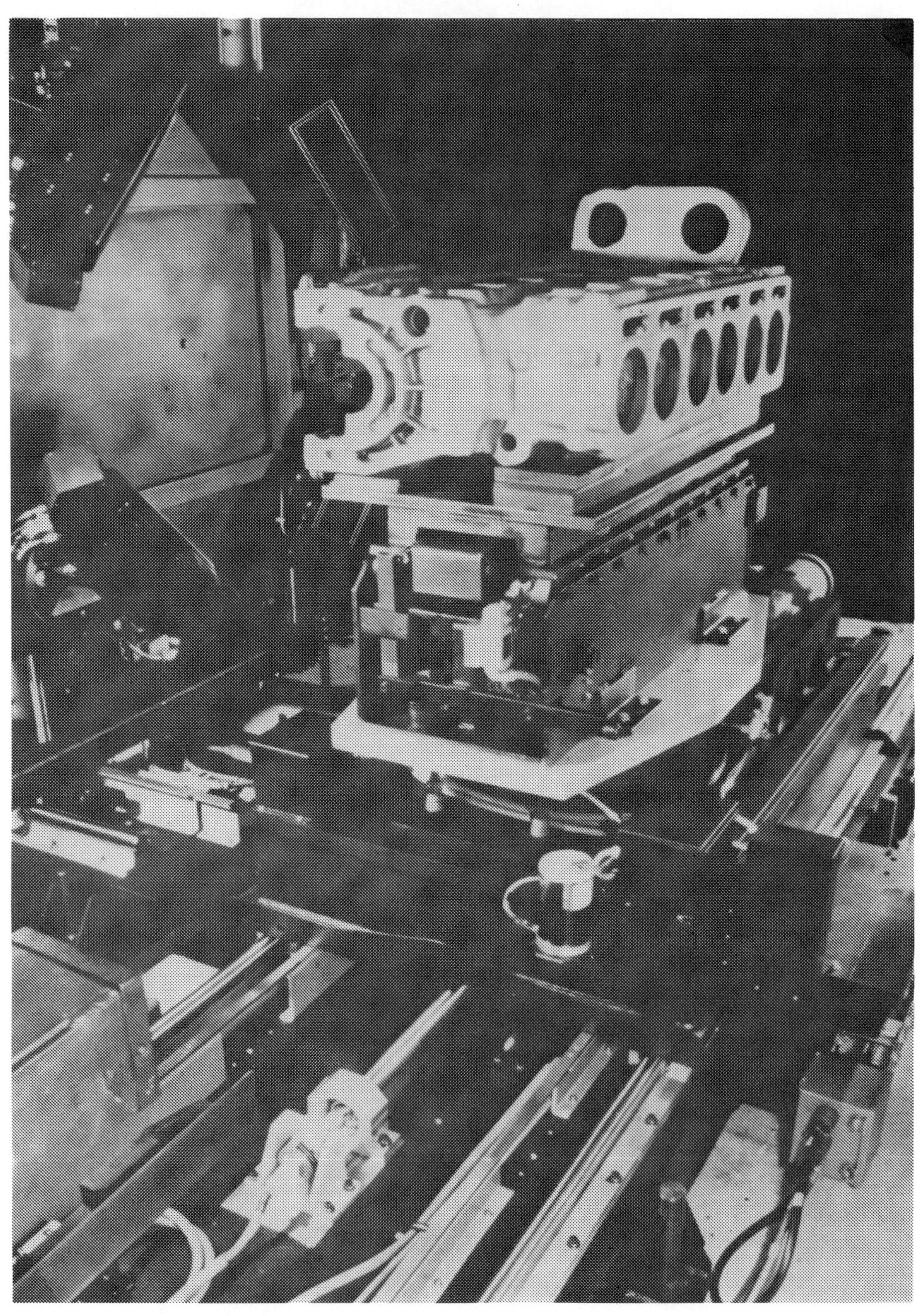

FIGURE 13
A SENSOR MOUNTED TO CUSTOM TRANSLATION UNIT

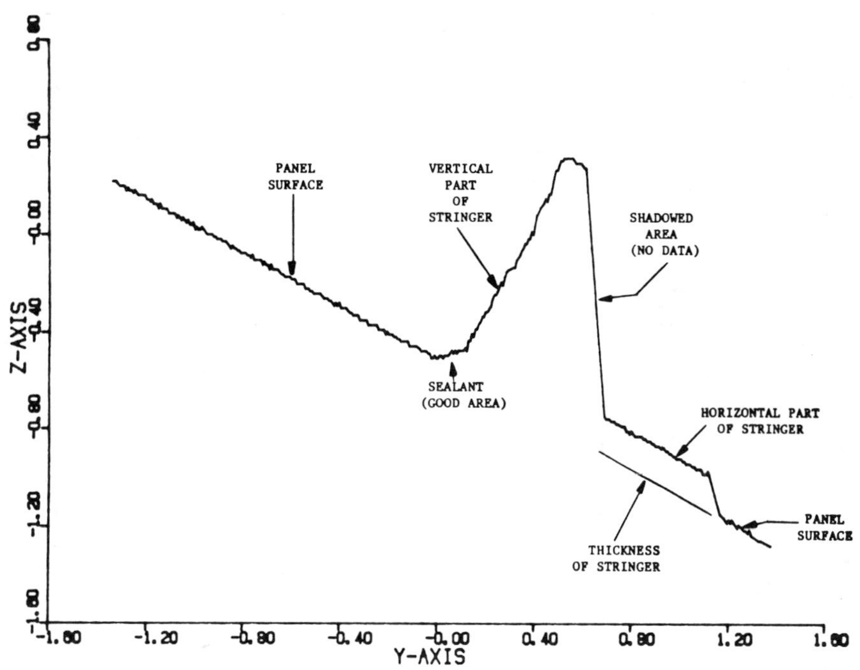

FIGURE 14

DATA FROM SAMPLE POINT 1

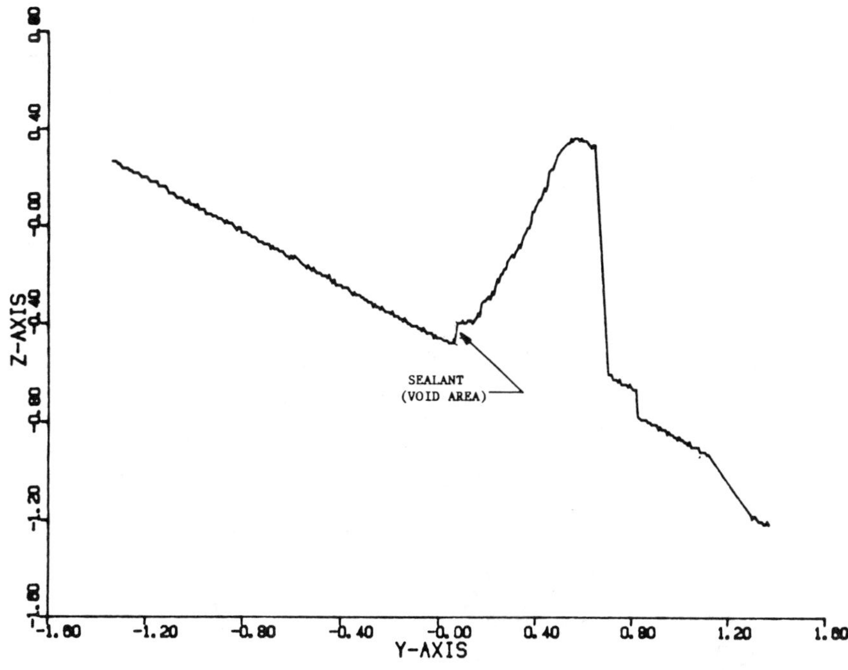

FIGURE 15

DATA FROM SAMPLE POINT 2

DATA BASE GENERATION

Although it is not an actual manufacturing or inspection process, the use of 3-D vision sensors to aquire data from surfaces can be a useful tool in developing tooling and robot programs. For example most parts of existing aircrafts are not on a CAD data base. There is a great cost associated with the conversion from standard drawing to CAD and in reality master parts or models often serve as the standard and not the drawings.

For example of a surface data acquisition for CAD processing is being developed under the AFWAL sponsored Honeycomb Repair Center program. The prime contractor is the Lockheed Georgia Co. RVSI is developing a large volume Optical Measurement System under a contract with Grumman Aerospace Corp. who is Lockheed's subcontractor for the Integrated Core Cutting cell of the HRC. The system uses a modified Cybotech electric G80 gantry robot shown in Figure 16 to translate a vision system over the part to be digitized. The system will be used to acquire data from F111 parts to generate NC tapes for core cutting. The parts will be scanned in a rater scan as shown in Figure 17a. The data will be automatically thinned and converted to IGES format for processing by a CAD system. The data processing will reduce the surface data to a set of constant axis slices as shown in Figure 17b. The orientation and spacing between the slices is selectables by the operator at the time of scan. In addition any steps will automatically be detected and measured and will also be reported. This system is presently being constructed and will be ready for testing during the third quarter of 1985.

ADVANCED TRUE 3-D SENSOR

Under AFWAL sponsorship, RVSI has constructed and evaluated a real-time version of a true 3-D vision system. The system is based on a film based 3-D copying system developed by RVSI in the mid 1970's. That system was able to make high quality portrait sculpture as shown in Figure 18. The Sensor was demonstrated during the summer of 1984. It is able to acquire and process in real time (less than 3 seconds) a complete 3-D data base of every resolvable element in a cubic filed of view of 2 inches (50.8 mm) on side. Since the system used optical triangulation, only those surfaces which are viewable to both the camera and the projector can be measured. A photograph of the accuracy module mounted on a Cybotech H80 robot is shown in Figure 19.

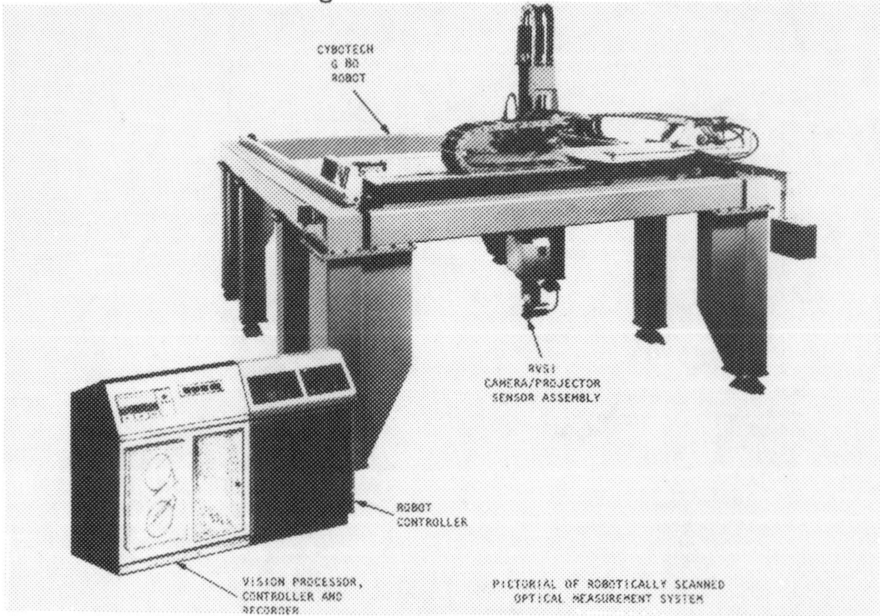

FIGURE 16

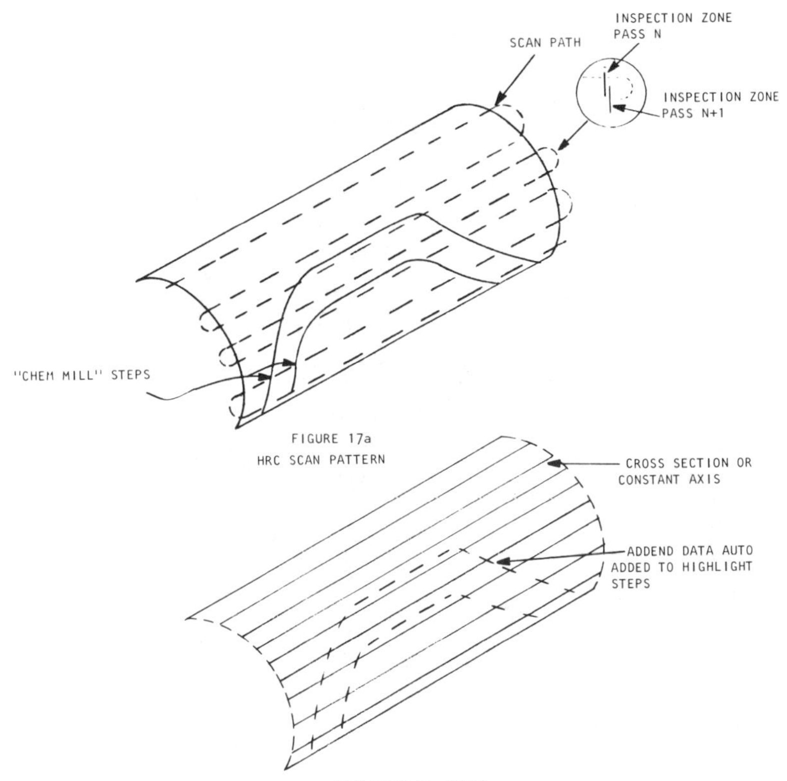

FIGURE 17b
HRC SCAN RESULTS

HONEYCOMB REPAIR CENTER SCAN DATA

FIGURE 18
PORTRAIT SCULPTURE

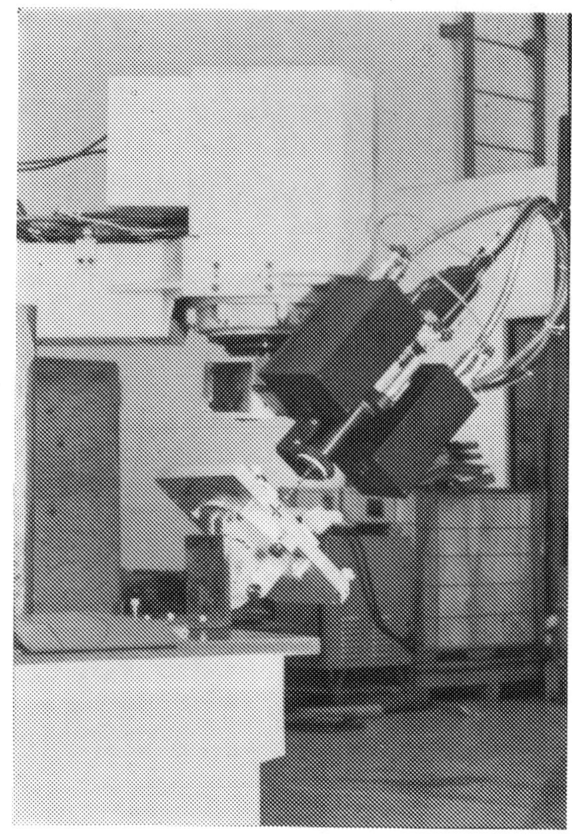

FIGURE 19

ACCURACY MODULE MOUNTED ON A CYBOTECH H80 ROBOT

The 3-D sensor has an accuracy of +/-0.005 in. (0.13 mm) for the 2 inch (50.8 mm) on side field of view. The field of view can be increased or decreased with the corresponding scaling in accuracy. RVSI is continuing work on this product and a production model will be available. The speed of data acquisition can be decreased to approximately 0.3 seconds in production units depending on improvements in a key component of the sensor.

CONCLUSION

Although the use of 3-D vision is a relatively new technology, the experience gained in other industries and the various tests performed to date, indicate that it has many applications in the aerospace manufacturing industry. As the technology matures, the developmental costs for new applications will significantly decrease and vision based robotic systems will compare favorably in cost, versitility and performance with specialized hard tooling.

Presented at the RI/SME Applied Machine Vision Conference,
February 1984

Vision Measurement Technology Using Photogrammetry for Quality Control in the Aerospace Industry

by Stephen M. Hall
Industrial Measurements and Controls

Photogrammetry is a new concept to many in quality control. Although new applications are being found in aerospace, the use of photogrammetry is quite old.

Photogrammetry is a measurement technology which provides three-dimensional coordinates of points in space, with input data provided by measurements from photographs of the object under inspection taken from several surrounding locations.

Photogrammetry can be divided into two application groups, Aerial, for the generation of topographic maps; and Close Range Photogrammetry, for the measurement of terrestrial objects typically with camera distances not exceeding a few hundred feet. This second branch of applications is the subject of this discussion. While lengthy texts are available dealing with entire subject of photogrammetry, it is the intent to present a brief exposure to aerospace applications and the elements of a simple system.

Close Range Convergent Analytical Photogrammetry

As objects of a few feet to several hundred feet in size can be physically difficult to accurately measure, a scale model of the object could more easily be measured. If the model were precise in dimension and the exact scale were known, measurement of the original object could be deduced. A series of photographs can provide this model. By using convergent camera locations, coordinates of discrete points can be determined in three dimensions.

The three dimensional coordinates of a target are calculated by means of two or more rays generated by each photograph. (See Figure 1)

As these rays approach being parallel, the perception of the intersection becomes very weak. As the angle increases toward 90° and as the number of rays increases, the solution is strengthened. If the number of photographs is increased above two, the ability to detect and reject a ray measured in gross error is possible as it will not intersect in agreement with other rays to that same target location.

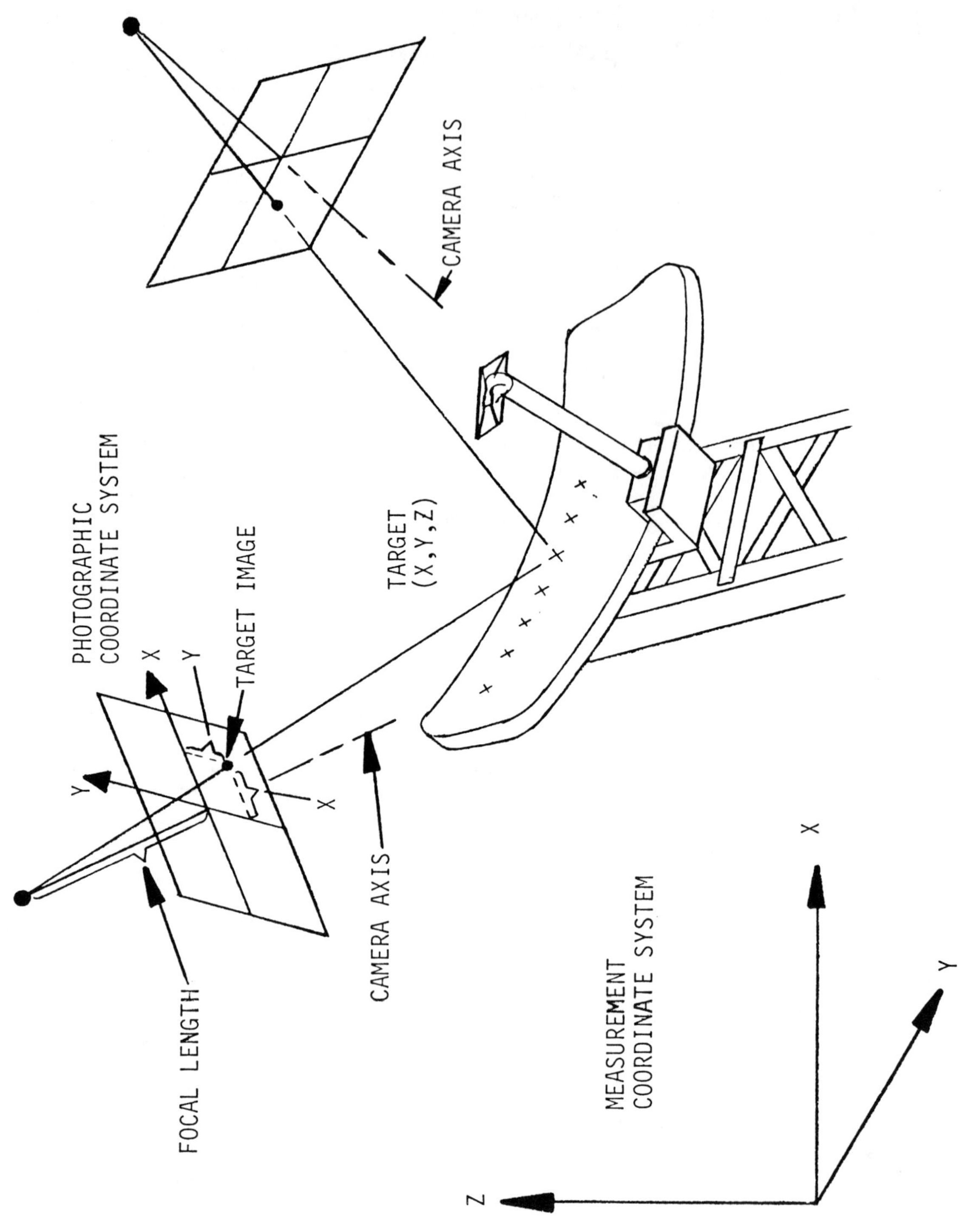

Figure 1
Photogrammetric Triangulation

Targeting

Normally measurements are desired between discrete points or a three dimensional coordinate is desired in relation to one or more additional points. To facilitate identification of these points on an object, visual targets are placed on the points of interest. Each target is assigned a unique ID number. Targets must be identifiable on each of the images in which it is visible.

Targeting may be simple temporary gumbacked paper dots on a surface or permanent metal target if repeated use is anticipated.

To ensure an easily recognizable target on a photographic image, size as well as lighting must be considered. Depending on the method of film reading employed, a specific target size should be selected to provide an optimum target appearance on the film plane.

The human eye has a specific preference in target size being measured for best results while automated film readers are tolerant of a wider target size.

Many shapes of targets are employed successfully under varying conditions but a circular dot or single ringed bullseye is generally a preferred shape particularly when an automated image processing film reader is employed.

Film Analysis and Automated Film Scanning

One of the fundamental steps in any photogrammetric survey is the accurate two dimensional coordinate measurement of targets appearing on a set of photographic images. The accuracy to which this can be achieved can be the limiting factor in overall system accuracy. State of the art film analysis requires a measuring device capable of measurement accuracy of between 5 and 0.5 micron accuracy with the greater accuracy preferred. These accuracies must be available over a measurement area equal to the format size of the camera being used.Accuracy numbers quoted are entirely dependent on the application and many opinions prevail as to the current state of the art and how it is best achieved.

Most photogrammetric cameras employed for static surveys use a format size between approximately 4" x 5" and 9" x 9". Normally, bigger is better, but larger film reader format size increases cost significantly. In astronomical applications, 20" x 20" plates are producing excellent results but the difficulties in photographic plate handling and storage would be impractical for many aerospace applications, to say nothing of the camera size required.

Advances in image processing and microcomputers have allowed recent improvements in photogrammetric film analysis. While the human eye and brain show excellent adaptability to quality control inspection requiring inovative decision making, endless repetitive quantitative measurement is best accomplished by automatic systems. With the introduction of digital image processing, measurement points on photogrammetric film can be easily recognized by computer algorithms. This has shown to greatly increase throughput of photogrammetric film reading. Additionally, as each target is optically subdivided into hundreds of smaller optical elements or pixels (picture elements), an increase in target measurement repeatability has been proven over manual measurement.

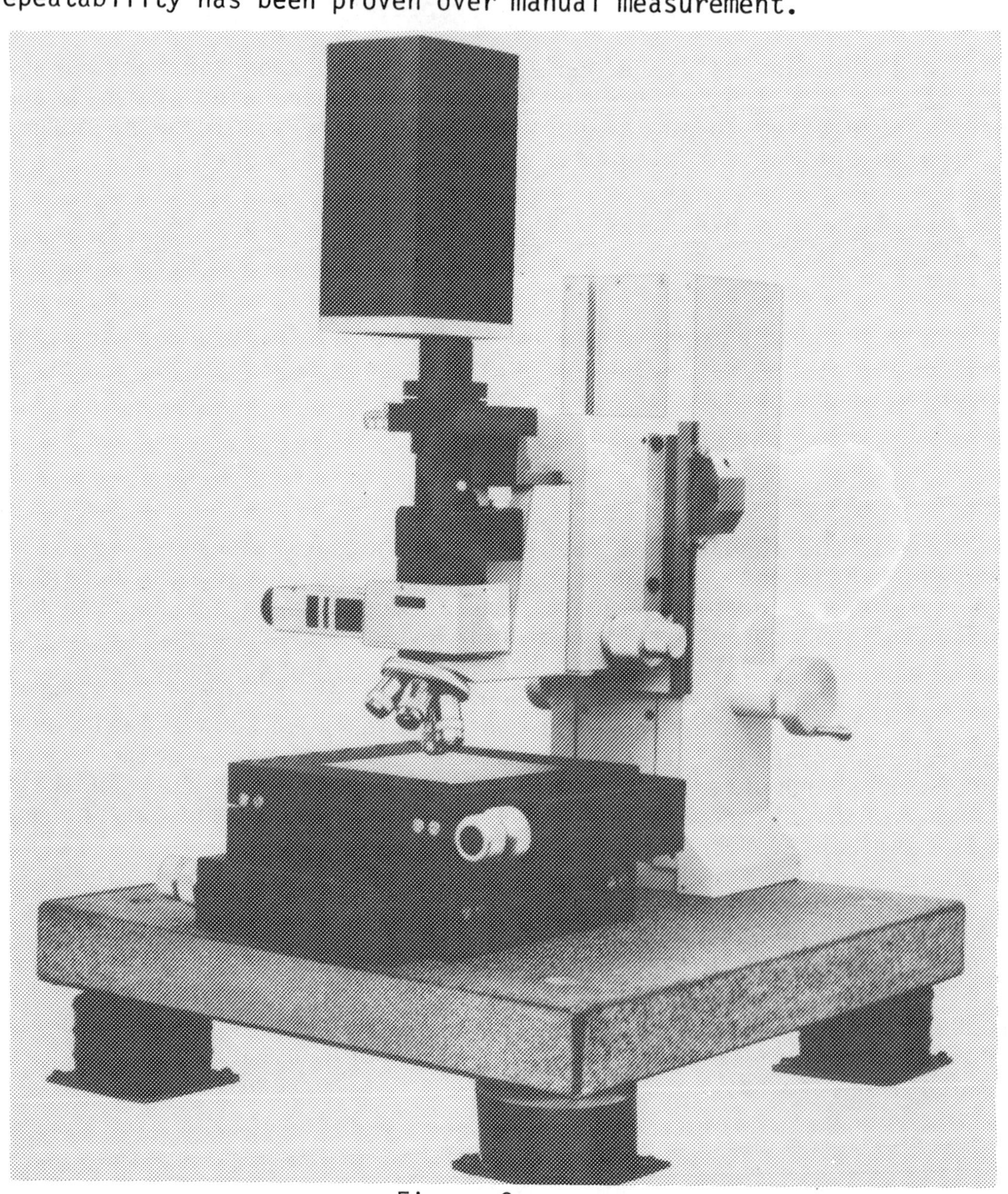

Figure 2
Industrial Measurements & Controls Automatic Scanner

In conjunction with this processing automation, film handling is also given to computer control, allowing the automation of alignment of film targets to optics. Industrial Measurements & Controls has developed such a system specifically designed for photogrammetry. It allows the use of all popular film format sizes including high speed motion picture film. This automated scanner is shown in figure 2.

Cameras

Due to the size or location of some objects to be measured it can be difficult or impossible to move to a measurement system. It then becomes a requirement of the measurement device to be portable if it is to be used in the field. The portability of a photogrammetric camera allows this convenience.

Cameras, depending on the application, can be hand held, tripod mounted or be supported by specialized structures. Normally for static measurements, a single camera is moved to each camera station in succession. This requires that the object under inspection be stable. If the subject of the survey is a study of dynamics, high speed cameras can be used in synchronization; with each set of simultaneous frames providing discrete measurements. Analysis of motion can then be extracted as needed.

As photogrammetric cameras currently in use are in the $30,000 range every effort is made to use a single camera as a matter of economics.

No more controversial subject can be mentioned in the presence of photogrammetrists than cameras preferences. In the aerospace field three particular models should be mentioned, as they are most commonly seen.

First is the CRC-1 built by Geodetic Services. It uses roll film of 9" x 9" format. The camera has a detachable film magazine with a 250 foot capacity. Film advance is achieved with a motor winder which allows remote operation. The camera suffers no loss of accuracy through the use of film, as a vacuum reseau platen is incorporated.

Second is the aus JENA model UMK 1318. This is a popular camera using a 5 x 7 format. It is quite rugged and fairly heavy for its size. It features the capability of using both glass plate and roll film with a 60 exposure capacity.

The smallest and lightest of the three is the WILD P31 using 4 x 5 inch glass plate or cut film. No roll film attachment is offered. This camera fails to use an internal light source to expose fiducial marks as proviced in the CRC-1 and 1318. These fiducial reference marks must be visible on the image to compute orientation of the image to the lens. Under certain lighting conditions they are lost with the Wild camera and this complicates film reading unnecessarily.

Photographically Induced Error

The fundamental accuracy of a photogrammetric system relies on an undistorted image for measurement. Any image distortion or displacement will present an altered subject for measurement. Care is taken within the camera to minimize or correct through calibration any such distortion. Each camera lens is individually calibrated with correction factors calculated to allow computer correction of the lens distortion. Even an image displacement of a few microns will degrade system accuracy.

An additional area of error introduction is the lack of film flatness in the camera. Many techniques are employed to assure flatness at the time of exposure. The use of glass plates is frequently used because of the stability of glass compared to flexible polyester based film. Glass has the ability to better resist distortion due to thermal and humidity expansion. Due to the increased demands of system accuracy required by current aerospace applications even the small instability of glass can limit system accuracy.

To overcome this problem, a technique of exposing a grid network of reseau marks on the film, overlaying the image, allows the detection of post exposure film distortion regardless of the source. This allows data correction. So effective is this technique that film can be used in preference to glass plates with a great cost savings and ease of film handling and storage.

Accuracies Available with Photogrammetry

Many factors influence accuracies achievable including technique and application induced limitations. Error expressed as a ratio to the largest dimension of the object photographed results in ratios of 1/300,000 or better. This equates to an inaccuracy of .001 inch measuring an object 25 feet in length under the optimum conditions. Such accuracy depends on the redundancy of many convergent photographs providing a strong target location solution. Relaxing of this accuracy simplifies a survey by reducing the number of photographs required.

A second technique not using convergent camera positions but using parallel stereo photography is available. The advantage is that complete targeting is not required. This is typically the technique used in map making as targeting of the earth surface would be impractical. The sacrifice made is loss of accuracy with stereo photography. The stereo technique has been attempted with unusable results where accuracy comparable to convergent photography was required. The best accuracy achieved by single stereo pairs is in the order of 1/10,000 to 1/20,000. This is about 1/10 the accuracy of convergent photogrammetry and requires a more highly skilled film reader operator. Shown in figure 3 are the camera positions of convergent photography vs. stereo. Note difference in the included angles. The larger angle contibutes to greater accuracy.

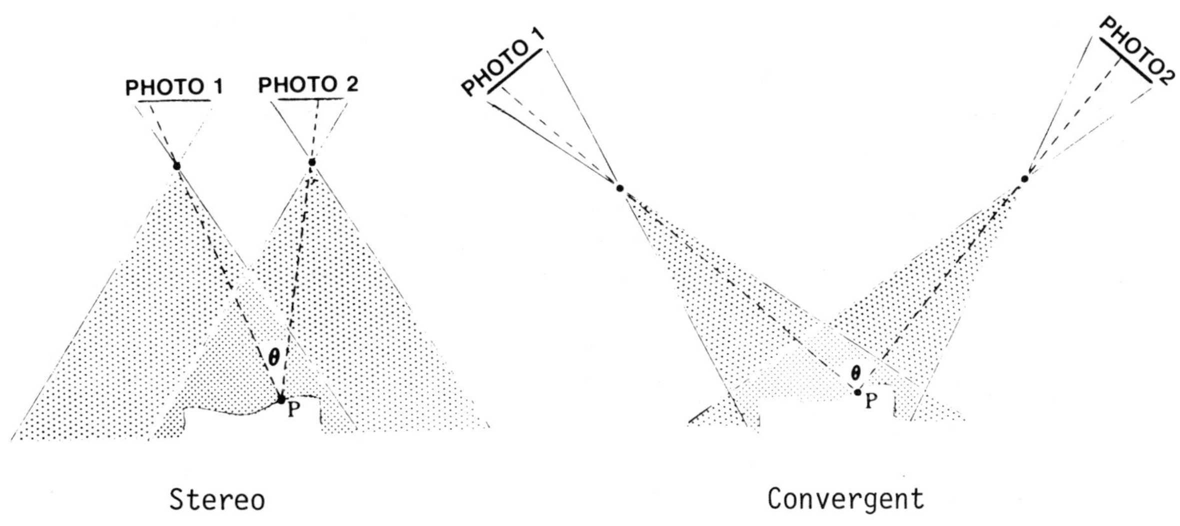

Figure 3

Aerospace Applications

Many uses have been found for photogrammetry in aerospace manufacturing and quality control. Following are some of the most successful programs.

General Dynamics, Fort Worth, TX

A study was made in cooperation with the U.S. Air Force for assembly tool inspection for the F-16 program. A requirement of airframe assembly necessitates periodic calibration of the tooling. Depending on quality control procedures and production rates this may be required several times a year. During inspection the tool must be taken out of production and will be unproductive from several days to a week or longer under conventional inspection techniques, using either optical transits or theodolites. With the introduction of photogrammetry into the F-16 program this non-productive period in some cases has been reduced to less than one shift, with a corresponding increase in productivity. Since an in-house photogrammetric capability has been in place, many other uses have been found at General Dynamics.

Northrop Corporation, Hawthorne, CA.

The aircraft division of Northrop investigated the cost savings achievable by using photogrammetry in tooling inspection. The payback period was sufficiently attractive to justify the purchase of a photogrammetric system which was integrated by their Manufacturing Technology Group and is being turned over the Quality Control Department.

Prior to the purchase of a photogrammetric system, a test fixture of known dimension was built to test the accuracy of the technology. Two surveys were performed with discrete, accurately controlled changes of target positions made by Northrop between surveys. The test employed was the photogrammetric systems ability to detect target movement, simulating shifts of locators in aircraft tooling. The results were excellent with average error of approximately .001 inch varified by Northrop.

A notable advancement incorporated in the Northrop system is the addition of an automatic film reader capable of full computer control and target recognition. A weakness of prior systems available was the manual measurement of target imagery. With a manual reader the operator is required to locate each target and using an eyepiece, center targets on a crosshair or other point locator. In a production environment accuracy quickly deteriorates as the operator may be dealing with hundreds of measurements with a required accuracy of a few microns. Northrop recognized this shortcoming of manual film readers. Using image processing software and a computer controlled X-Y table automation has been added. The film reader used is the Perkin-Elmer PDS Microdensitometer.

Grumman, Bethpage, NY

Grumman has introduced this technology into their quality control of space antennas. Conventional measurement techniques failed as conditions dictated measurements in a vacuum chamber. The large antennas required several thousand points to be measured quickly in a vacuum to simulate space conditions. Photogrammetry has allowed these measurements to be made in a most difficult environment. While all early work was done under contract by outside services, Grumman is now implementing an in-house system.

ITT Gilfillan, Van Nuys, CA

Quality control at this facility requires an ability to measure surface curvatures of radar antennas. With surface tolerances of a few thousandths of an inch and surface areas over one hundred square feet, Quality Assurance needed an effective means of inspection. Both the engineering and quality departments have investigated photogrammetry and found it a candidate for all future antenna surface measurements.

duPont Aerospace, Santa Ana, CA

A survey was conducted involving the study of shape and air flow within jet engine nacelle inlets.

Photogrammetry provided an ability to model the shape, giving an accurate contour. Many hundreds of temporary targets were placed at key locations and surveyed. By curve fitting to discrete points the surface was accurately mapped. Due to the portability and speed of the technique, the survey was performed in one day on an aircraft during a routine maintenance period without interruption of its schedule. See figure 4.

Figure 4
Engine Nacelle

Raytheon Company

An extensive number of surveys were made of portable communication antennas under various load conditions. The measurements were conducted in a wind tunnel to determine the deformation of surface, feed structure and supporting tripod over a velocity range from 0 to 100 knots. This required the use of three remotely operated and syncronized cameras.

Rockwell International

Under contract to Rockwell, Geodetic Services Inc. of Melbourne, FL performed several surveys in conjunction with the shuttle program. It involved the position measurement of approximately 200 points within the payload bay of the Columbia prior to Flights 3, 4, and 5. Other surveys in the Orbiter Processing Facility at Kennedy Space Center were performed on the launch pad. Accuracies of .006 inch were achieved over the 60 foot length of the shuttle bay. Feasibility studies were conducted showing the ability to perform surveys during space missions. Figure 5 shows the placing of targets within the bay.

Figure 5
Space Shuttle Payload Bay

The future

A future advancement currently under development involves the elimination of photographic film cameras and the introduction of live video cameras at each camera station. With live images feeding analytical equations, real time positional information could be available.

Photogrammetry offers measurement solutions in many unexplored applications. The inovation of aerospace companies will dictate the speed in which it will become common practice in quality control departments. This same inovation will dictate the direction of future system developments involving increased speed, accuracy and photographic techniques.

As additional systems are installed in diversified aerospace companies, additional uses will be found for the capability it supplies.

Many current limitations are due to the state of the art in image processing speed, video camera resolution, photographic camera and film resolution and the mechanics of film reading. As each of these areas progress new opportunities will open for photogrammetry.

Presented at the MVA/SME Vision '86 Conference, June 1986

The EndSpector™ System for Automated Inspection of Beverage Can Ends

by Robert L. Jackson
Mechanical Technology Incorporated

Russel S. Demuth
Mechanical Technology Incorporated

Boyd Eldridge
Adolph Coors Company

INTRODUCTION

Technical advances in machine vision equipment in recent years have resulted in more widespread application of the technology to real-world industrial inspection problems. Dramatic improvements in the cost and performance of this equipment, particularly in camera sensors and in electronic hardware to perform image processing functions at high speed, are chiefly responsible for the advances made. Availability of these systems has also stimulated innovation in various application techniques, such as specialized lighting methods and unique software approaches.

This paper focuses on a machine vision application involving a number of precision, dimensional measurements on a high-volume aluminum part, the top to a beverage can. The EndSpector™ system was developed specifically to address the requirements of can end dimensional inspection. State-of-the-art sensors and machine vision hardware, along with a number of specialized illumination and image processing techniques, make up a system that performs over a dozen types of dimensional measurements. A robotic loader and a precision, motorized stage are used to provide fully automated handling of parts during the inspection process. Benefits of the system are that inspection operations can be performed with considerably less labor and with improved consistency. Among the unique features of the system is a differential measurement technique for measuring metal thickness in the score and rivet areas of the can end. Two high-resolution video cameras, special structured-light projectors, and a series of specialized image processing operations provide the basis for this technique. Noncontact differential measurements can be performed with accuracies to 100 μin..

CAN END MANUFACTURING AND QUALITY CONTROL AT COORS

The Adolph Coors Company, a leading U.S. brewer, produces over 14 million barrels of beer annually. Approximately 70% of the product is packaged in aluminum cans. Coors manufactures virtually all its own packaging, including the cans and can ends. Typically, the cost of the packaging is greater than the cost of the beer it contains. The can end represents the highest cost element of the packaging. Coors strives to maintain a leadership position in beverage packaging technology through the production of the most advanced and most economical containers. This goal creates a constant demand for high technology improvements in the product and the manufacturing and inspection process. In addition, there are significant demands for more timely and more cost-effective quality control procedures.

Can ends are complex, precision items that must satisfy a variety of demands. They must provide reliable, leak-proof containment of the product, and yet be easy to open. They must meet ecological and safety demands and requirements, be satisfying to the customer, and be economical. Coors has always placed a high priority on state-of-the-art manufacturing systems and processes to achieve these requirements. Recently, emphasis has been placed on a closer association between production and quality control, because it is recognized that more timely and effective results from the quality control function are vital to improving the overall production process. However, the extremely large product volumes create a highly labor-intensive operation in adopting statistical process control (SPC) methods.

Coors conducted an evaluation of existing SPC methods, equipment, and results which revealed a number of problems. First, labor requirements are prohibitive if a number of parts representing a true statistical sample is inspected. Second, all current inspection devices are manual, contact-type gages that require ends to be transported to various bench locations to complete the 13 inspection operations. Third, data collection is handwritten on charts for eventual statistical processing, which causes significant delays in corrective action. Fourth, inaccuracies occur in the data, despite the efforts of conscientious quality control technicians.

As a result, Coors is actively pursuing the implementation of numerous improvements to the quality control operation. Much of this effort is directed at implementing state-of-the-art systems that can provide more meaningful and consistent inspection results in a timely fashion. A major objective is to improve the process of performing dimensional inspection on sampled can ends. Since this process typically requires up to 20 full-time technicians who perform up to 13 types of dimensional measurement operations using contacting type sensors and manual operations, a fully automated inspection and data acquisition system is desirable.

INSPECTION REQUIREMENTS

The can ends are produced in a continuous process on high-speed, multiple-station presses. Large rolls of aluminum stock fed in at one end become finished parts that exit each of the presses at rates up to 18,000 parts per hour. Total production volume is on the order of 8,000,000 parts per day. Sample parts are pulled from each press several times a day for routine QC inspection. These parts require measurements of up to 13 critical parameters; these are:

Score Residual	Curl Diameter	Panel Deboss
Score Width	Curl Opening	Panel Depth
Rivet Residual	Curl Height	Countersink Depth
Rivet Diameter	Emboss Height	Stock Thickness
Dome Height		

The most critical parameters are those associated with the tear-open score and the tab anchoring rivet. The thickness of the metal at the bottom of the score is referred to as the "score residual." Similarly, the thickness of the metal at the center of the rivet is referred to as the "rivet residual." These parameters, as well as the rivet diameter, must be controlled precisely to ensure that the end will open easily, while also ensuring that leakage is never a problem.

Several sizes and styles of parts must be measured. Additionally, since part designs are likely to change, an inspection system must have the flexibility to accommodate modifications to future inspection parameters, procedures, and tolerances. To meet the goals for reducing labor cost in the inspection operation, a fully automated parts handling capability that accommodates a stack of ends taken from many different presses in the plant is preferred. All loading, handling, and inspection should occur automatically, with operators notified immediately of any out-of-tolerance condition. Another requirement is for all measured data to be stored on hard disk with an interface provided to an existing computer system which conducts SPC analyses.

SYSTEM CONCEPT

In evaluating potential solutions for the can end inspection requirement, several approaches were considered. Various standard inspection system products were ruled out because of the inability to meet all system requirements. Coors elected to pursue development of a specialized system dedicated to this specific task.

A review of relevant measurement technologies indicated that machine vision offered the most promise for providing a flexible solution to the varied requirements presented by the 13 parameters. Contact gaging methods were judged unacceptable because of the need for a wide variety of unique probe arrangements, problems of probe wear, and inflexibility in accommodating future changes in requirements. Various noncontact techniques, while applicable to many of the measurement requirements, failed to offer the complete solution that machine vision offers.

Figure 1 shows the EndSpector™ system that was developed to meet these requirements. The major system elements are illustrated in Figure 2. Five video cameras acquire the image information at various stations in the system. Four of the cameras are high-resolution vidicon devices which are required to achieve the resolution needed in many of the measurements. The fifth camera is a solid-state CCD device selected for its small size and price advantage and is used for some of the less rigorous measurement requirements.

The vision processor consists of five Multibus™-compatible boards that enable high-speed digitizing, storage, and processing of the image data. Any one of the five cameras can be selected under computer control and its output digitized at video rates. Image data are stored in the three frame-buffer units, each having the capacity of storing 512 x 512 pixels with 256 gray-scale levels. A high-speed numeric processor performs a variety of low-level image processing operations needed for the dimensional measurement algorithms. The processor also provides a real-time video output to display either live or processed images to the system operator.

The host computer, which provides overall system control, is based on an Intel 80286 processor. The system has a 750-kilobyte memory, a 320-kilobyte floppy disk drive, and a 19-megabyte hard disk storage device. The RMX™ operating system is used. Major functions provided by the computer are:

- Direct interface to and control of the vision processor operations
- Operator interface
- Control of all system mechanical functions, including the part stacker, loading robot, and five-axis positioning stage
- Control of illumination devices
- Interface to existing QC computer.

The three mechanical part-handling elements are the stacker, the robot, and the positioning stage. The stacker unit was designed to allow an operator to load up to 100 parts of a similar design. The function of the stacker is to present can ends, one at a time, to the robot for automatic loading into the system. It accomplishes this function through a single slide mechanism at the base of the stack. This slide is operated by an air cylinder under control of the system computer. The mechanism momentarily lifts all parts of the stack, except the bottom part which is translated laterally toward the robot. After the robot grasps the end, the slide mechanism returns to the home position. Various inserts are used to allow the stacker to accommodate can ends of different size.

Figure 3 shows a close-up of the inspection station. Three high-resolution vidicon cameras are visible, as well as the loading robot on the right and the positioning stage and fixture in the center. The loading robot is a commercially available, air-operated, five-axis, pick-and-place device equipped with a special end effector to enable it to hold can ends. Figure 4 shows the robot lifting a part out of the stacker. Figure 5 shows the end effector.

A key feature of the system is that all cameras and light sources are rigidly fixed on a granite base, with the can end being presented to each camera station by a multi-axis positioning system. This five-axis positioning stage is comprised of three linear stages and a rotary stage assembled from commercially available products. The fifth axis is a specially designed rotary fixture that accommodates a range of can end designs. This fixture accepts an end presented by the loading robot and then clamps and centers the part with a three-element pawl arrangement. Once locked on the fixture, the part remains precisely centered and is supported on a three-point reference surface.

All five axes are operated by stepper motors that are under control of the system computer. For all but one of the parameters, measurement accuracy is not dependent on precise positioning of the stages and part. As long as the desired feature is in the camera's field of view, the measurement can be performed. Accuracy is dependent on the camera and vision system. This approach minimizes the need for extreme accuracy in positioning the part.

MEASUREMENT APPROACHES

To perform all of the various measurements, a variety of techniques were required, most utilizing specialized illumination. This section describes several of these techniques. Many of the parameters to be measured exist only at certain locations on the end, and it is necessary for the EndSpector™ system to know precisely where to look for these features. Therefore, one of the first steps in the inspection sequence is to identify the angular orientation of the end with respect to the positioning stage. Since the angular orientation of the ends as placed in the stacker is arbitrary, when a part is placed in the fixture by the robot, the system does not know the exact location of the features to be measured. The initial vision operation is directed at precisely defining the angular orientation. This operation must be accomplished with sufficient accuracy to allow automatic placement of the score feature in front of one of the cameras to an accuracy of several mils.

First, the end is viewed by the solid-state camera which has a relatively large field of view, on the order of 1.5 in. After the stage has positioned the end so that the center rivet is precisely in the center of the field of view, a rotational pattern recognition algorithm is executed. This sequence enables a gross determination of angular orientation, to within one angular degree. Once this procedure is completed, approximate location of all the features is known with respect to the rotary fixture position. Second, the score feature is placed in front of one of the vidicon cameras, which has a 0.25-in. field of view. A second pattern recognition algorithm then defines the score location precisely and allows an accurate identification of the angular orientation. This information is used in most of the subsequent measurements to ensure that the desired feature location is imaged properly.

Curl Height and Countersink Depth

Figure 6 illustrates curl height and countersink depth, designated H and D, respectively. The measurements are performed in the EndSpector™ system by first presenting an edge of the can end to a camera. The camera has a field of view of approximately 0.25 in., which is adequate to view the entire edge and provide sufficient resolution for the measurement. Once the end is properly positioned, three backlight sources (A, B, C in the figure) are moved into position by computer-controlled actuators. These light sources consist of fiber optic bundles having beveled plastic end terminations designed to direct light toward the camera. The camera then sees the two desired part features as dark bands against light backgrounds. The desired dimensions are obtained in the vision processor through simple edge detection routines. Figure 7 shows the backlight sources in their retracted position in front of the camera.

Curl Opening

This measurement requires a determination of the width of the opening into which the can is ultimately inserted during final packaging. Figure 8 illustrates the approach to this measurement. A collimated light source is projected from the side of the end, casting a shadow onto the inside wall of the curl opening. The camera viewing the curl opening from an angle of about 45 degrees sees an image as shown in Figure 9. The right side of the image is the curl edge. The dark region in the center is the curl opening. The region on the left shows the light projected on the inside edge of the curl opening.

Figures 10 and 11 show the results of two operations in the vision processor that enhance the desired edge transitions. The first of these is a threshold operation which displays all gray levels above a certain intensity as white and all others as black. The second figure is the result of a high-pass filtering operation. In both cases the edges are more clearly defined, although not with sufficient accuracy and repeatability to meet the measurement requirements. This is due to the effects of rough surface texture on the part which are clearly visible in the figures. The roughness essentially adds a random effect to the standard edge measurement approaches. Figure 12 illustrates the method implemented for this measurement in the EndSpector™ system. First, a vertical filtering operation is used to average image information in a preferential direction, minimizing errors from random surface texture. Next, a scan is made across the center of the image as indicated in the figure. The gray level intensity along the scan (shown plotted at the bottom of the image) is then analyzed with an edge detection routine. The number of pixels between the two edges is then converted into dimensional information.

Score Residual Measurement

Measurement of the score residual, which is one of the more critical parameters in the can end, presented the greatest challenge in the EndSpector™ design. The score is the portion of the end that fractures when the can is opened. The thickness of the metal in that region (residual) must be carefully controlled. If it is too thick, the can may not open properly; if it is too thin, a leak may develop. Current techniques using contact probes often give inconsistent and questionable results. An important requirement is to provide a true, differential noncontact measurement of score residual to the desired accuracy of approximately 100 μin. The solution, illustrated conceptually in Figure 13, involves the use of two high-resolution cameras and precision, structured-light sources. The cameras are rigidly mounted coaxially, each having a field of view of approximately 0.02 in. Each of the structured-light sources projects a line pattern onto the end in the field of view of the camera. With the camera and light source held fixed with respect to each other, locations of the line pattern in the camera image can be used to determine the position of the can end target. By combining such information from the two camera systems, a differential thickness measurement can be derived.

Figure 14 shows an example of such an image with a camera having a field of view of approximately 0.1 in. The score region can be seen as the horizontal feature in the center of the image. Projected vertically down the center is a black line pattern on a white field. Where the line pattern passes over the score, the offset of the line to the left can be clearly seen. The dark line on light background is shown here to enable the score to be visualized. In the EndSpector™ system, a slit of light is projected on a dark background. Figure 15 shows a greatly magnified view of the score region, again showing the structured-light stripe passing vertically through the image. The measurement process requires that the position of this stripe in the image be detected and measured with a repeatability of one or two pixels. This presents a considerable challenge, because the relatively rough surface texture of the metal seen at such high magnification makes the structured-light image appear very imprecise. Figure 16 is the result of a threshold operation and illustrates more clearly the difficulty in defining the "center" of the stripe to within a pixel or two.

In the EndSpector™ system, the vision processor implements a number of operations to characterize the structured-light feature precisely. Figure 17 illustrates one of the results of these operations in which the reference image and the actual image were correlated. The magnitude of the correlation is displayed as intensity, with a plot of the actual correlation value superimposed. The region of interest is the bright band between the two dark stripes. The maximum value in this region represents the "best-fit" value for the correlation and is used as the definition of the "center" of the structured-light stripe. Similar information is obtained from the image on the back side of the score.

Press Identification

To utilize EndSpector™ data for trending and early identification of problems with a particular press, it is necessary to correlate the measurement data for a particular end to the press that made it. The system is able to automatically read a special press identification code on each part. The format of the code allows for future expansion of the information provided so that other identification parameters can be added, if required.

Progression Measurements

The presses that manufacture the ends use a five-step sequence to implement all the forming operations. Periodically, a press is stopped and parts are removed from each of the five stages. These parts are called "progression ends" and have special parameters to be measured which are unique to their position in the progression. The EndSpector™ system provides additional inspection routines for these progression measurements.

SYSTEM OPERATION

As part of the routine quality control procedure in the Coors can end manufacturing facility, samples are taken from each of the presses on a periodic basis and brought to the QC area. The ends are grouped in lots based on several size and feature characteristics. For example, all ends having a stay-on type tab and a diameter of 2-9/16 in. are grouped in a single lot, and the lot is then loaded into the EndSpector™ part stacker. The operator enters various setup parameters into the system via the CRT terminal. The data include descriptive information about the part size and type, as well as information about the time at which the sample was pulled.

Once all setup information is provided, the inspection sequence can be initiated by a command at the CRT. Parts are then loaded automatically into the system, one at a time, and cycled through the complete inspection procedure. During the measurement cycle, all measured data are stored on the system hard disk. In addition, a single-page inspection report is produced for measurements of each part. The measurement sequence is roughly as follows:

- Stacker presents a part to the robot
- Robot grips part and inserts it into the fixture and positioning stage
- Part moves to first camera position to identify angular orientation, locate the emboss and read press identification code; part moves to next camera station to complete accurate angular identification
- Measurements made at second camera station
- Part moves to third camera station for additional measurements
- Part moves to differential measurement station for score residual, score width, rivet residual, and stock thickness measurements
- Part returns to loading position, robot removes part and drops it in discharge area.

As each measurement is performed, it is compared to the expected value for that part type to determine whether the actual value is within tolerance. If a measurement is determined to be out of tolerance, the system interrupts the inspection operation and sounds an alarm to notify the operator. At this point, the operator can repeat inspections on the part and confirm the out-of-tolerance measurements so that the manufacturing function can take the appropriate corrective action. If desired, this automatic interrupt feature can be overridden to enable continuous operation of the system, despite out-of-tolerance measurements. This may be desired at times to obtain additional data from lots having known defects.

In some cases, it is necessary to run special lots of parts to obtain selected measurements rather than the full set of 13 parameters. This might be done, for example, if one parameter such as rivet residual had been out-of-tolerance on a particular press, and a special set of sample parts needed to be run to evaluate a corrective action. To accommodate these special cases, the operator has the option of defining various subsets of the total inspection sequence to be run. During the entry of setup parameters at the CRT, a menu of measured parameters can be selected and various parameters eliminated from the sequence, so that, for example, a lot could be run to measure only rivet residual.

Special calibration masters are provided for periodic recalibration of the vision measurement hardware. There are two basic calibration methods. The first method uses precision circle targets placed in front of the cameras to provide a two-dimensional distance reference. The targets are precision chromium plating on glass optical masters. These targets are fixtured in a part which simulates a can end so that they can be loaded into the part fixture and positioning stage and positioned in front of each camera. The second calibration method uses precision gage blocks, also mounted on simulated can ends. These masters are placed at the two-camera, differential measurement station to provide calibration of the score and rivet residual measurement algorithm.

The system hard disk provides for storage of approximately 30 days of inspection results. This information must be transferred periodically to another computer which performs SPC operations for the overall QC function. Transfer of data is accomplished through a serial RS-232 interface. The external computer appears as a terminal device to the EndSpector™ system and receives blocks of inspection data that are stored on a separate disk device for later analysis.

The system also contains diagnostic and self-check routines that verify the correct operation of major system elements. Some of these routines are executed automatically each time the system is powered up. Others can be invoked at the request of the operator.

CONCLUSIONS

The EndSpector™ system is a unique example of specialized machine vision techniques applied to the inspection of a precision packaging part produced in very large quantities. The system illustrates how various equipment and techniques in precision positioning, robotics, structured light, machine vision, and image processing can be combined to address a specific functional need in an industrial quality control environment. The system provides the capability of performing a full set of dimensional inspections on a family of parts at considerable labor savings and with major improvements in the credibility of the results and the ability to document, analyze, and trend the data in a timely manner.

In the future, it is anticipated that several of the measurement techniques developed for the more critical parameters will be considered for in-process application in the manufacturing environment. Ultimately, the application of these techniques on or near the manufacturing equipment could provide even greater payback and the possibility of real-time monitoring and closed-loop control of the manufacturing process. Clearly, application of these techniques in the manufacturing environment presents a formidable challenge. The EndSpector™ system, however, is an important first step in applying these specialized machine vision techniques to precision, dimensional measurement.

ACKNOWLEDGMENTS

The authors wish to acknowledge the contributions of several individuals who have been instrumental in the development of the EndSpector™ technology. Dr. R. Kraft developed most of the image processing approaches and algorithms; L. Hoogenboom developed the optical and mechanical aspects of the precision differential measurement approach used for score residual; and S. Fallek made major contributions to the mechanical design of the system.

Figure 1 EndSpector™ System

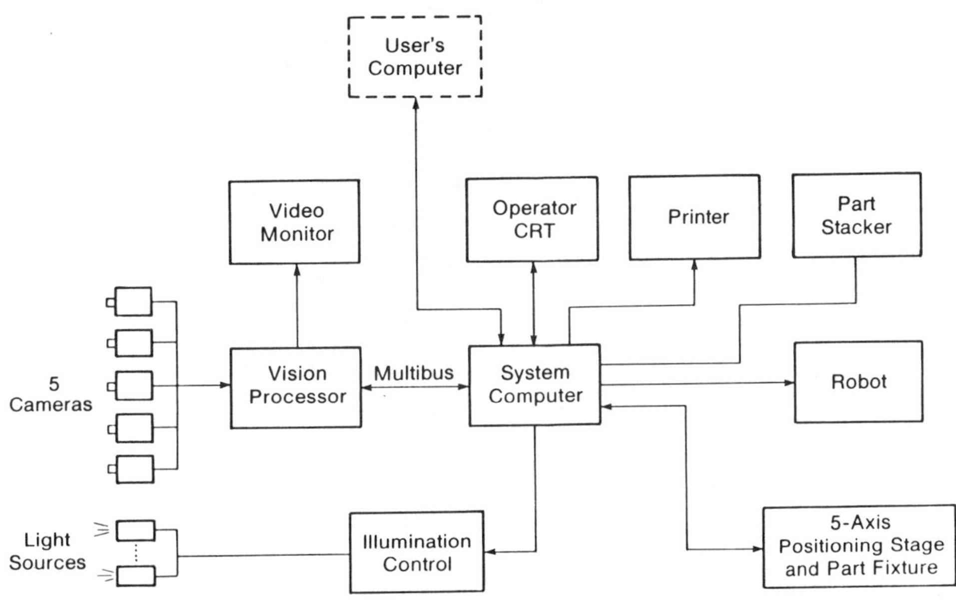

Figure 2 EndSpector™ System Configuration

Figure 3 Inspection Station Showing Cameras, Loading Robot, and Part Positioning Stage

Figure 4 Loading Robot

Figure 5 Robot End Effector Holding Can End

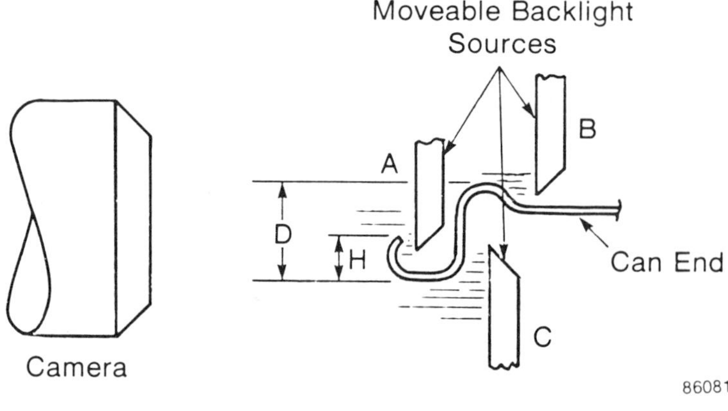

Figure 6 Curl Height and Countersink Depth Measurement Approach

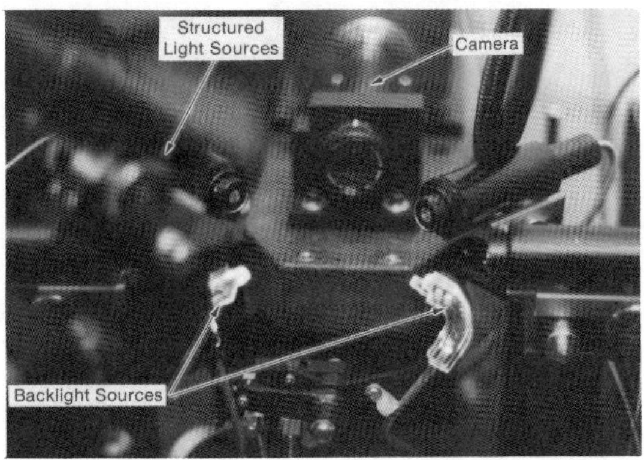

Figure 7 Measurement Station Showing Camera, Backlight Sources, and Structured-Light Sources

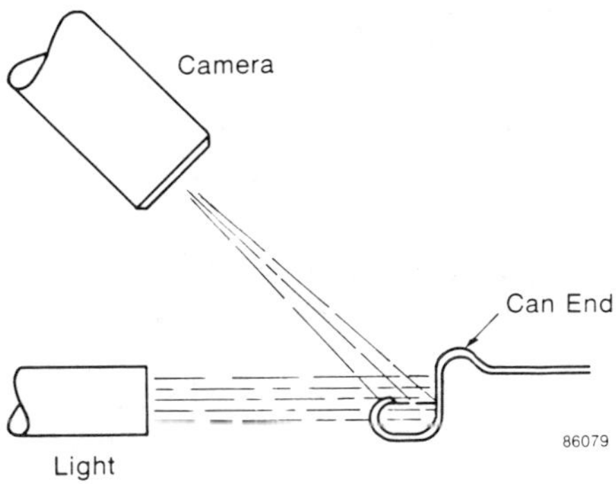

Figure 8 Curl Opening Measurement Approach

Figure 9 Image of Curl Opening - Un Processed

Figure 10 Image of Curl Opening after Threshold Operation

Figure 11 Image of Curl Opening after High-Pass Filter Operation

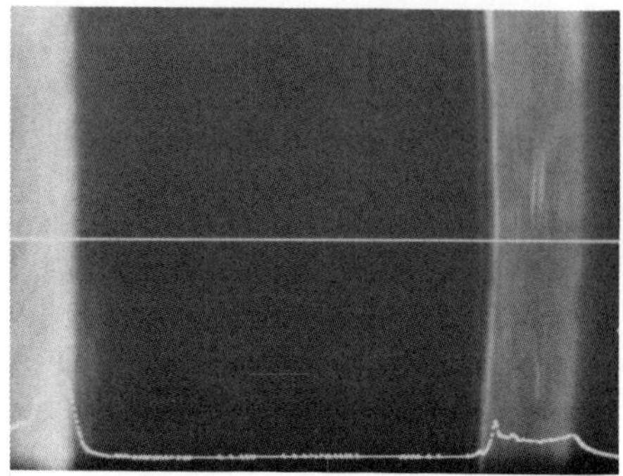

Figure 12 Image of Curl Opening after Vertical Smoothing and Edge Measurement

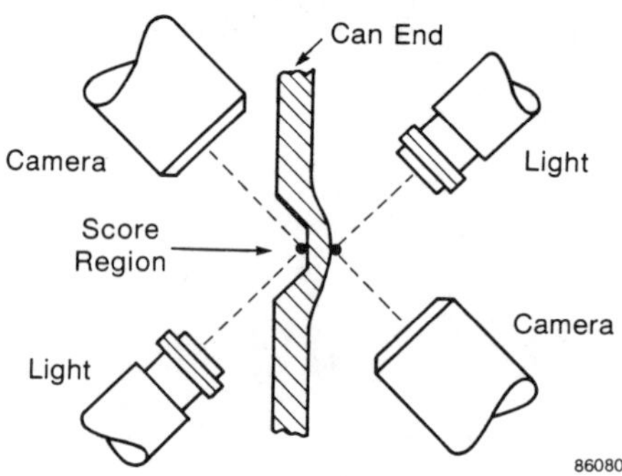

Figure 13 Score Residual Measurement Approach

Figure 14 Image of Score Region

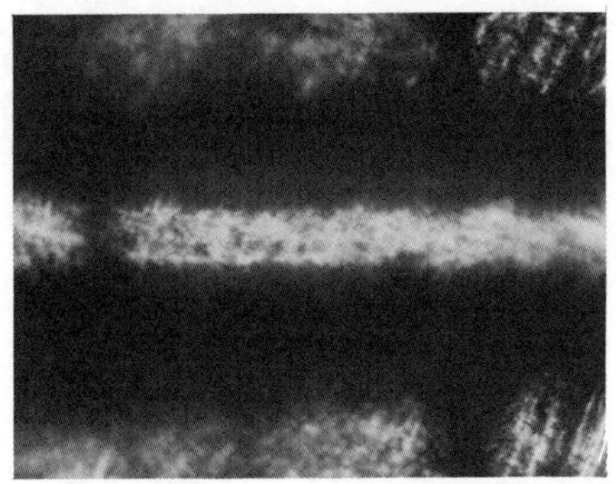

Figure 15 Image of Score Region with High Magnification

Figure 16 Image of Score Region with High Magnification and after Threshold Operation

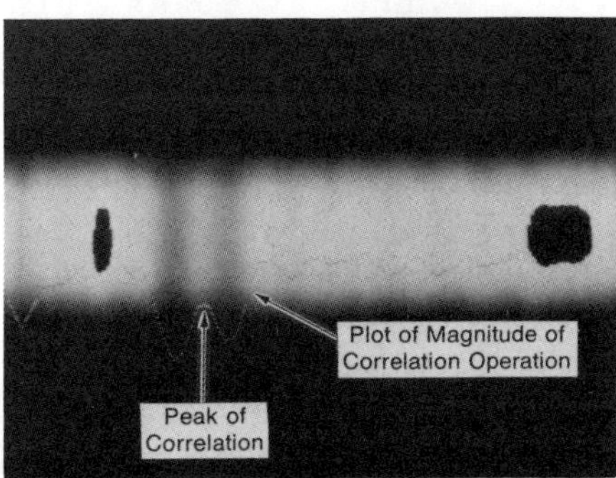

Figure 17 Result of Score Residual Correlation Operation

Presented at the CASA/SME Sensors '85 Conference, November 1985

Noncontact Dimensional Gage for Automated Applications

by Victor B. Burke,, Ph.D.
Current Technological Systems, Inc.

Introduction

A noncontact displacement-measuring system utilizing reflected light is described. The noncontact gage is suitable for making dimensional and geometrical measurements on a wide variety of materials. Since the measuring system is operated through a microcomputer, various levels of automation are possible. As illustrated below, in figures one, two and three, the gage may be manually positioned, very much like a dial indicator, or automatically positioned by a multi-axis CNC system. Many benefits follow when a noncontact gage is mounted on a CNC precision positioning table. Two non-obvious benefits are: the measurements become absolute, not relative; very large dimensions can be measured by adding the noncontact gage reading to the table's offset from a reference location.

With a noncontact system, rapid measurements are possible for two reasons. The incident light beam may be regarded as a "probe tip" which readily follows sudden changes in the surface of the part being measured. The "probe tip" has high mechanical bandwidth. There is no need to interrupt the measurement to lift the probe tip over an obstacle, or to slow down the motion of the part because the probe is "bouncing". Second, the noncontact gage is a transducer which converts motion of the part to an electronic signal via reflected light. The electronic circuitry has high bandwidth. During a measurement the gage is stationary, there are no moving mechanical parts; only the light beam moves as it follows the shape of the part.

High speed combined with automatic recording of measured values overcomes limitations of traditional measuring devices, such as dial indicators and electronic displacement gages. Furthermore, since the noncontact gage is controlled through a microcomputer, measured values may be analyzed and displayed graphically, GO/NO-GO criteria may be evaluated, and concise reports may be provided.

Industrial metrology is concerned with the science and art of measuring dimensions, geometry, and surface finish of manufactured parts. A truly remarkable array of measuring devices and techniques have evolved to answer the following recurring questions. Are specified dimensions within tolerance? Is total runout within acceptable limits? Is surface smoothness within spec? Are related surfaces flat and parallel; round and concentric? Is a particular region of a surface properly located spatially?

Within many manufacturing plants, previous achievements have created significant knowledge bases. Productivity and quality may be enhanced when available information is made accessible to intelligent systems which unite a microcomputer with a measuring device. Some capabilities of a measuring system which is operated through a microcomputer are: faster, more thorough measurements; immediate and objective interpretation of the measurements; GO/NOGO decisions; error reports; concise "final" reports characterizing the essential features of the measured part(s). An automated measurement system which provides little more than a tabulation of measured values is not acceptable in a production environment; it would more likely be disruptive rather than helpful. A carefully integrated system, on the otherhand, blends in with existing operations and provides answers which are consistent with previous experience.

Noncontact measuring systems are inherently attractive because they do not perturb the surface being measured. Specifications vary so widely for different applications that it is helpful to have some insight into the main factors which influence the design of a noncontact gage head. For this reason, relationships among the resolution, standoff distance, range, and aspect ratio are discussed later in the paper. Another complex of ideas is concerned with the rate at which measurements may be performed. That discussion centers around the exposure of pixels in the detector.

A measurement problem

Consider for a moment the cross-section of a slotted two-section part shown in Fig.1. The step "s" between adjacent sections is to be less than 0.005 inches. Dimension "b" has a tolerance of +0.000/-0.010 inches and a total indicated run out (T.I.R.) of 0.005 inches. The average and standard deviation of the measured "b" values are to be determined. It is illustrative to contrast three methods for measuring the part:

1. Plunger-type dial indicator, height stand & gage block.

2. Manually-positioned noncontact measuring system, height stand & gage block.

3. CNC noncontact measuring system.

Dial indicator

Low cost, small size, ease of use, and outstanding versatility are characteristic of dial indicators. They have been employed throughout industry for decades. Measuring the step shown in Fig.1 is straightforward. The gage would be placed on one side of the step and adjusted for a "zero" reading. Sliding the height stand to position the gage on the other side of the step gives a direct indication of the step size.

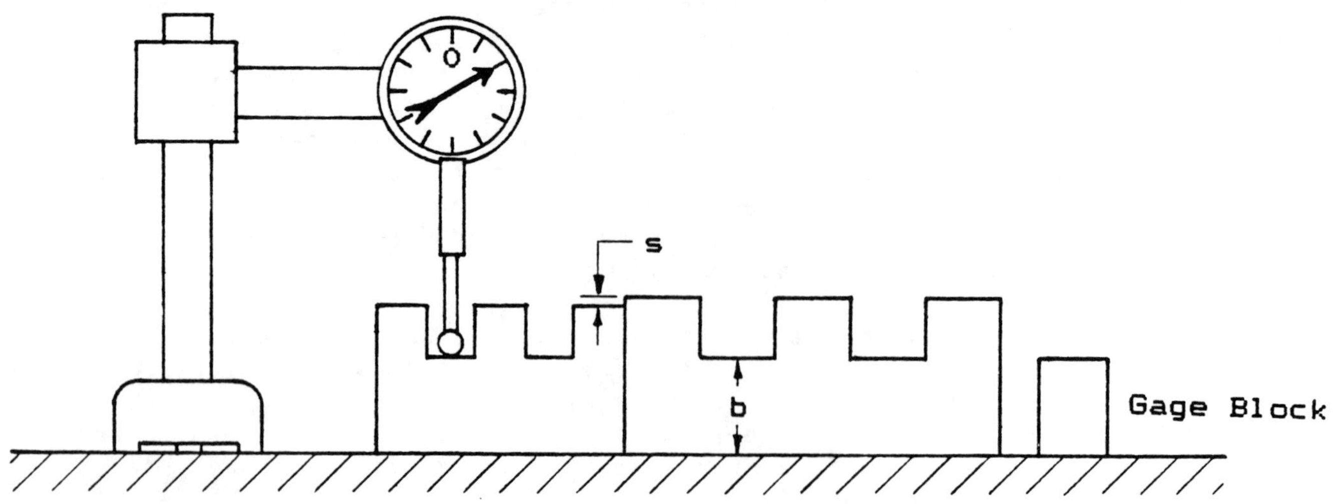

Fig.1 Dial Indicator

To measure the "b" dimension for each slot, the indicator could be "zeroed" against the gage blocks - built up and wrung together to form an accurate dimensional reference - and then shifted to the part. Time is not a friendly ally though when it comes to measuring the bottom of each slot, since it may be necessary to lift the indicator's tip before moving to the next slot. Measured "b" values are not recorded automatically, nor is the data processed automatically. So additional time is required to write down the measurements and to calculate an average value and the standard deviation. T.I.R. may be obtained by scanning the recorded values or by remembering the high and low deflections of the indicator.

Noncontact measuring system

Manually-positioned

Let's suppose that the dial indicator is replaced by a noncontact gage and that the gage is operated through a microcomputer, as shown in Fig.2. An important consideration with a noncontact measuring system is the standoff distance - the minimum distance between the part and the gage for which a measurement is possible. Let's suppose further that the standoff distance is greater than the slot depth, allowing the gage to glide above the part. The step height may be easily determined. Whether or not the bottom of the slot can be measured depends upon its aspect ratio, which is the slot depth divided by slot width.

$$\text{SLOT ASPECT RATIO} = \frac{\text{SLOT DEPTH}}{\text{SLOT WIDTH}} \quad (1)$$

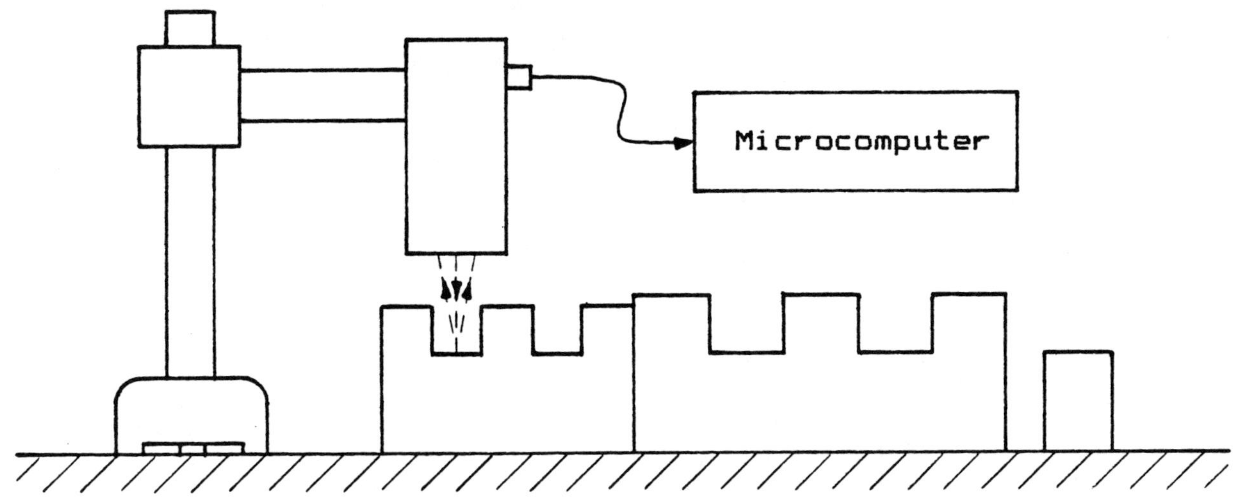

Fig.2 Noncontact Gage

Noncontact gage heads also have aspect ratios. For a given slot width, the greater the aspect ratio of the gage, the greater is the depth of the slot which can be measured. If the noncontact gage can "see" the bottom of the slot then rapid measurements are possible because the gage can glide above the surface at a constant level. An omnidirectional noncontact gage has the further advantage that its aspect ratio is independent of the orientation of the slot. In other words, the slot could just as well be a circular hole.

As with the dial indicator, the noncontact gage attached to a height stand would be manually positioned over each slot and a reading taken. Because of the ease in positioning the noncontact gage, measurements could be performed much faster. By automatically recording each measured value in the microcomputer, the operator, on command, could determine the T.I.R., average, and standard deviation of dimension "b".

Three modes of operation are available to initiate a measurement: manual trigger, free-run, and external trigger. Measured values are stored in memory and displayed on the monitor as numerical values and optionally in graphic form as a line or polar chart. Displayed points may be unconnected or they may be connected by an interpolation routine.

Generally, many dimensions on a part would be checked. GO/NO-GO determinations could be made by the microcomputer if it had a list of tolerance specifications for the various dimensions being checked. In the event that an out-of-spec condition is detected, any of a variety of responses could be initiated. One would be to generate an error report which identifies the part and the offending dimension(s). In simple cases an alarm could be sounded or a red light turned on.

When the part is within spec, a concise report could include the part identification, step size, T.I.R., average value, and standard deviation. Whether the part is in spec or not, in most cases, it is unnecessary and even undesirable to print out all of the measured values. Of prime concern are the significant results from the analysis of measured values.

CNC noncontact measuring system

Shown in Fig.3 is a noncontact gage mounted to a vertical linear table which, in turn, is mounted on a horizontal table. Accurate placement of the part on the reference surface may be obtained with a fixture. Typically, a CNC controller directs the motion of one or more precision tables, including rotary as well as linear tables. Incremental motions with linear resolutions of 0.0001 inches and angular resolutions of 0.002 degrees are available. Linear travels up to 36 inches (and more!) are also available. Large complex surfaces may be inspected.

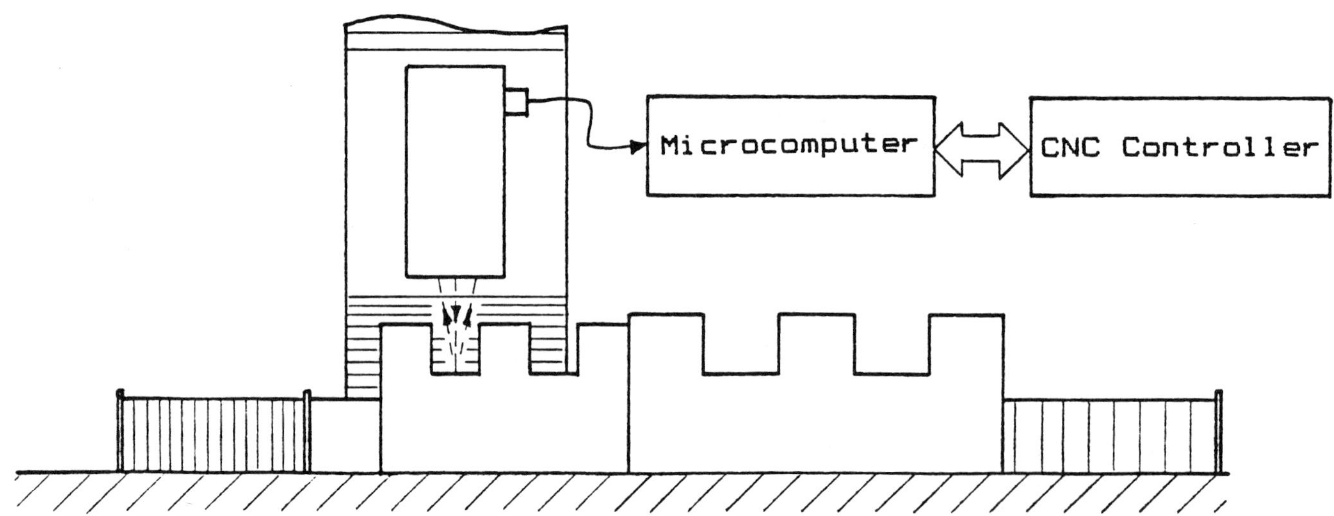

Fig.3 Two-axis CNC System

Since a given part may be measured at various stages of production and assembly, different measurement tasks may be necessary. It is possible for the microcomputer to program the CNC controller based upon a minimal description of the part and on a description of the measurements to be performed during the current task. With two degrees of freedom, the microcomputer can position the gage at any height and transverse position consistent with the range of the tables.

In Fig.3, note that the reference gage block has been removed; the positional accuracy of the tables is usually sufficient to obtain absolute measurements. Dimension "b" would be the sum of the noncontact reading and the offset of the table from a reference position.

Let's assume that the top and bottom of the slot are both within the dynamic range of the noncontact gage. In one horizontal scan spanning the region of interest, it is possible for the gage to measure dimension "b" for each slot and to measure the top on either side of the step. Measured values may be automatically recorded, analyzed, and GO/NO-GO determinations made. In the event that an out-of-spec condition is detected, any of a variety of responses could be initiated. One would be to generate an error report which identifies the part and the offending dimension(s). In simple cases an alarm could be sounded or a red light turned on.

When the part is within spec, a concise report could include the part identification, step size, T.I.R., average value, and standard deviation. Whether the part is in spec or not, in most cases it is unnecessary and even undesireable to print out all of the measured values. Significant results of the analysis of the measured values are of prime concern.

Palletized operation

Completely automated measurements are possible when the part is mounted on a pallet. Fig.4 illustrates a CNC system for measuring inner diameter, outer diameter, and axial dimensions of cylindrical parts of various diameters and heights. In the figure, an assembled part is mounted on a pallet which, in turn, is clamped to a rotary table. Not shown are the CNC controller and the microcomputer. Interchangeable lenses for measuring deeper cavities and for axial measurements are sketched in Fig.5. In operation, the pallet arrives and is identified; the rotary axis spins the part at constant RPM; measurements are made; rotation stops; and the pallet is released.

The microcomputer, by referring to a data file for each part, programs the CNC controller, which then moves the linear and indexing tables to position the noncontact gage. In between measurements, the microcomputer determines average diameters, T.I.R., and concentricities. Out-of-tolerance conditions are reported to a repair station. When all measurements are within specifications, a final report is generated which documents the essential features of the part.

In an integrated manufacturing environment a means of communication among hierarchial computer systems is available. To use the available network as little as possible, it is recommended that unnecessary and highly redundant data not be transmitted. This leads quite naturally to a guiding philosophical principle: send a minimal description of the job to the measuring system's microcomputer; let it develop the full set of data and instructions necessary to complete the assigned measurement tasks.

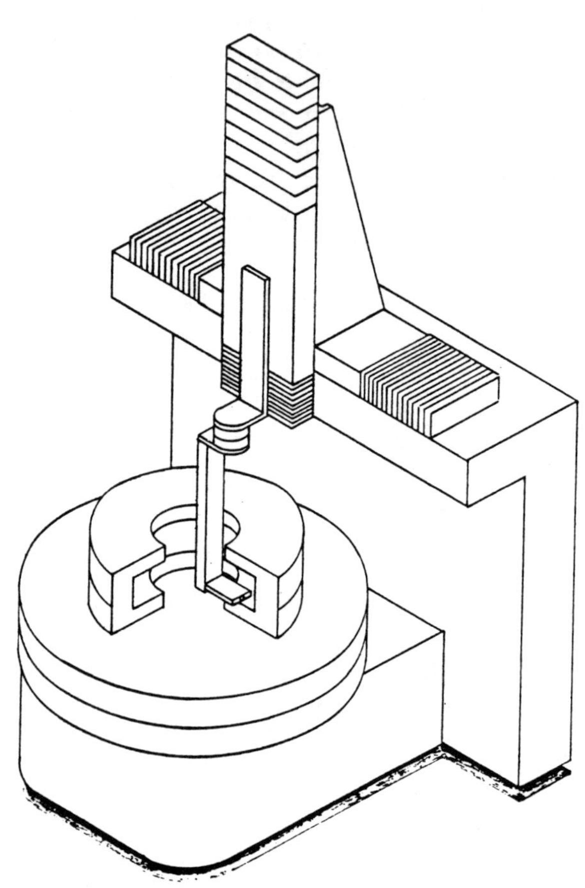

Fig.4 Palletized CNC Noncontact Measuring System

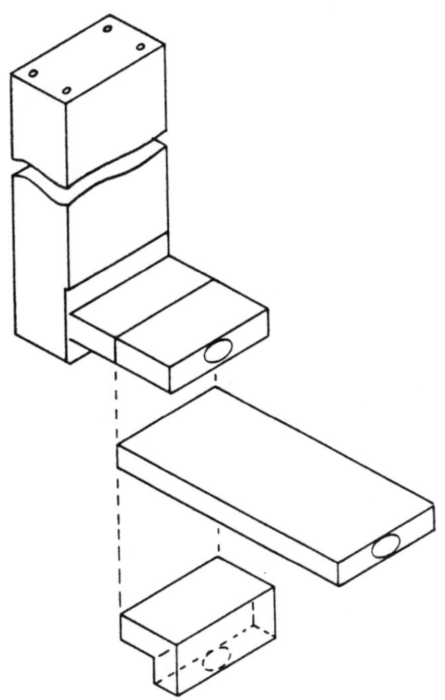

Fig.5 Interchangeable Lenses

Surface effects and reflected light

All noncontact gages which operate on the basis of reflected light have fundamental limitations imposed by the condition of the surface of the part being measured. Important surface effects are the absorption of incident light, surface finish, and the tilt of the surface. Two ways to overcome the absorption of light are to select a brighter light source or one with a different wavelength. The role played by surface finish and surface tilt are discussed in the following paragraphs.

Specular vs diffuse reflection

Suppose that a small spot on the surface of a measured part is illuminated. If the surface is mirror-like, then the angle of incidence of a light ray equals the angle of the reflected ray. This familiar type of reflection is called specular reflection. As illustrated in Fig.6, the detector would have to be at a particular angle to see the reflected light. Naturally, if reflected light never enters the detector then displacement measurements would be impossible.

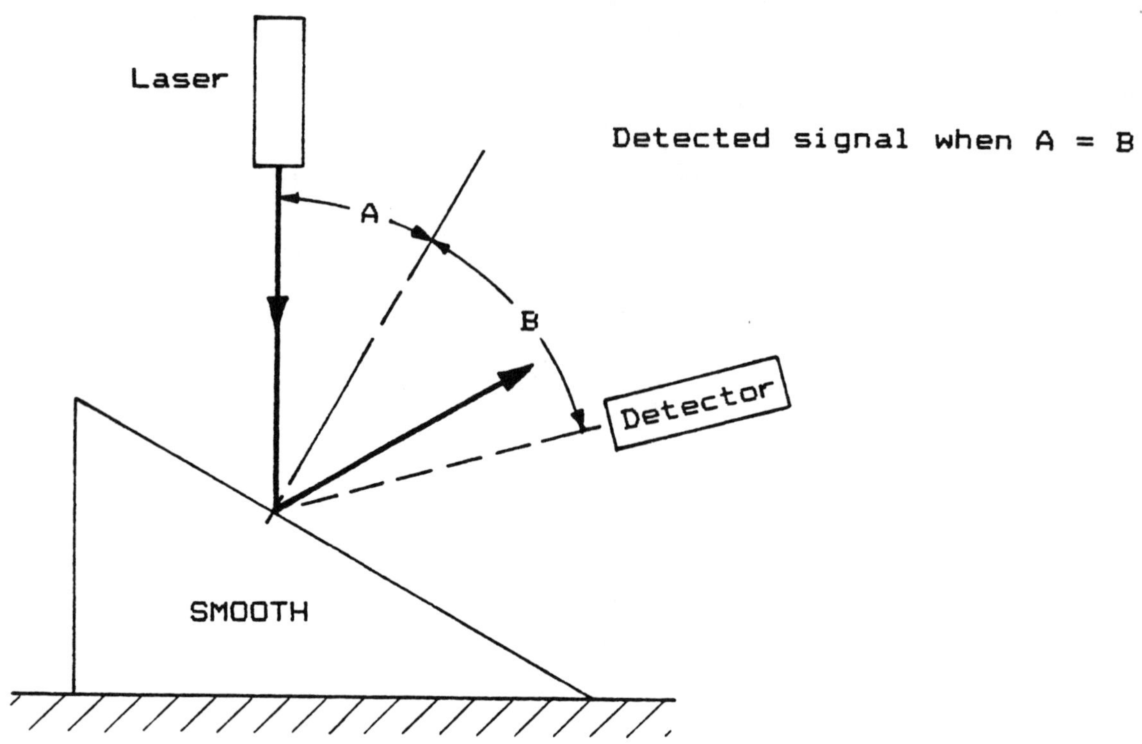

Fig.6 Specular Reflection

On the other hand, a "rough" surface results with diffuse reflection. Any surface which does not give a reflected image is "rough". Common examples are parts with a satin finish, such as the barrel of a micrometer, cast parts, even a piece of paper. As illustrated in Fig.7, an incident light ray is scattered in all directions in the hemisphere above the part. Brightness of the reflected light is not uniform at all angles. For a Lambertian surface the intensity of reflected light is proportional to the cosine of the angle of the normal to the surface.

Light reflected from the surface of a manufactured part is partly specular and partly diffuse. Most of the light from the source enters the detector when specular reflection dominates. From the detector's viewpoint, the small illuminated spot looks very bright. In contrast, with diffuse reflection the detector receives a small part of the available light,.so the spot looks relatively dim. If, however, "dim" light is sufficient for the detector to operate, then, when reflection is mostly specular, a smaller, lower power, more economical light source may be employed.

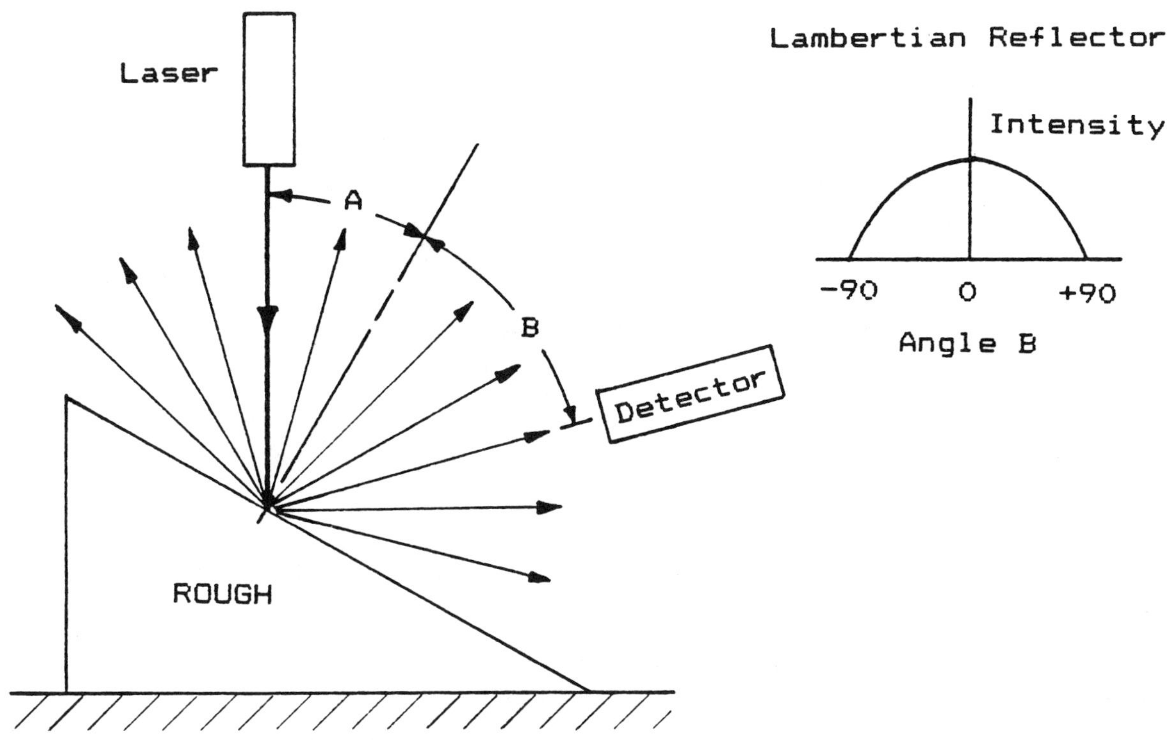

Fig.7 Diffuse Reflection

Curved/tilted surfaces

When it comes to measuring curved or tilted surfaces, diffuse reflection is very forgiving while specular reflection is not. Tilting the surface of a diffuse reflector (without changing the distance) alters the amount of light which enters the detector. Accurate displacement measurements are still obtainable if the detector is relatively insensitive to intensity variations and as long as the incident light exceeds a minimum threshold. The maximum tilt angle for which the distance of the part may be measured depends upon the degree of diffuse reflection from the surface of the part.

Consider the curved surface in Fig.8. When the noncontact gage (NCG) is positioned opposite point "a", the incident beam is at +30 degrees to the surface normal; at "b" the incident beam is normal to the surface: at "c" the beam is at -30 degrees to the surface normal. With diffuse reflection, light will enter the detector in each case and measurements may be obtained. With specular reflection, only point "b" could be measured.

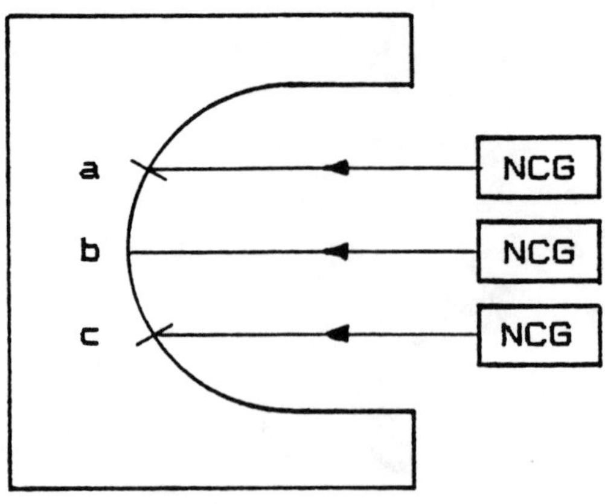

Fig.8 Tilted Surfaces

Surface conditions vary markedly from one application to another but may be fairly consistent within a family of similar parts at the same stage of production.

Alignment of the surface normal with the incident beam is much more critical with specular reflection. Since the reflected beam is narrow and directional, a small variation in the tilt of the surface may cause the beam to completely miss the detector. It should be pointed out, however, that a disadvantage in one application may be an advantage in another. If the spatial orientation of a surface is critical, then specular reflection may be quite useful.

Noncontact gage head

Principal components of the noncontact gage head are shown in the block diagram, Fig.9. A light source and beam optics contained within the gage project a small bright spot onto the surface of the part being measured. Some reflected light is collected by a lens which is coaxial with the incident beam, and is processed optically. The conditioned light illuminates a detector and is converted to an electrical signal. Real-time processing of the electrical signal results with a digital value for the distance between the part and a "zero" reference location.

An important set of interrelated parameters are the resolution, range, standoff distance, and aspect ratio. Three other parameters are the spot size, the rate at which measurements may be performed, and the brightness of the light source.

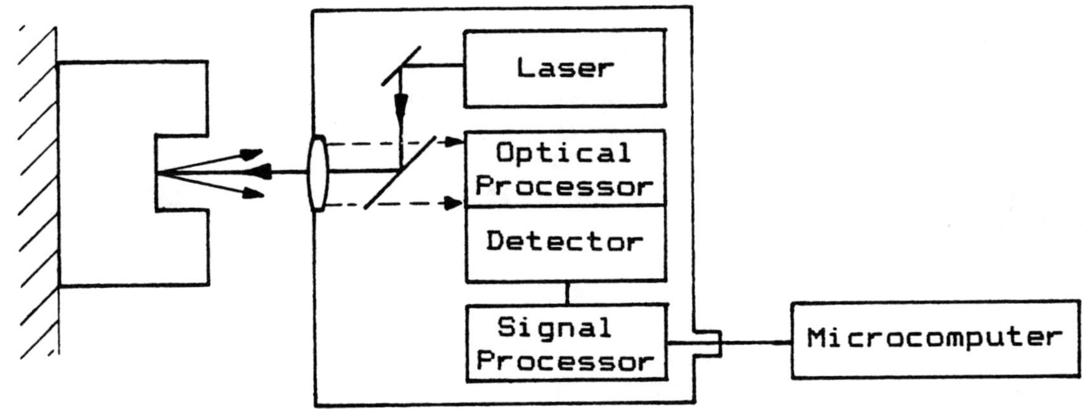

Fig.9 Block Diagram

Representative values

Before discussing relationships among the fundamental parameters of the noncontact gage, it may be helpful to list representative values for three different applications. Please refer to Table 1 below. Resolution of 0.0003 inches and a minimum standoff distance of three inches were the prime concerns in the design of system A. Design B required a range of four inches and a resolution of 0.001 inches. Both system A & B were designed to measure fairly large parts. System B, in fact, is for tire dimension measurements. Small spot size and moderate resolution were the overriding concerns in system C. Measurements at the rate of 1000 per second was more than adequate in each of the three applications. Acquiring measured values at faster rates is straight forward.

Table 1

Representative values for three applications

	A	B	C	Units
Resolution	0.0003	0.001	0.0005	inches
Standoff distance	3.00	16.00	0.300	inches
Range	0.5000	4.000	0.2500	inches
Aspect ratio	10	30	2.5	inches
Spot size	0.020	0.040	0.005	inches
Rate	1000	1000	1000	measurements/sec.

Basic curve

In the discussion which follows, the individual ideas are reasonably simple. But in combination it is easy to get confused. Keep in mind that one way to regard the noncontact gage is as a dial indicator with a light beam instead of the mechanical plunger. When the surface is displaced and "pushes against" the light beam, the gage responds by indicating the new position of the surface.

Fig.10 shows a basic relationship between resolution and distance separating the gage and part. Resolution is the least displacement of the part which can be detected by the gage. As shown in the diagram, resolution is given in thousandths of an inch; it improves the closer the gage is to the part being measured. Better resolution is at the lower region of the graph.

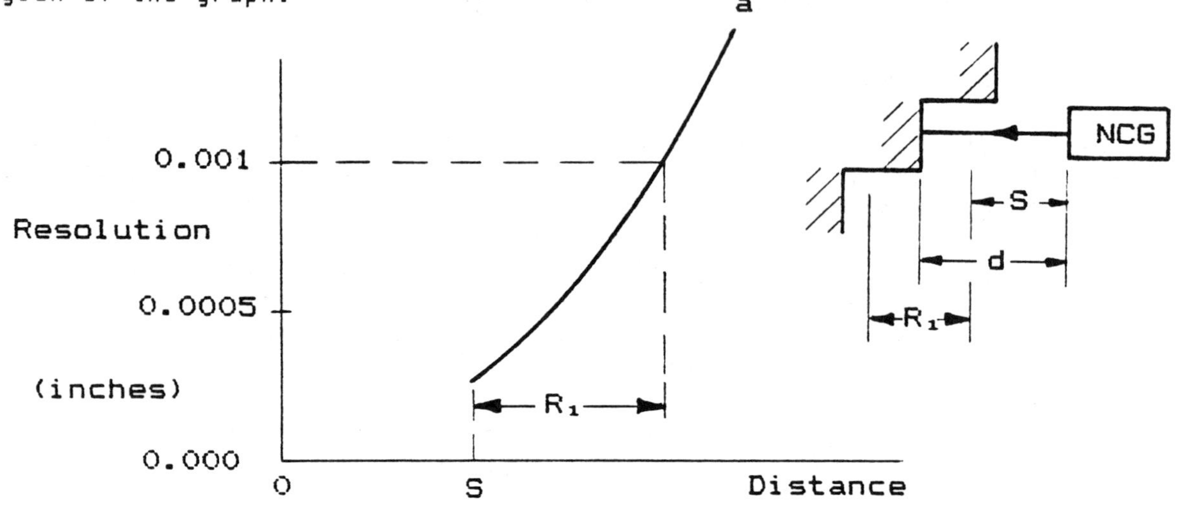

Fig.10 Basic Curve

Each noncontact gage is limited in how close it can be to the surface and still measure the position of the surface. This minimun distance to the part is called the standoff distance (S). If there are any irregular features on the surface of the part, such as with molds and castings, then the standoff distance should be large enough to avoid physical contact between the part and the gage. When a protruding surface reduces distance to a value less than S, an out-of-range value is indicated.

The active, or dynamic range of the gage may be determined by specifying the largest resolution which is acceptable for the application. To illustrate, let's specify a resolution of 0.001 inches. In Fig.10 project a horizontal line at the 0.001 level until it strikes the basic curve, from there drop a vertical line to distance axis. R_1 denotes the dynamic range of the gage. Within this range the resolution of the gage is 0.001 inches or better. What happens if this range is not large enough for the application? Fortunately, there are several ways to extend the range. These will be considered shortly.

Aspect ratio

Reflected light is collected by a lens of diameter D. The aspect ratio (a) of the gage is given by the following equation:

$$\text{GAGE ASPECT RATIO} = \frac{S}{D} \tag{2}$$

Do you recall the earlier discussion of the slot aspect ratio? The gage aspect ratio indicates how feasible it is to measure the surface at the bottom of a slot or hole. For example, with system A the aspect ratio is ten and the range is 0.5 inches. So that particular gage can "see" the bottom of a hole which is 0.050 inches diameter and 0.500 inches deep. Or 0.025 inches diameter and 0.250 inches deep, and so on. The gage aspect ratio must be greater than or equal to the slot aspect ratio; also, the depth of the slot or hole must not exceed the dynamic range of the gage. An important feature of the gage aspect ratio is that it is omnidirectional, that is, the same at all angles. Otherwise, to measure the narrowest possible slot, there would be a preferred orientation of the slot relative to the gage. With an omnidirectioal gage there is no preferred orientation.

Extending the dynamic range

Suppose that the aspect ratio is larger than necessary but that the range is too small. It is possible to trade aspect ratio for range while maintaining the same resolution and standoff distance. A family of curves with decreasing aspect ratio and with increasing range is illustrated in Fig.11.

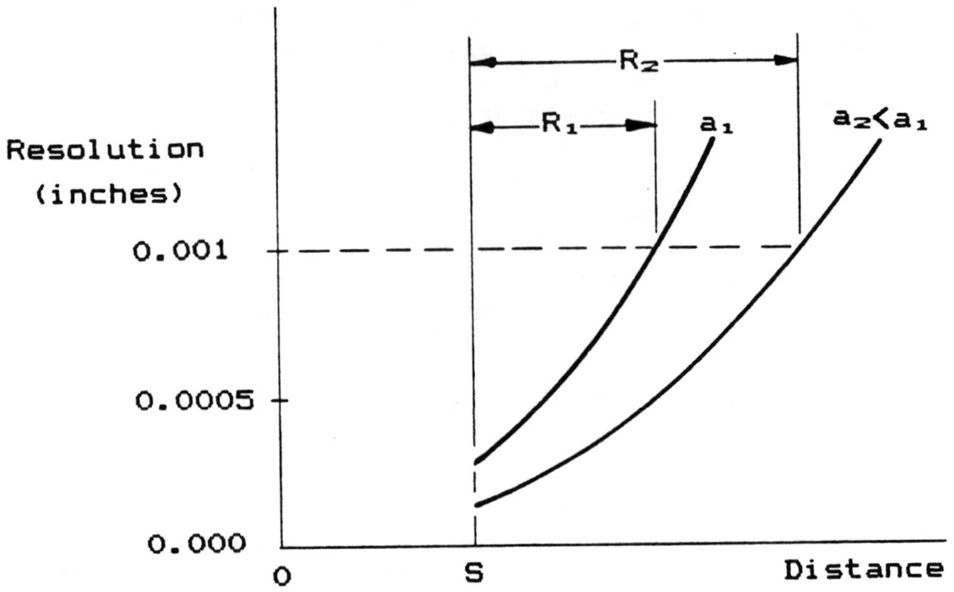

Fig.11 Trade Aspect Ratio for Range

If the aspect ratio cannot be traded for increased range, there is a subtle alternative. In simplest terms, the region of the curve above a specified resolution (0.001 inch in our example) may be traded for improved resolution and greater range below the 0.001 inch level. A family of such curves is shown in Fig.12. The new element in the discussion is the upper limit of the dynamic range imposed by the gage itself, not by a resolution criteria. By reducing the gage-limited range, resolution improves and, remarkably, the dynamic range below 0.001 inches increases. Since two opposing factors are competing, there is an optimum condition. In any event, neither the aspect ratio nor the standoff distance are affected.

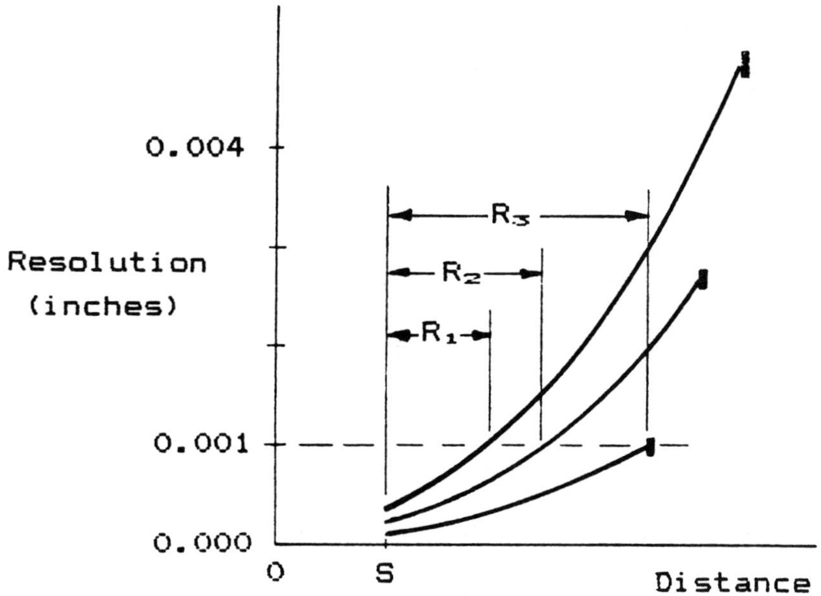

Fig.12 Trade Max. Range for Useable Range

Adapting to an application

This considerable flexibility shown in Figures 10 through 12 allows the noncontact system designer to configure a suitable system for a wide range of applications. Furthermore, the elements which comprise the gage head may be assembled in a form which is determined primarily by shape and size restrictions imposed by the application.

Measurement rates

It's helpful to think of the detector as a collection of cells or pixels. The word "pixel" is a contraction of "picture element". Fundamentally, measurement rates are limited by the rate at which light enters the pixels. Exposure of a pixel is the product of light intensity and time, just like film in a camera. To be useful, the pixels must be neither underexposed nor overexposed.

Low light levels mean slow measurements. It may be that the surface of the part absorbs much of the incident light or that the gage aspect ratio is too large for the application. Refer to Fig.9 and to equation (2). If the standoff distance is large and the lens diameter is small then comparatively less light enters the detector. There are two things that can be done to boost the amount of available light and thereby make more measurements per second. Use a brighter light source and use the smallest acceptable aspect ratio. A more sensitive detector may also be necessary.

Limitations

Because of the available electro/optic and electronic components, distinct classes of performance are obtainable. One figure of merit for displacement measuring systems is the ratio of range to resolution. Noncontact system ratios are listed in Table 2.

Table 2

Range/Resolution Limits

Class	Range/Resolution Ratio
R1	100 - 1000
R2	100 - 2000
R3	100 - 4000
R4	100 - 8000

Even with highly reflective surfaces, the rate at which measurements can be performed is limited by available components. A particular noncontact measuring system is also a member of one of the speed classes tabulated in Table 3.

Table 3

Speed Limits

Class	Measurements/second
S1	0 - 1000
S2	0 - 2000
S3	0 - 4000
S4	0 - 8000

Doubling the speed means that the time available to collect the light is halved. To maintain a moderately uniform exposure the amount of light entering the detector should be doubled. Similarly, to double the range/resolution ratio means that twice as many pixels are available. If the exposure is to remain constant then, once again, the amount of light entering the detector should double. As more light is needed a brighter light source may be necessary.

Spot Size

Spot size is a design parameter; it can be tailored for the application. It seems contradictory, but larger diameter beams have less divergence and may be focused to a smaller spot. For this reason it is common practice to insert a beam expander and converging lens in the path of a laser beam to obtain a small focused spot. With quality optics to preserve the phase of the wavefronts, diffraction limited performance may be obtained.

Another consideration with regard to spot size is a consequence of the relative transverse motion between the part and the noncontact gage. Suppose that the spot size is 0.010 inches, that 1000 measurements per second are being taken, and that the relative transverse speed is 50 inches per second. With an exposure time of one millisecond, the spot illuminates a strip on the part which is 0.010 inches wide and 0.050 inches long (0.25x1.25 mm). The gage, in effect, is not measuring the distance to a single point on the part but is measuring the average distance to a collection of points. The degree to which this is of any consequence depends upon the application. Decrease the transverse speed or increase the measurement rate, either one will reduce the length of the strip associated with each reading.

Conclusions

The noncontact gaging system discussed in this paper may be employed to measure dimensional and geometrical features of manufactured parts. There are a variety of design tools which may be employed to accommodate the requirements of very different measurement applications. Principal features of the measuring system are the following:

1. High measurement rates
2. High range/resolution ratios
3. Automatic recording of measured values
4. Automatic processing of measured values
5. GO/NO-GO determinations
6. Generates error reports
7. Generates final reports
8. Graphic display
9. Manually - positioned systems
10. Multi-axis CNC positioning systems
11. Palletized CNC positioning systems

CHAPTER 6

ON-LINE METROLOGY

Reprinted courtesy of Marc Bothel
Everett, Washington

Linear Array Cameras In Two and Three Dimensional Tasks

by Marc Bothel
Opcon, Inc.

The Linear Array Camera is among the most elementary forms of Machine Vision Systems. As the name states, it is merely a single row of pixels. The obvious application of a single line is to do point to point measurement and simple presence/absence type analysis. In fact, it has been productively employed in many such applications.

The purpose of this paper is to go a step further and suggest some techniques which will extend the Line Imager's range of use into 2 and 3 dimensional analysis. In one case, the use of "structured light", that is lines projected at an angle, allows the measurement of height or thickness. In another technique, the camera's grayscale facility is used to integrate light from a moving object, resulting in an efficient shape analysis.

Advantages of Line Imager -

The very simplicity of the Line Imager is one of its chief advantages. One frame of data, at most, contains 1024 bytes of grayscale data. Operating in binary mode it may yield no more than the pixel addresses of 2 or 3 edges. With this small amount of data, a simple microprocessor is all that is needed for data analysis. It also follows that data analysis can be very quick, allowing the Line Imager to keep up with high speed processes. The advantages continue with less complex electronics, high reliability, more flexibility and especially a much lower cost than other vision products.

Structured Light -

The concept of determining distance by triangulation is well known. Figure 1 illustrates a simple way this can be used in vision applications. A line of light is projected onto an object while the camera observes from a different angle. The displacement of the line will be directly proportional to the thickness of the sample. Using the Line Imager, we are now able to measure both the width of the object and its thickness. This technique has been applied to accomplish a wide variety of tasks; among them sorting packages by size, assuring fill level and checking for part presence.

If we go a step further and use several parallel lines together, we can use this technique to determine the contour of an object. As you see in Figure 2, the offset of each line tells us the thickness at that point.

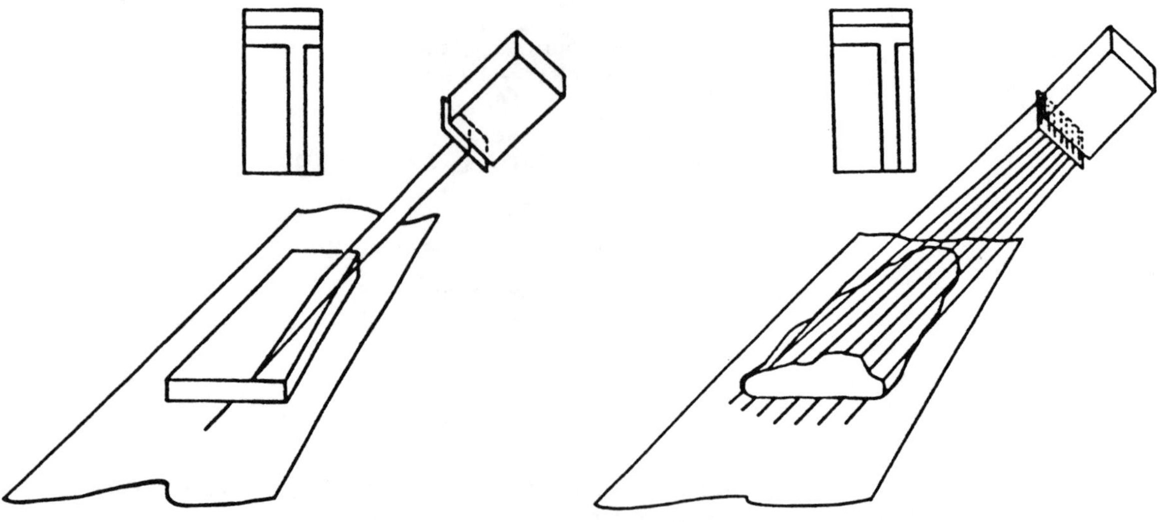

Figure 1 - Structured Light   Figure 2 - Contour Detection

Fish Portioning Example -

This use of structured light was recently applied to the task of portion control in cutting fresh salmon. The problem is as follows; restaurants, institutional food services and packagers of frozen meals all want uniform portions of fish for their meals. Typically, they want 4, 6 or 8 ounce servings. Until now, cutting the fish to these sizes was a totally manual task and accuracy was hard to attain. A fillet of fish is a very irregular shape and cutting it to a predetermined portion size is difficult, even for the most experienced person.

A Line Imaging System has now been applied to this task with excellent results. As described above, lines of light are projected onto a conveyor belt carrying fish fillets. A rotary encoder is attached to the belt and used to trigger a picture every tenth of an inch. For each exposure, the thickness contour is determined and the volume thereby integrated over the length of the fillet. This information is then used to calculate where the cuts should be made to produce portions of the desired weight. A few feet down the line the fish is automatically cut into the desired portions. In this case the vision controller directs the actuation of a high pressure water knife.

The result of this development has been a dramatic improvement in operation. Labor productivity has more than tripled. Accuracy of the system is consistently within 1/4 ounce and usually better than that. For comparison, an experienced person is considered to be doing well when he stays within one ounce.

While this application is interesting by itself, our purpose here is to illustrate a concept which has broad application potential. That is, by using structured light with the Line Imager in this way, contour and volume can be measured for many purposes.

We do want to give credit to the developers of the Fish Portioner. Opcon was pleased to participate as a co-developer with Seafreeze and Design Systems Inc. in this project. The Portioner I™ is now marketed by Design Systems Inc. of Auburn, WA.

Grayscale Shape Imaging -

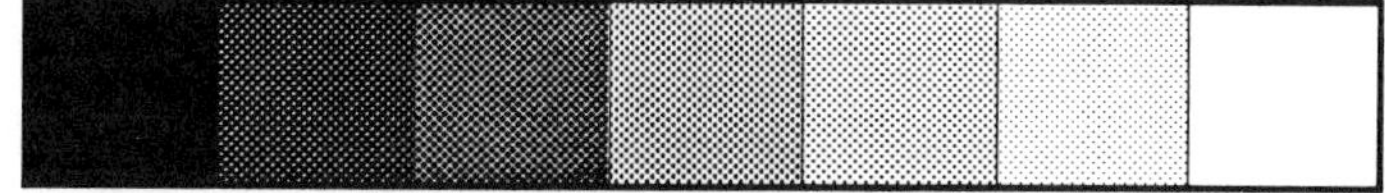

Figure 3 - Grayscale

Up to this point we have not been concerned with using the grayscale capability of the Line Imager. In Opcon's case, the system will differentiate 256 levels of gray. It is clear that if we were to take a picture of the illustrated gray bars we could produce the grayscale image. From that we could locate the edges of the bars and make whatever measurements we wanted.

Taking the use of grayscale a step further, we can also relate it to an object's shape. Figure 4 displays a grayscale image which we will use to demonstrate this point. We normally use an oscilloscope to display the image obtained by the camera. The image is, in essence, a plot of the grayscale for each pixel. If you look closely you can see that the display closely resembles the profile of one half of a bottle laid on its side.

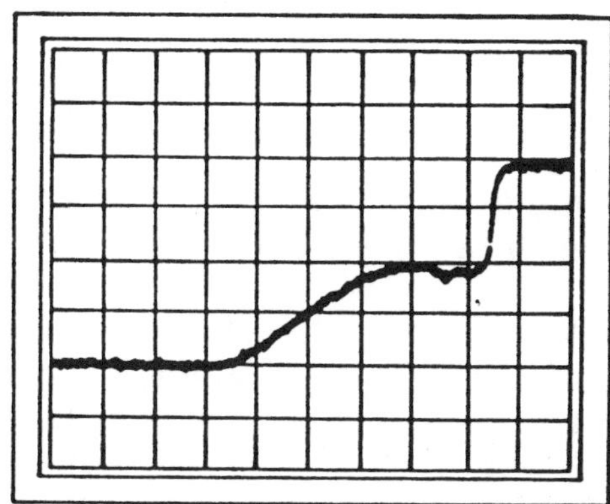

Figure 4 - Grayscale Image of Bottle

The image is created by taking a time exposure as a bottle moves along a conveyor in front of the camera. The exposure begins when the bottle gets to the scan line and continues to the middle of the bottle. The effect is that we integrate the backlight over time, the thinner the bottle (as at the neck) the less time the light is blocked and hence the higher (whiter) grayscale value. The bottom blocks the light most of the time - almost 0 graylevel. With bottles coming by at say 800 per minute, the exposure time will be roughly 25 msec. Using a 1 x 256 array, the result of the exposure is a vector of 256 elements representing the amount of time the light was blocked at each point along the bottle. Or in terms of our desired analysis, each pixel has a value which is directly proportional to the width of the bottle at that point.

Now, here is the advantage gained from the Line Imager; this "shape vector" was obtained with no digital analysis. The grayscale image from the camera directly yields a compact digital profile of the bottle without any data processing or computation. Because of this efficiency, this technique can be used in applications requiring high speed analysis such as the following example.

Bottle Sorting Example -

Opcon has used the above technique to develop a high speed bottle sorting routine. A special concern for many bottlers is to make sure that all the bottles coming into their processing lines are the right brand. The first level of our sorting routine does just this; what we call "Foreign Bottle Detection". It checks all bottles to be sure that they match the desired standard.

By storing several templates this concept can also be extended into a sorting algorithm. Each bottle is compared to several templates until a match is found. As you would expect there is a trade-off between the number of templates to check and the throughput speed. The chart shows the processing rate for different numbers of templates.

# of Bottles To Recognize	Production Rate (Bottles/Min)
1	1800
2	1100
3	900
4	700
5	575

These numbers assume the lack of any major differentiator such as height. A preliminary sort on such a feature can often speed up the process.

Again, our purpose is not to emphasize the Bottle Sorter but rather the technique. The Line Imager is not the obvious tool of choice when the task at hand is a shape related problem. However, the concept of light integration makes an effective tool. The profiles of many objects can be recognized and measured by using graylevel as an indirect representation of width.

In the above example we developed the light integration using a time exposure of a moving object. We have also obtained similar results with shorter exposures and on stationary objects by using a cylindrical lens.

Conclusion -

As we discussed at the beginning, the Line Imager has some attractive features in its simplicity, speed and low cost. When it is capable of doing the task at hand, it is usually the most cost effective and reliable tool to use. The above discussion is offered in the hope that these techniques will suggest additional applications where the Line Imager can be used to your advantage.

Presented at the MVA/SME Vision '85 Conference, March 1985

Optical Gaging—Crankshaft

by Donald W. Whitney
General Motors Corporation

I. INTRODUCTION

The introduction of interchangeable parts by Eli Whitney caused an immediate need for means to compare those parts to a standard. Advances to meet these needs have occurred in step with demands to improve part precision and accuracy. Micrometers, calipers, and snap gages gave way to open-air orifice gages, which, in turn, have given way to combination air-electronic and pure electronic gage systems.

These gages all rely on contact to perform their measurement. This contact also creates trouble for the gage. Wear of the gage head occurs; part geometry variations reduce measurement accuracy. Part wear or damage can occur if contact pressure is excessive. If contact area is increased to lessen pressure, surface characteristics may cause more variation. Compressed air blowing on the part carries with it moisture and/or oil, attracting dirt. Part temperature is also affected by this air; expansion and contraction of both the part and the gage must be considered. Contaminants, like dirt, tend to jam linkages, LVDT's, and air orifices, resulting in frequent maintenance. These contaminants may even be detrimental to the part itself.

The opposite of contact gaging is non-contact gaging. The effects of the gage on the part are obviously gone. Similarly, the part cannot wear or contaminate the gage. When dealing with desired measuring accuracy and precision (repeatability) at levels of 1 micron (40 millionths of an inch) or less, the attributes of non-contact gaging are most attractive.

A type of non-contact gaging is the venerable optical comparator ("shadowgraph"). Its drawback has been lack of repeatability among operators and slow usage speed; its strongpoint is the ability to measure a wide variety of dimensions without special fixture requirements. A "shadowgraph" operator, assigned to measure parts on a typical mass production schedule, would soon fall behind, make errors, have a nervous breakdown, or, most likely, all three! By adding light sensors to read data and a computer to manage that data, this "shadowgraph" evolved into a crankshaft inspection system capable of measuring and tracking 54 separate characteristics on each crankshaft produced.

II. OPTICAL GAGING SYSTEM

The gaging system is completely automatic. Just prior to entering the gaging stations, part temperature is measured. The first gaging station rotates the part to measure journal diameters, keyway position, pilot bushing bore, and threaded holes. The second station gages the part in a stationary mode to provide rod pin journal ovality data. Parts gaged reach the operator station, where any defects found are displayed on a CRT screen for operator attention. The defective parts are segregated from acceptable parts at this point. For the purposes of discussion, the system shall be divided into several parts:

1. Gage sensors and light source.

2. Characteristics monitored.

3. Data and system management.

Figure 1 (page 11) shows system layout.

1. Gage Sensors and Light Source

The earlier description of this system as a computorized "shadowgraph" is fairly accurate, although an oversimplification. A "shadowgraph" projects light past an object. The shadow cast by the object is focused on a screen at a selected magnification. Measurements are made by comparing the particular feature to a grid on the screen.

For this crankshaft gaging system, light is projected past a part. The shadow cast is focused on a diode array, which is a device for converting light presence to an electrical output. The diode array, through its arrangement as a grid system, "counts" how much light is blocked out by the part. The count is then compared to the master shadow count. The unit comprised of the diode array and the focusing lens is referred to as a sensor unit.

The light source is placed directly opposite each sensor unit. A simple incandescent light bulb is utilized; positioning and focus are critical to project the required light to the sensor. It should be noted that with careful alignment of the sensor unit, light source, and part, the resultant measurements are along line-of-sight or true tangents, eliminating error due to geometrical properties. See Figure 2 (page 12) for illustration.

2. Characteristics Monitored

The crankshaft is probably the most complex machined part in an automotive engine (rotary engines excepted). Added to the complexity of shape is the need to fit bearing clearances and to maintain piston stroke and timing in the engine assembly. These needs result in the large number of characteristics being monitored. See Figure 3 (page 13) for chart.

A partial list of these characteristics includes:

a. Main bearing journal diameter.
b. Rod pin bearing journal diameter.
c. Pilot bushing bore diameter.
d. Oil seal diameter.
e. Gear fit diameter.

The above characteristics are measured directly by the sensor units. Bearing journal taper and ovality, main bearing journal alignment, crankpin/keyway index, pilot bore runout, and oil seal runout are among the characteristics requiring part rotation and/or calculations in order to be checked. A computer system was married to the sensor units to perform this task. In all, 54 characteristics are checked and monitored by the system.

3. Data and System Management

The computer system devised consists of a master microprocessor and four slaves. Each microprocessor performs specific tasks related to various measurements. The slaves communicate with the master unit, which, along with its own unique measuring tasks, compiles and reports the data as a measurement record.

In conjunction with the above units, a Gould Modicon programmable controller is used to control physical movements of the parts within the gage. These movements include station-to-station transfer, part rotation, rotation drive synchronization, and reject marking. Part rotation and rotation drive synchronization are required only in the first gaging station.

This data management system performs the task of making over 10,000 individual measurements in order to evaluate each crankshaft. The following chapter will describe the dynamics involved.

III. MEASUREMENT DYNAMICS

The development of the data management system has resulted in unique measurement evaluation. Many automatic gages accept/reject parts based on measurements taken only in one plane, or, if rotated, based on passing over limit lines. Any dirt present on the part which could affect the measurement is ignored. Human operators, on the other hand, actually average their findings by checking dimensions in several places. The system used was developed to reflect that human habit and to take into account the presence of dirt or other contaminants. A great advantage of this averaging is to ensure that characteristics are evaluated equally from part to part. Measurements which do not meet specifications are always noted for the operator; the coding of the defects also allows the operator to evaluate the seriousness of the defect.

Five sections will be used to briefly explain the measurement evaluation and calculation:

1. Main bearing journal, oil seal, and gear fit.

2. Rod pin bearing journal.

3. Pilot bushing bore and threaded holes.

4. Temperature variation.

5. Data management.

1. Main Bearing Journal, Oil Seal, and Gear Fit

As the crankshaft is rotated, 5 flash sensor readings are taken at every 10 degrees of rotation, beginning with the top-dead-center position of the #1 rod pin. The average, in raw counts, is recorded for each interval. Following completion of 1 revolution, the raw count data is filtered by comparing each record with the preceeding and following record. If the record in question differs from its neighbors by a value exceeding a specified window, the preceeding value is substituted. In this way, dirt, dust, and other contaminants have no effect on the measurement. Once filtering is completed, the data is converted to linear measurements using preset scale factors.

2. Rod Pin Bearing Journal

Rod pin bearing journal measurement requires special treatment because the offset of the journals from centerline moves the journal in and out of the view of the sensors. In order to measure the journals, flash sensor readings are taken only as the pins pass through their respective top-dead-center position. As above, the raw counts are averaged and filtered. Scale factors again convert the raw counts to linear measurements.

A second gaging station is needed to supply data for calculating pin journal ovality. This station takes flash readings in a stationary mode. The averages of these measurements are similarly converted to linear measurements.

3. Pilot Bushing Bore and Threaded Holes

The measurement of the pilot bushing bore and the threaded holes requires different hardware, although the light sensing technique is the same as previously explained. Carbide inserts attached to lever arms follow the bore sides. These arms pivot at midpoint, allowing the other end to reflect the movement of the inserts. This reflected movement is read by the light sensor unit; again, filtering and conversion to linear measurements are performed.

The threaded holes are checked for thread depth and thread form. Light is projected into the hole. The amount and pattern of the light reflected back to a sensor unit determines the measurement. The presence of any contaminants is so critical to this measurement that the limits used were modified so that only presence of threads and hole depth are effectively measured.

4. Temperature Variation

Part temperature variations can account for a significant portion of measurement variances. In the load station immediately preceeding the first gaging station, a transducer senses part temperature. This temperature is compared to that of the gage master at the last calibration. Scale factors are used to convert the measured data to account for expansion or contraction.

Sensor focus can also be adversely affected by temperature. Contraction/expansion of the sensor unit mount varies the focus of the sensor unit with respect to the part. Such movement is prevented by maintaining a constant temperature in the sensor box. In this case, a heater is used to maintain temperature at 40 degrees centigrade (+/-1 degree). An added benefit is the increased life of electronic components which do not experience cyclic temperature variations.

5. Data Management

Once all the diameter measurements are taken, the data management system takes over. Average diameter, taper, and ovality are calculated as needed. In addition, for each bearing journal, an overall average of the two diameters measured is found. The results are compared to the preset limits. If any resultants fall outside the limits, the offending characteristic value is displayed along with a reject code to indicate reason and severity.

Main bearing journal, oil seal, gear fit, and pilot bushing bore runout is calculated relative to the centerline of the end main bearing journals. Ovality of the rod pin bearing journals is calculated by comparing the average values found in both gage stations.

These characteristic values of each crankshaft are all stored in the computer memory until reaching the operator station. At that point, defective characteristics are displayed for the operator. The entire measurement display or portions of raw data can also be off-loaded to a recording device if desired. This feature is very useful during runoff and during quality checks of the manufacturing process. Work which can take a week of painstaking manual activity can be accomplished in an hour.

Another important use of the computer system is for self-diagnosis and troubleshooting of the gage system. Understanding of the measuring sequence and observation of the data will pinpoint trouble spots in a minimal time period.

IV. PERFORMANCE

This crankshaft optical gaging system is presently in use on the plant floor as an integral part of the production process. The first step in this integration was to pass a runoff, both on the vendor's floor and on the plant floor.

In the case of gages, primary consideration is given to repeatability (precision) and accuracy. These parameters must be statistically stable over time. Pontiac Motor Division Manufacturing Engineering Standards specifies a sequence of measurement in a given time period using approved masters to determine gage acceptability. Even more stringent tests were undertaken for this system because of the newness of the technology for this application:

1. Five gaging cycles for each of the three masters (minimum, mean, maximum) for both mains and pins.

2. Five production samples gaged five times each.

3. Repeat steps 1 and 2 above at 30 minute intervals over four hours (total 9 repetitions).

4. Acceptability based on data stability, repeatability within 10% of the tolerance band width, and accuracy within +/-5% of the tolerance band width from the known value.

These criteria are checked by using X-bar/R control charts and 6-sigma capability. Finding these values is important, since they establish the amount of gage drift over time, the amount of variance in the measurement due to the gage itself, and that the variance that is present is predictable. Putting it simply, the risk of gaging error is reduced to nil.

The repeatability of this gage was found to be less than .5 microns (20 millionths of an inch). Accuracy was found to be .3 microns (12 millionths of an inch). These values are only 1% to 2% of the tolerance band widths. Even more remarkable is that through these tests involving nearly 20,000 characteristic measurements, only 6 individual thread measurements exceeded the established limits. A portion of the test was repeated after 8 months of system usage; identical results were found.

Production rate of the gage itself is around 250 parts/hour (net rate is somewhat less due to shuttle system constraints). Calibration is needed only weekly; this interval was found by observing master drift over time. Calibration is performed automatically following loading of the appropriate master. Essentially, this calibration is a gaging cycle run so that the computer can memorize the master shadow and relate this shadow to the master's known value. Only the mean masters are required to perform the calibration.

Maintenance required consists mostly of cleaning the glass shielding for the lights and sensors. At one time, this cleaning was thought to be a potential headache; experience has shown that cleaning is usually needed only weekly. Without the oil/water contamination found in most compressed air systems, dirt does not accumulate very fast. The precision part rollers are cleaned and checked periodically. Replacement has been needed only once following a "wrekcident". Many cycle safeguards have been built into the system to minimize these occurrences.

Some failures of sensor units have been encountered since installation. The sensors that failed had all been in use during the entire development period (almost 3 years). Many electronic components fail in industry in much shorter time frames. The light sources have proven to be suprisingly durable. After a full year of operation, the light bulbs were replaced; even at that point, no bulb had failed, although some were a little dim. At various times, sensor sensetivity was adjusted to account for this light variation, enabling the continued function of the gage.

The system is installed on the plant floor. There is no special floor vibration isolation; no controlled environment enclosure; and no special washer for parts cleaning. Employes have readily adapted to the system operation, giving them more oppurtunity to track process quality trends and to solve quality problems which may occur.

V. SUMMARY

The system described is an operational vision gage. Many advantages have been pointed out:

1. Non-contact measurement.

2. Computer-driven to allow management of the enormous amount of data in real time.

3. Self-diagnostics.

4. Measurement variation is nil (less than .5 micron repeatability; .3 micron accuracy).

5. Minimal calibration drift.

Disadvantages are also present, as compared to conventional gaging already in place:

1. Considerable training is required to develop understanding of the system. Skeptics like to see the measurement on a column; they believe they are able to visualize the measurement itself as it takes place and are thusly verifying its existence. This system only shows the final result as a linear measurement. It takes some length of time to convert these people.

2. Considerable training is required to develop in-house service. Along with knowledge, real interest must be developed. It is difficult to generate such interest in many new systems, let alone one that is "different".

3. Component life is still somewhat unknown. Much of the technology used can be viewed as a "black box"; it either works or it doesn't. Factors beyond the control of the system may be responsible for some failures. The sensor units and controller boards in use have low failure rates so far; some have been in almost continuous usage for the last four years. Experience will show if component life is a problem in the long run.

Although these negative factors are small in number, they point out that this gaging system is merely a part of the whole manufacturing system. Much thought and preparation is required to bring any such unit on-line. For those willing to adapt, plan, and learn, this system will perform beyond their expectations.

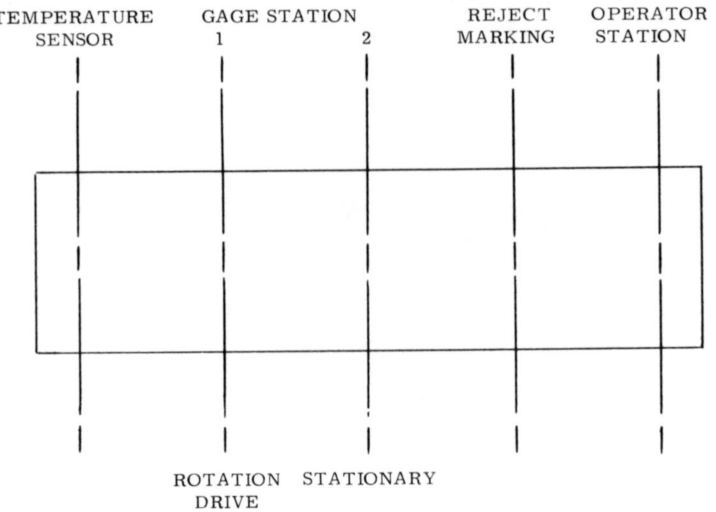

FIGURE 1 - CRANKSHAFT OPTICAL GAGE

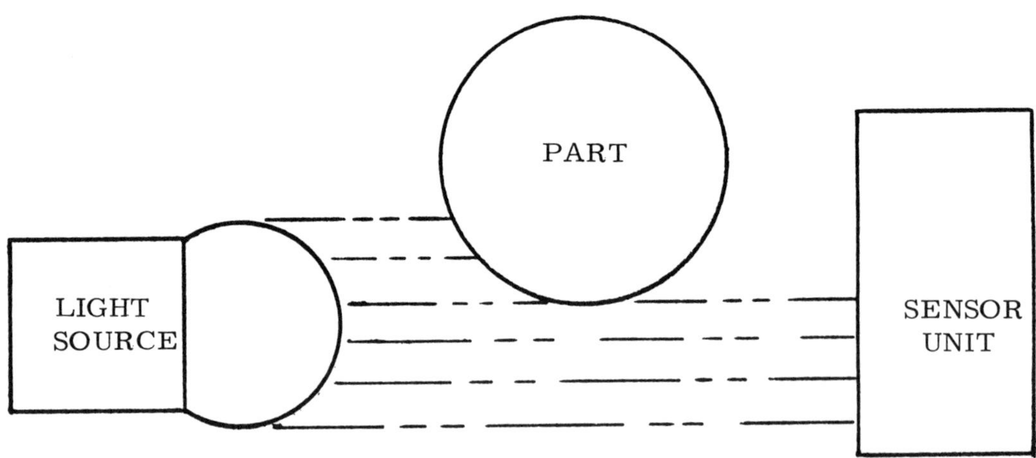

FIGURE 2 - SENSOR UNIT and LIGHT SOURCE

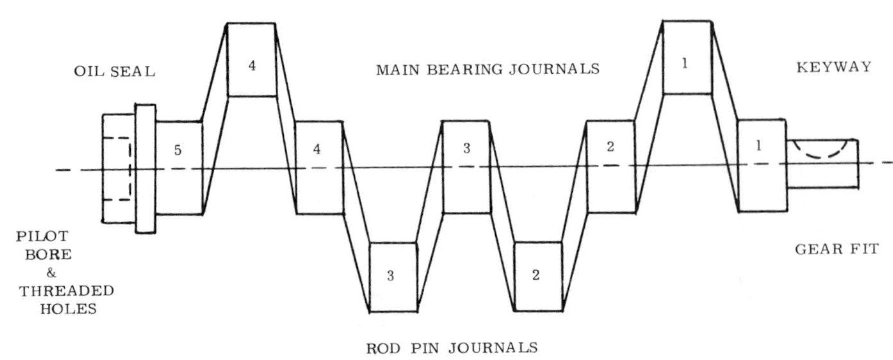

FIGURE 3 - CHARACTERISTICS MONITORED

Presented at the MVA/SME Vision '85 Conference, March 1985

Crop Shear On-line Software Directed Automatic Position and Speed Control

by Peter W. Loose
and
Peter J. Hunt
European Electronic Systems, Ltd.

INTRODUCTION

I think it is important at the outset that we understand how a crop shear controller fits into the scheme of a modern integrated steel works. Why is it needed at all? What are the economics of employing high technology in such a heavy industry as steel making? (Those who are familiar with the process may like to skip this section!)

A hot strip rolling mill receives slabs as its raw material. Each slab will weigh upwards of 10 tons. The slab may be typically 10 inches thick, 40 inches wide and 12 feet long. The mill end product will be a coil of similar width to the entry slab but thickness reduced to 0.050 inches up to 0.2" - depending on product destination and purpose. Slab entry temperatures will be about 2000°F for rolling through successive "stands" and finishing at around 1450°F prior to controlled cooling with water and then coiling. For process reasons the hot strip mill divides into two sections, roughing and finishing. The roughing mill reduces thickness to 1.25 inches. The rolling process deforms each end of the bar much as is experienced when rolling out pastry!

This deformed end shape has to be threaded through the finishing stands at 1000 ft/min. and greater. As the material gets thinner so its speed increases. It is important that threading is effectively carried out. Symetrical and square ends are essential to avoid mill roll damage and encourage a straight material path. A failure here is spectacular - the notorious cobble - in which red hot steel piles up or spills out from the mill.

Accordingly, there is placed a rotating shear just at the entry point of the finishing mill. The shear works on "the fly", that is a moving shear hits a moving bar of steel. The task is to hit and cut in the correct place. Too little cut off and the cobble risk becomes dangerously high, conversely too much cut off and cobble risk falls but yield loss goes up. It is the purpose of this paper to describe one way in which use of solid state non contacting cameras and associated visual signal processing has given rise to a new crop shear control system.

TODAY, many crop shears are not automated.

Those which are automated use a pair of position sensors, hot metal detectors or nuclear switches to time the bar end travel between two points, deduce its speed and position, and issue a cut signal past the last trigger point. The shear then accelerates without feedback, until the cut is made. Thus, a nominal fixed length, as an operator preset cut length, is measured from the head (or tail) and cut. The length cut is a safe average to insure square ends in the finishing mill.

One variant uses a single position sensor and a contact wheel for velocity. There are practical difficulties in maintaining tachometers in contact with the hot steel, as well as problems of wheel slip which contribute to cut errors - crop loss errors of as much as 1.0% have been found with such automated systems.

Another variant uses the same system as a basis of applying width optimization but includes a width camera, typically located at the roughing mill exit to take advantage of the time necessary for the bar end to travel to the shear mouth. During this relatively long period (in computer terms), width data is processed and a cut offset from the leading or trailing point is calculated. The initialization time of the shear position/velocity system is adaptively modified by the shape based data. This system was not adopted in the European Electronic Systems (EES) development for the following reason.

With a bar travelling several hundred feet from the roughing mill to the shear, the bar can be displaced sideways over the distance. The transfer of the trigger point will then be offset due to the gradient of bar end shape.

In the EES development, a camera is arranged to scan across the bar at the shear entry to overcome the displacement problem (Fig. 1). However, this approach poses the question - why use two cameras?

Exit speeds from the roughing mill are typically 80 to 100 ips whereas entry speeds to the finishing mill are approximately half of that value, 40 to 50 ips. Thus, for a given camera scan time, the resolution along the bar is twice as good at the finishing mill. On a head end of the bar, interest is only in the first 6 to 16 in. which corresponds to 60 to 160 ms and places severe constraints on the camera exposure. At the shear entry, a position camera with a 22 ft field of view (FOV) can be used to prime the width/shape camera exposure.

A psychological problem in placing two cameras close to the shear is the limited time, 3 or 4 s to view the bar and make decisions about the cut. In practice, this is not a problem and operator confidence is increased as the real-time cut point can be observed on a graphic display unit.

DEVELOPMENT

In the mid-1970's when EES became involved in the application of solid-state image sensors to the steel industry, a decision was made that all video data would be processed as a self-radiation digital 8-bit stream and that no external light sources would be used.

This decision was made because the inherent performance capability of solid-state arrays with regard to on-line performance potential was not being realized. The reason was signal processing, and particularly, cost effective 10 or 11 (1024/2048 grey levels) analog to digital converters were not available.

However, although limited to 256 grey levels for each picture element (pixel), significant progress had been made in the application of solid-state images to a family of instruments, i.e. the stereoscopic width gauge, thermal profile gauge and, most recently, the width optimized crop shear controller. The mill location of the two cameras which provided the data input for the position system and shape system is illustrated schematically in Fig. 1.

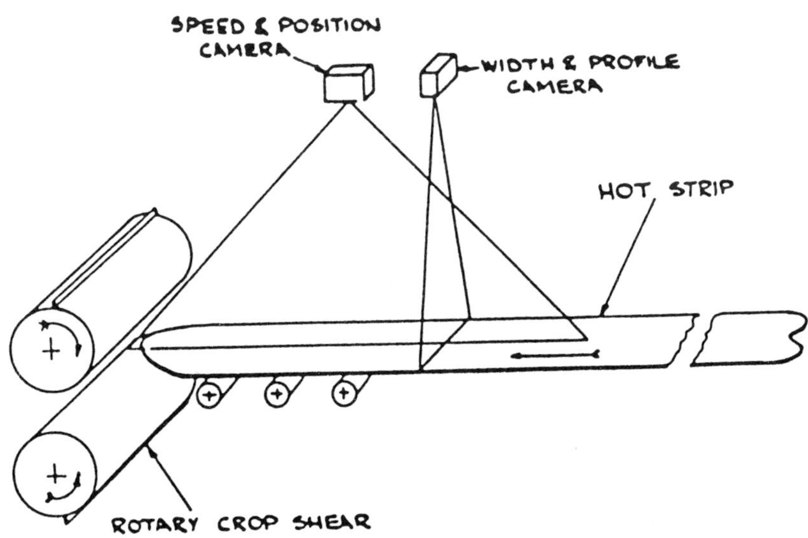

Fig. 1 - Crop Shear Control System

There are five key elements to the EES crop shear control system (Fig. 2): the system processor: a crop shear position velocity control: a position system: a shape optimization system: and an engineering control/diagnostics unit.

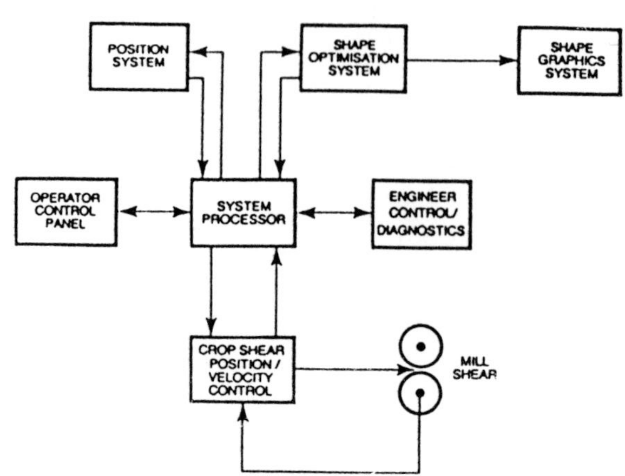

Fig. 2 - Key Elements of Crop Shear Control System.

The objective is to produce an optimized cut on the fly of both head and tail ends without contact devices to achieve high system availability times. Ideally, the system should be self-contained, including the shear position feedback within the one computer system. The goal of the system is to identify the optimum crop line (the line on a bar which has just reached constant width) to track it and control the flying shear such that the tracked line is cut.

SYSTEM DESCRIPTION

Mill data is collected by two cameras each consisting of a 2048 pixel linear CCD array with 22 and 8-ft FOV. Camera data is digitized to 256 grey levels/pixel (2 kbytes/scan) in the camera assembled and transmitted over balanced pairs to the associated computer. The purpose of the position system, with a 22-ft FOV, is to collect primary video data about the position of a head or tail end along the delay table. The viewing axis is along the mill centerline so it is that part of a head or tail which is tracked. Typical track points for a variety of bar ends are depicted in Fig. 3. There is no direct link between the tracked point and the desired cut line.

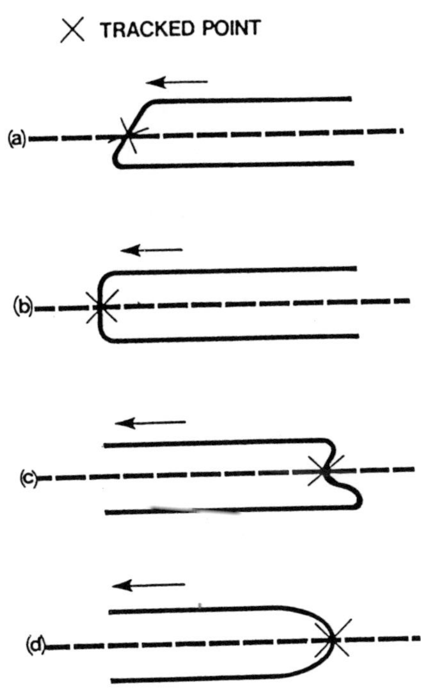

Fig. 3 - Tracked points for a variety of bar ends.

A typical video signal for a head end is shown in Fig. 4. It is a representation of the temperature distrbution along the bar centerline. Position measurement is effected by one of two algorithms. In the first case, a simple threshold is used based on the peak video per scan. The threshold ratio is software programmable and has inbuilt default or limit values. Although this simple level edge detector has adequate position resolution (one pixel is approximately 0.125 in.) it is inherently weak because of its dependence on a peak energy value which, as the bar progresses further into the FOV becomes hotter. Thus the peak used may be, and usually is, derived from a point several feet from the relatively cold leading edge. Accordingly this technique is used as a hardware pointer to reduce software data processing time. Edge detection is essentially performed independently of local temperature variations and takes into account edge blurring due to bar motion.

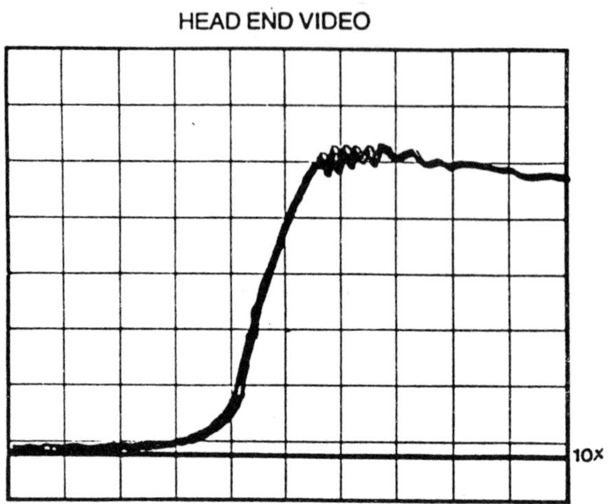

Fig. 4 - Typical Video Display of Head End.

Position information is filtered and fitted to a curve using a least squares fit method from which bar position trends are established. Since the purpose is to cut at an optimum width, data from the position camera are collected and labelled with a time tag at which it originated. A time resolution of 128 uS has proved sufficient. Similarly, shape data is time tagged using the same real time reference clock.

Consequently, data on bar position and shape are correlated retrospectively to predict a cut time and then to start the cut cycle on the shear.

Position processor - In practice, bars are variable both with respect to nominal entry temperature and scale. Entry temperatures vary between steel grades and types by several hundred degrees Fahrenheit. Scale build up, particularly at the leading edge, is a combination of speckle scale, which has the appearance of black drops of rain, and flake scale. Flake scale originates from at least two sources; scale from the current bar, and scale pick up on the head which was dropped in the roll gaps of the delay table from earlier bars.

Speckle scale is normally not a problem unless there is a single strip along the bar centerline. This occurs in the roughing mill and is due to minor malfunction of the roll cooling sprays or entry descaling sprays. In the case of a continuous scale line condition, other tests are performed on the video to test for valid tracking.

The most common condition experienced is head end scale pick-up. Scale acting as a thermal insulator, has a substantially lower emissivity than a clear surface and results in an artificially depressed camera signal. However, the effect is not uniform. A typical camera trace is illustrated in Fig. 5.

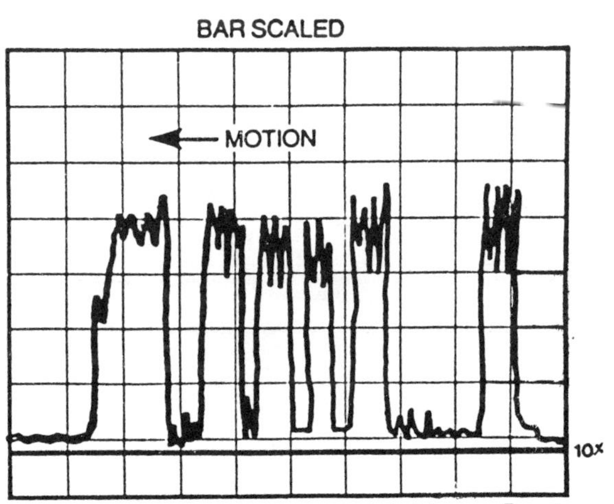

Fig. 5 - Typical Video Display of a Scaled Bar

The main problem resulting from scale is the potential identification of multiple head end track points. The worst conditions can be eliminated by simply checking the distance between successive video transitions of the threshold. However, the marginal case remains which produces a "hopping" between one edge and the other. This ambiguity is resolved by the mathematical processing in the system computer because it has access to recent historical data.

A bar status feature is also built into the position processor which determines whether there is no steel in view, a head, all steel, a tail or a tail/head pair. Tail/head pairs potentially resemble scale (Fig. 6). The criteria for distinguishing between these two conditions is the distance between the falling edge (tail) and the rising edge (head). If this distance is greater than approximately 16 ft. it represents two bars.

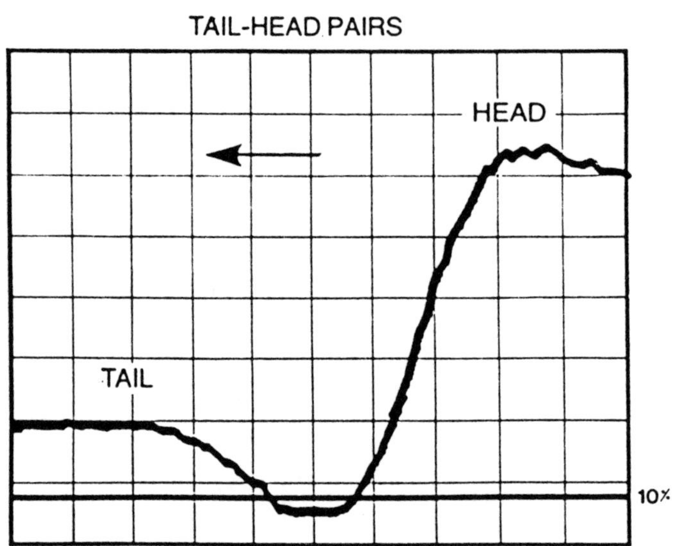

Fig. 6 - Video Display of Tail & Head Bar Ends

The position camera has a narrow track line (<0.10 in.) along the centre of the bar. If cyclic effects are present due to earlier descaling difficulties, substantial gaps of energy occur which appear like a no-steel condition. Potentially, this was the most damaging aspect of automatic noncontact shear control.

Bar status is calculated by reference to the direction of the video change at the leading and trailing edges. If the first transition is positive, a head is declared; a negative transition indicates a tail. This type of status information is classified as raw information because, on its own, it is not rigorous enough to make operational decisions. Through a bar, status may flicker several times because of scale and water which depresses the video response below the threshold level. All final decisions are made by the system processor.

EDGE DETECTION - The critical feature of edge detection is to identify the edge region to subpixel resolution. This is necessary to achieve width accuracies of 0.01 to 0.02 in. which are largely independent of strip position. With bar movement at approximately 40 ips, considerable blurring occurs. At a scan rate of 25 ms, the bar moves 1 in. or 8 pixels, scan. This is compounded by temperature effects. In practice a modified width gauge edge detection algorithm is used to take advantage of the bar motion through the application of a statistical technique.

An enlarged thermal energy profile of a bar leading edge is illustrated in Fig. 7. There are three regions of interest: A, B and C. Region A is a defect in all linear arrays when exposed to radiation bands 700 to 1100 nanometres (close to infrared). It is charge spillage arising from the body of the array and leaking to nonilluminated parts of the array. On some devices, this effect makes them unsuitable for infrared applications. However, in the EES approach this is overcome by the introduction of special features in the design of the arrays. The intersection of Regions A and B is at approximatey 10% of the peak level and, it is in this transition region, that optical effects compound the inherent array limitations. It is the location of Region B which is of interest.

By differentiation of the video in the design of software, limits are set for the A-B and B-C regions, based on slope and slope ratios, pixel to pixel. Limits are set which govern the correlation of each element within region B so that the edge position can be established by the system processor from position/time information derived from each camera scan.

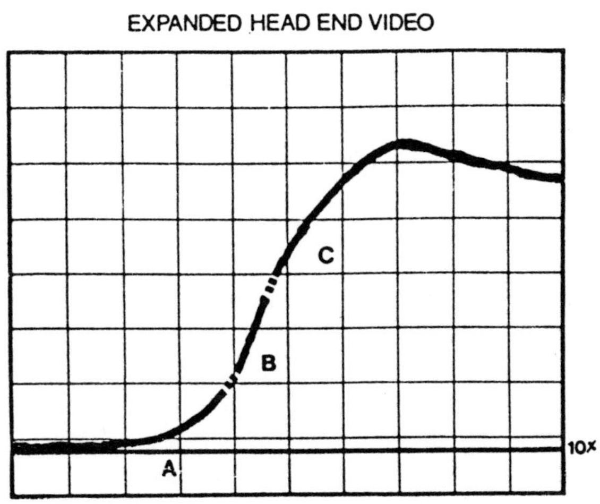

Fig. 7 - Enlarged Head End of Bar.

SYSTEM PROCESSOR - The system processor is the control core of the system. It directly supervises both cameras for sensitivity and receives data for both tracking and cutting. Exposure control plays a significant role in the tracking system. It is not uncommon to experience a head to tail temperature run down on a bar of 100 to 200°F. Mill delays can lead to further cooling resulting in a working range approaching 500°F. In energy terms, at the camera wavelength, this temperature range represents a ratio of approximately 10:1. The total transit time for a bar end through the FOV of the position camera is approximately 4 s. However, a typical shear will take 2 seconds from go to cut. Therefore, within 2 s of a head appearing within the FOV tracking data must be stable, width optimized and a cutting decision made. Time is critical. One function of the camera hardware is to determine the peak video level in each scan and to pass this information to the system controller. Additional reference signals in the array establish dark and nominal white levels. Thus, the required exposure can be calculated. The shape camera exposure is preset from the position processor. A programmable offset and gain scaling factor links the two cameras for a preset number of shape camera samples to insure a correct exposure. So the head end shape data are accurately collected.

Position tracking is established after the exposure has been stabilized.

The raw bar status is then verified by a system check for consistency. (Status validation counts are site programmable as part of commissioning.) With a valid head end being declared, position trending now commences by allowing raw position/time pairs to "stack" for curve fitting. A small number of initial, successive, position samples are allowed to the stack to prime the curve fit. As this number is exceeded, successive trend position limits are reinforced as an envelope around the core line produced by the least squares fit algorithm. A long memory buffer is maintained which holds the position/time pairs for subsequent width optimized correlation. In commissioning, a trend position error limit is set by reference to a monitor function which, at the completion of each potential cut, reports the number of raw sample positions available and the number used to perform the curve fit.

A wide limit will approach 100% utilization but with attendant dilution of accuracy. In practice, a limit of 1 in. from the predicted position yields a sample utilization of approximately 80%.

Assuming that no steel is in view by either camera and that the crop shear is ready for the next cut, the camera is set to a background sensitivity reading for receiving the head of the next bar. When the bar appears in the FOV of the camera at the end of the roughing mill, the system position processor real-time clock is reset and the system and shape processors are synchronized. All three processors are now synchronized within approximately 10 us and count in 128 us steps. Synchronization is critical because it is the link to all the processors. With synchronization complete, the exposure is established for the position camera and primed in the shape camera. Thus, head tracking is established and bar speed is successively recalculated by reference to the slope of the fitted curve.

SHEAR CONTROL - The shear must effectively accelerate from a known park position and cut the bar end at both the desired position and speed. For a head cut, a shear speed a few percent above bar speed at impact is desirable to aid in the separation of the crop end. The reverse is true for tail ends. Since shears vary from site to site, shear control is characterized in the following manner.

SHEAR CHARACTERISATION

An optical encoder with a 10-bit resolution (approximately 1/3°/step) is coupled to the shaft of the shear. Care is taken to mechanically isolate the encoder from mill shock. The optical encoder generates a digital signal directly. The optical encoder was selected on signal to noise grounds because of the high motor starting currents and high voltage in the immediate vicinity. There is an encoder fail alarm which is capable of aborting a cut should difficulty be experenced.

The encoder is mounted and connected to a test computer. The output of the test computer is a software controlled analog drive voltage connected to the motor drive control. Speed demands are then forced into the shear motor to determine its movement time (a lag in the system) together with acceleration data for a range of terminal velocities.

Encoder data is then logged and the control program optimized through loop factors adjusted to give an overall impulse response which is critically damped. Loop gain and velocity feed forward are gauge set functions.

CONTROL STRATEGY

There are at least two control strategies which can be adopted. One method is to always trigger the shear when the cut point is a fixed distance, i.e. 130 in. from the shear and to accelerate linearly for the cut. For a tail end this method would be acceptable because of the relatively constant speed. Clearly if mill speed is zoomed then this method has obvious shortcomings. However, head ends are more variable, especially when delay table speeds are not automatically matched to the entry speed of the finishing mill.

The method adopted in the EES system is based on established servotheory which was originally applied in analog systems and is now applied to a digital and intelligent control strategy (Fig. 8).

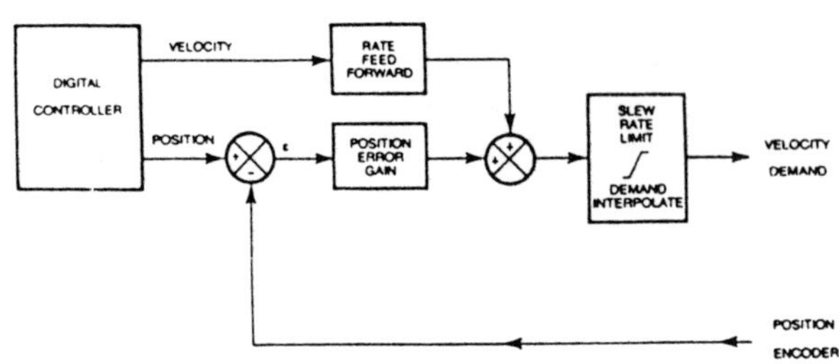

Fig. 8 - Cropping Control Strategy

There are three aspects to the control strategy which are illustrated in a position-time plot (Fig. 9).

The shear is parked up to time t_1 but the control program, in a wait mode, is continually calculating the pretrigger distance, p_t. When the bar speed and position conditions give a go signal at time, t_1 the control loop switches to a lag control phase. Lag is the time for the shear to begin rotating, i.e. a greater than 1 LSB change in encoder feed-back position.

This time is in the order of 0.3 s but varies from shear to shear. It is a preprogrammed value determined from predelivery tests. It is also a commissioning variable!

At time t_2 the shear moves to a second-order position prediction loop. (If this is not done, the early calculations of position error could cause the shear to go backwards.) From stored data, the controller develops a position error, not by direct reference to bar speed but by comparison with the theoretical expected position at any given time. The shear is constrained and accelerates in a known way to the point where shear and bar are in position and velocity synchronism at the earliest possible moment.

At time t_3 the controller switches to a track mode, during which errors in shear to bar position are progressively corrected. Plots of position error though the entire cut cycle have shown that the system errors are significantly less than 1.0 in.

A bar position demand update every 50 or 60 mS provides a measure of system deadband and in the case of more modern responsive shears can result in "chatter". The control program module is however, accessed at a higher rate and hence interpolating the demand based on the last pair of bar values can infill between samples to give a smoother output. The output to the shear drive is rate limited to ensure that basic current trips are not exceeded in any control phase.

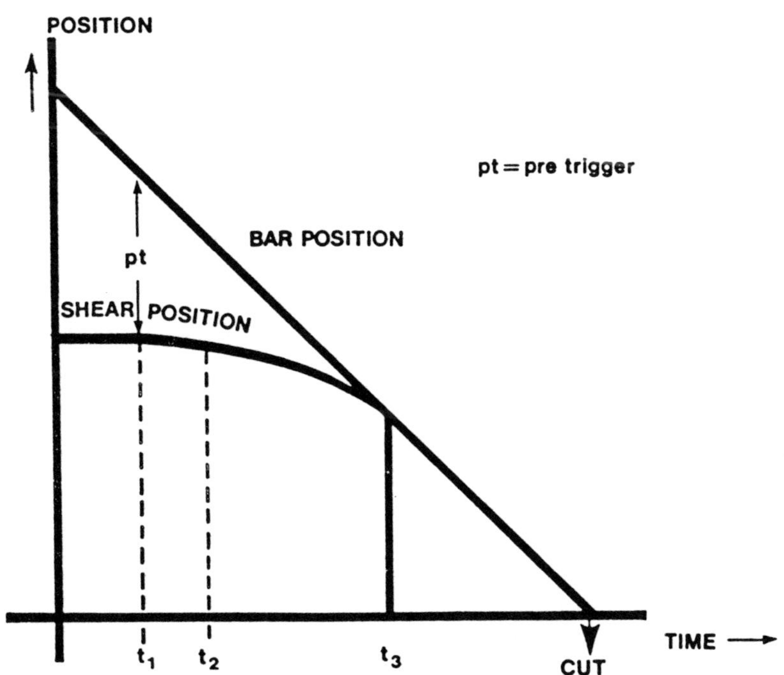

Fig. 9 - Position-time relationships for Bar and Shear

WIDTH OPTIMIZATION

Mounting a camera transversely to the motion of a bar is in concept a simple method determine the optimum width to cut a bar. However, there are, in practice, a number of problems to be overcome. Bar speeds of 50 ips are common so that a program loop time of better than 20 ms is required to give a longitudinal resolution between than 1 in. Since the difficulty of sealing and maintaining backlights has been overcome by using self-radiation, programming must take into account the optical effects of water and scale which can potentially lead to incorrect decisions.

SCALE FILTERING

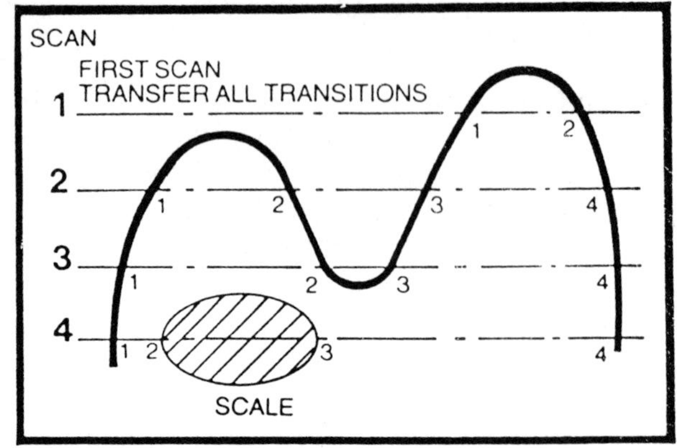

SCAN 2 TRANSITIONS 2 & 3 TRANSFERRED BECAUSE THEY ARE OUTSIDE BOUNDARY OF FIRST AND LAST IN SCAN 1.

SCAN 3 TRANSITIONS 2 & 3 TRANSFERRED BECAUSE MEAN OF TRANSITIONS 2 & 3 OF SCAN 2 FALLS BETWEEN TRANSITIONS 2 & 3 OF SCAN 3.

SCAN 4 TRANSITIONS 2 & 3 IGNORED BECAUSE MEAN OF TRANSITIONS 2 & 3 OF SCAN 3 FALLS OUTSIDE TRANSITIONS 2 & 3 OF SCAN 4.

Fig. 10 - Recognition of Head End in presence of scale patch.

For example, a recognition problem arises if a patch of flake scale is carried on the head end (Fig. 10) that can be interpreted as a fish tail. In this instance, by storing successive transitions, it can be determined if the shape is enclosed or open. A check is kept on the number of transitions per camera scan and, to a limited extent, the edge detector threshold can be adjusted to minimise the number of transitions. Successive transverse transitions are paired and each mean is compared to the mean of the adjacent scan pairs to provide a basis for distinguishing between a true end condition vs scale or descaling water runback. A comparison of raw and filtered data is illustrated in Fig. 11 together with the calculated cut point.

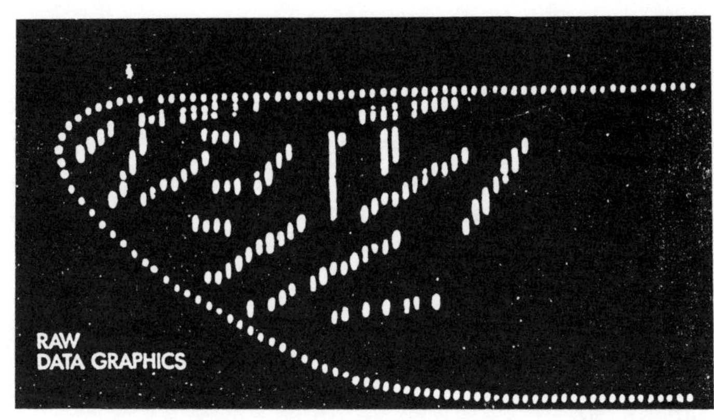

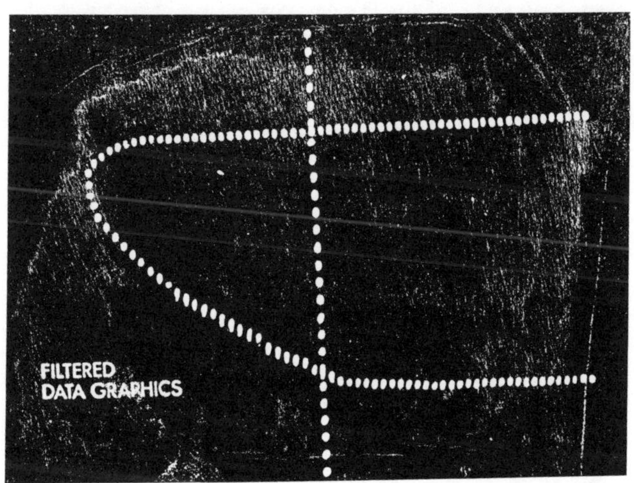

Fig. 11 - Visual display of raw and filtered data together with calculated cut point.

Cut point determination - Operating on filtered scan data, two groups of widths are setup in memory. For the position processor, each width has its associated time tag creating a data pair.

The optimum cut point is defined as the point where the width is greater or equal to a defined fraction of the mid-bar width:

$$W_c > XW_a$$

where
W_a = Measure of mid-bar width (approximately constant)
W_c = Width at cut point
X = optimum cut requirement

The mid-bar width is determined in the following manner.
A length of bar is examined (10 consecutive scans, i.e. approximately 12 in.

bar travel) starting at a point 32 scans further into the main body of the bar than the proposed cut point (Fig. 12). An average width is calculated over this 10 scan section. The width of each of these 10 scans is compared with the average width and a maximum deviation calculated. The maximum deviation should not exceed 3% of the average width if constant width is to be declared. The following relationships define the condition which must be met before the cut point test is performed, i.e. $D_{max} < 3\% \; W_a$

where

$D = W - W_n$ for n = 1 10
D_{max} = maximum value of D
$W_a = W_n/10$

The 3% value was chosen empirically from on-site tests.

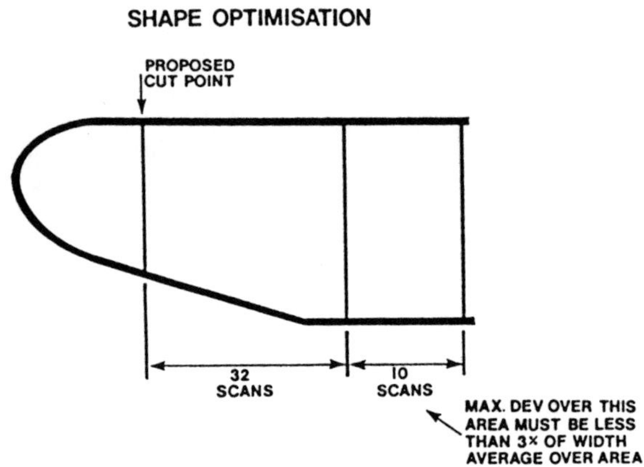

Fig. 12 - Determination of bar width.

In actuality, when the cut point is so found, it is the time at which the cut point passed under the shape camera that is transferred to the system processor. Fine tuning of the cut point is possible by an interpolation of the width change around the proposed cut point. The cut point offset is calculated by the system processor using prior tracking position time information.

Because of the time required for interpolation in the shape processor, it is likely that the system position time will not match that of the shape processor. This is handled by interpolating between the adjacent position time pairs which encompass the match to produce a final cut offset. In the rare event that no match is found, the system reverts to fixed length cut which is reported to the operator.

Crop shear optimization and control hinges on measurement of time. Unless care is taken with the data handling the control regime will always be substantially late. Thus, the bar position may not cause the shear blades to react for several milliseconds. Although several microprocessors are used to increase the flow of information data does flow, nevertheless, sequentially, one instruction at a time. This effect is minimized by realigning any time critical program phase, so that time-tagged data closely corresponds to the current real-time clock. The microloops are forced to execute significantly faster than the main loop so that time errors do not lead to significant position errors.

MILL CONSIDERATIONS

Key mill considerations are the delay table, water run back and steam, mill skids and camera protection.

Delay table - Bars are often jogged on the delay table until the desired entry temperature is reached. With an automatic crop shear, it is important that the head end of the bar should not be jogged into the FOV. of the speed camera. If this should happen, the reverse velocity of the bar, as it is jogged out of the FOV of the speed system, will indicate a velocity error and abort the cut. Alteratively, the entry speed into the FOV may be so high that the system will initiate the shear. The result is that the correct head cut will not be permitted for a preprogrammed number of seconds.

The delay table serves also as a region where the exit speeds from the roughing mill can be reduced to match the finishing entry speed. If the deceleration profile is not optimally set, significant speed variations can occur on head ends and the shear motor may be inadequate to respond to the rate of change of velocity. Pretriggering can again be a problem. If the entry speed into the FOV of the position system is high, the pretrigger distance (dynamicaly calculated by reference to the stored motor characteristic and bar speed) may not provide sufficient time to activate the shear and make a successful cut. For the optimum performance of the automatic shear control system it is important to maintain stable conditions over the last 20 ft. of bar travel. The camera/computing system is capable of establishing high velocity tracking but the shear may be incapable of responding in the time available.

Water run back - In the early stages of development, water run back from the high pressure descaling sprays could depress the video to 10% or less of the clear view condition. In the worst condition, with a sustained water flow in the reverse direction to the bar movement, the position tracking system responded as if the bar was going back in the roughing mill.

The computer control system is now equipped with filters that reduce the water effects on the position track.

Associated with water run back is the problem of steam in front of the cameras. When water run back is controlled, the sources of steam are also reduced. Some form of skip will be arranged to collect crop ends as they fall away. Steam originating from a water-cooled crop skip will, under certain mill conditions, drift back across the shear entry. This condition can be reduced by restricting the pit opening or by preventing steam from rising up the crop chute. A second source of steam can be the roll and shear drum cooling water: jet repositioning may be required.

Mill skids - In some mills, forks or skids are inserted between the rolls on the delay table to prevent a down-turned head end from jamming between the rolls. These inserts, however, collect scale which is heated by the bar and glows after the bar has passed. This effect may be sufficient to create a tail end velocity error and for a cut to be aborted. This condition has been rectified by using forks or skids that have an open design.

Camera Protection - The cameras are mounted directly above the hot bar and are exposed to radiant and convected heat. Because the FOV of the position camera is large, the lens front element is also large and initially, problems were experienced in designing an adequate cooling system. The compromise reached is to position a reflective N.D. filter between lens and steel and to use an appropriate substrate and deposit which withstands the high temperatures. This filter is located in a water-cooled assembly which envelops the camera and lens. In addition, there is a second outer water-cooled case. The camera unit also contains a temperature sensor which is monitored by the system computer which, in turn, controls the water flow to the cooling elements to maintain a set temperature within 5°F for all rolling rates and ambient temperatures. Condensation problems are avoided by a temperature control system which maintains the temperature of the camera and lens above the dew point.

ABORT STRATEGY - A set of test conditions, which are site programmable, is designed into the system to reduce the incidence of false cropping. The abort conditions are:

* Velocity too high. A preprogrammed number of successive velocity samples have been logged which would otherwise put the shear into an overspeed condition.

* Velocity too low. A bar condition that results in a shear velocity which would have too little kinetic energy to reliably shear the steel or stall the bar shear.

* Excessive scale. Spurious attempts to start a cut are halted by a position/shape camera test. A cut will not be made unless the width camera is declaring no steel for a programmed number of scans.

* Tail head too close. The system reverts to a manual mode when it is determined that the shear will not be able to complete both a pending cut and restart for the second cut due to inadequate bar separation.

* Bar temperatures too low. The system reverts to the manual mode when the tail end of a bar is too cold, as measured by the position camera exposure control.

* Encoder failure. System reverts to the manual mode because an excessive position error will lead to cut failure.

Operator panel. A small pulpit-mounted panel provides the shear operator with keyboard entry and alphanumeric display. The system status is passed to the operator from the main computer together with an abort and the reason. The operator can direct the shear cut mode, i.e. fixed length or width optimized, and can also set basic cut parameters, such as the current boost to shear at cut, percent overspeed, heads, etc. As the panel is a small microcomputer, a variety of operator requirements can be programmed into the system to conform with existing mill practice. A VDU based graphics package is provided as an extension of the shape processor. It is a real-time display, built up when the width computer has calculated a cut offset together with the cut point. Status messages are also shown to indicate the mode of operation, i.e. auto, optimized or fixed length.

YIELD IMPROVEMENTS

It is not uncommon to be unaware of the actual losses due to cropping or to maintain records for performance checks. Potential savings can be very large. The following example illustrates the savings that can be achieved by optimizing and controlling crop shearing. Actual performance in any given mill will also depend on other factors, e.g: continuous cast or ingot slabs: reversing or continuous roughing trains which affect bar shape: mean bar width: and annual production.

Example calculation - The determination of cost saving using an EES crop shear control system is based on the following hot mill data.

* Annual output, 1.5 million tonnes

* Slab weight, average 15 tonnes

* Slab width, average 1500 mm

* Slab thickness, average 25 mm

* Initial yield loss at crop shear, average 0.8%

* Cost of slabs $250/tonne

* Scrap value of cropped material $85/tonne

* Density of cropped material, average 7.6 gm/cc.

Loss in value of cropped material is \$165/tonne.
Weight saved per millimetre reduction length of crop through use of crop shear control system = $(1500 \times 25 \times 7.6)/10^3$ = 285 gm/cut or 570 gm/slab.

Thus, the increase in sheared product weight per millimetre reduction in crop length per year = $(1.5 \times 10^6 \times 0.285 \times 10^3)/15$ = 28.5 tonnes, when cutting head ends only, or 57.0 tonnes, when cutting head and tail ends.

The cost saving per millimetre reduction in length per year is \$4702, when cutting head ends only, or \$9405, when cutting head and tail ends.
To pay back an investment of approximately \$315,000 (equipment costs, commissioning and mechanical structures) the following average reduction in length to cut would be necessary:
315,000/13.110 = 33.5mm. However, the average weight of cropped material is $(15 \times 0.8 \times 10^3)/100$ = 120 kg/slab.

The average volume of cropped material is $(1.20) \times 10^3/7.6 = 1.58 \times 10^4$ cc/slab. And, the average surface area of cropped material = $(1.58 \times 10^4)/2.5 = 6.32 \times 10^3$ cc/slab. Thus, the equivalent length of cropped material = $(6.32 \times 10^3 \times 10^2)/1500$ = 420 mm/slab.

Assume, conservatively, that 50% of the surface area of cropped material is not recoverable (i.e. the optimum crop shear yield loss is 0.4%) and that the equivalent head and tail crop lengths are equal i.e. 210 mm.

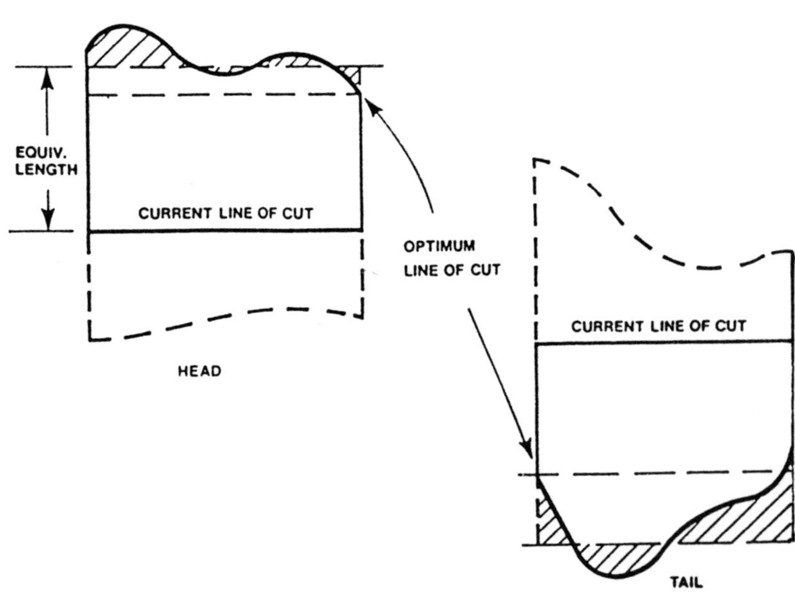

Fig. 13 - Comparison of current and optimum line of cut for head and tail end bar ends.

Probability plots of the crop length achieved currently with an EES crop shear control system follow a normal Gaussian distribution and show that the EES system significantly increases the probability of reduced crop lengths.

The achievable reduction in crop length is 105mm.

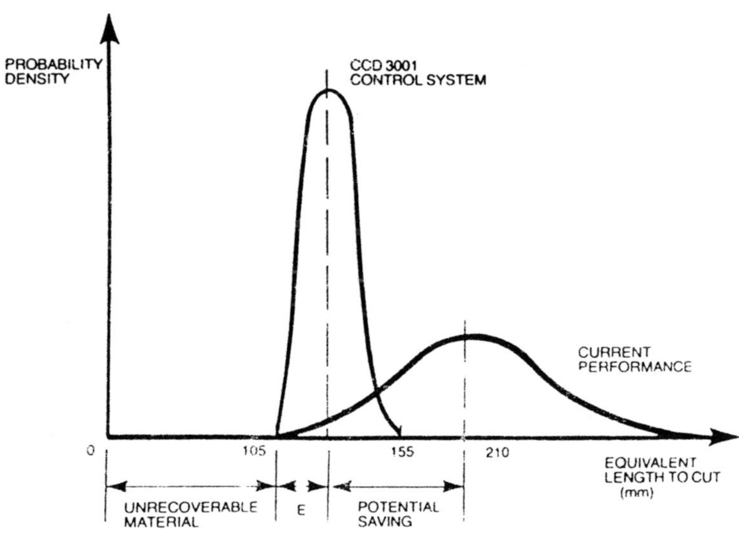

Fig. 14 - Probability distribution of lengths for EES control system and current practice.

Since the optimum cut point precision for the crop shear control system is within ± 10mm and the overall cut point precision of the combined control system and crop shear is expected to be approximately ± 25mm, the reduction in crop length should be significantly in excess of the 24mm required for a 1-year payback.

The system performance implies potential savings of approximately $62,700/month if both head and tail ends are cropped; equivalent to a packback period of approximately six months.

If only 25% of the cropped material is not recoverable, i.e. an optimum crop shear yeild loss of 0.2% the savings would exceed $103,000/monthly or a payback time of approximately three months.

While this example is based on an actual European hot strip mill, payback calculations for individual mills should be made because annual output, slab dimensions, cropping practices and shear performance vary between mills.

SUMMARY

A new noncontact sensor and control system has been developed. The system is fully digital, based on two solid state linear CCD arrays. A bar position camera scans longitudinally for approximately 22 ft from the entry to the crop shear. A transverse camera is mounted approximately 15 ft from the crop shear to optimize the cut point based on end shape. Video data from each camera are digitized to give 2 kbytes of grey levels per camera scan of typically 10ms. Only hot bar infrared radiation is employed.

Water and scale problems are resolved by using two 16-bit microcomputers dedicated to each camera, to extract position and shape data. A third microcomputer supervises camera exposure, validates camera-derived bar position data using curve fit routines and combines position/width information for shear control. The crop shear motor is controlled by a low-voltage analog output. As input, an optical absolute position encoder provides position and velocity feed back. The servoloop, a subroutine in the system computer, is user programmable to optimize stability margins. Basic data on the motor/shear performance is held in memory for each installation and is used to calculate the ideal crop speed/position profile on a bar end by end basis.

Presented at the MVA/SME Vision '86 Conference, June 1986

Using Machine Vision to Inspect Automobile Forgings

by James T. Barczak
GM-New Departure Hyatt
Tonawanda Forge

and

Jeanne Temesan Merchant
GM Technical Center
Advanced Engineering Staff

BACKGROUND

During the late Sixties and early Seventies, some said that the forging industry was a dying business. However, with the oil embargo came the down-sizing of vehicles and the need for high strength, lower weight forgings to reduce vehicle weight. This resurgence in the automobile forging industry brought about new processing techniques, producing forgings at high production rates with cost effective, near net shape parts.

The selection and installation of new equipment at General Motors New Departure Hyatt Tonawanda Forge provided the opportunity to be innovative...on the leading edge of technology. The new equipment (Figure 1) included four hotformers, each capable of producing forgings at the rate of 80 to 100 pieces per minute, an Automatic Guided Vehicle System for material handling, and a machine vision system to inspect the forgings.

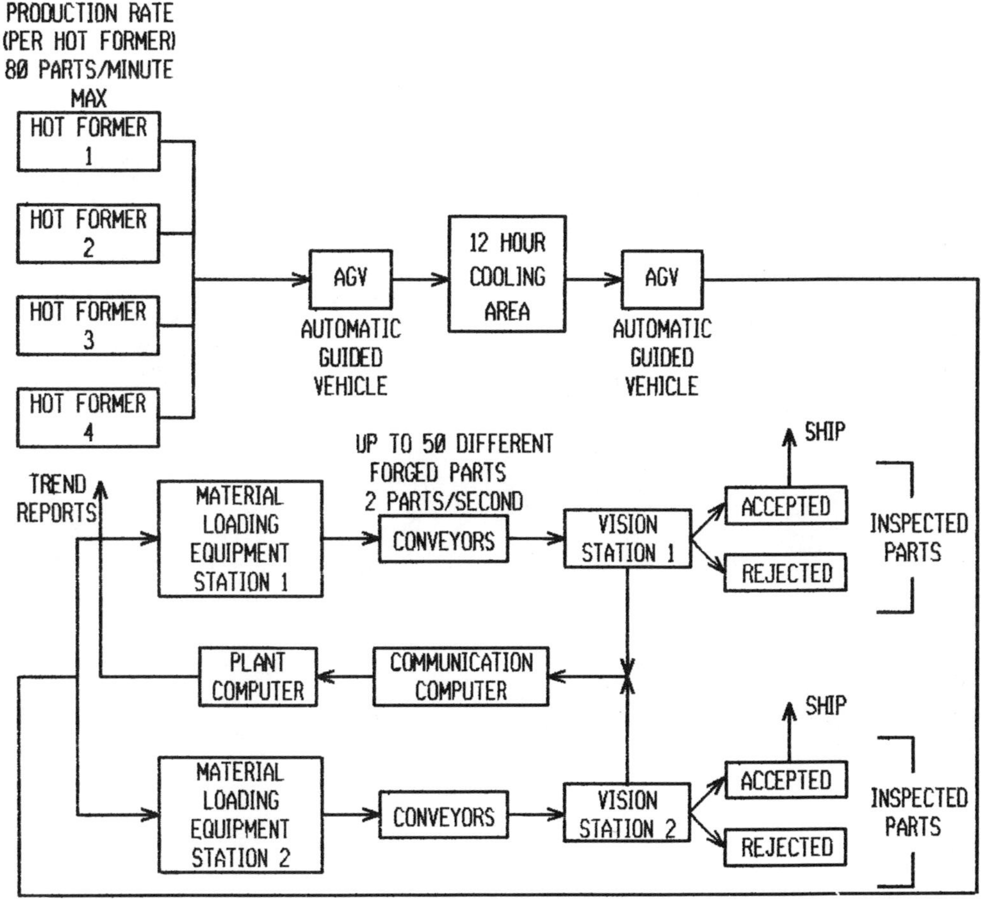

Figure 1 Material Handling and Inspection System

The vision system is necessary to locate any non-conforming parts that may enter the system for any reason. This technology was selected based on the logistics of 100% inspection of millions of parts.

When the decision was made to add vision to the material control system, it provided a challenging project to develop a system based on Tonawanda's needs. Since what was proposed had not been done before in the vision field, it presented the problem of finding a supplier capable of building such a system. General Motors Advanced Engineering Staff, Machine Intelligence Department saw the potential, and agreed to develop the vision system. Two identical vision systems were required, for left- and right-hand conveyors.

THE INSPECTION TASK

Parts are inspected as a batch process, at a rate of two parts per second. The parts travel on a conveyor at twenty inches per second. Thirty different parts are currently in production. They may be classified into six families, as follows:

1. Speed Gears
2. Side Gears
3. Spindles
4. Hubs
5. Sleeves
6. Trunnions

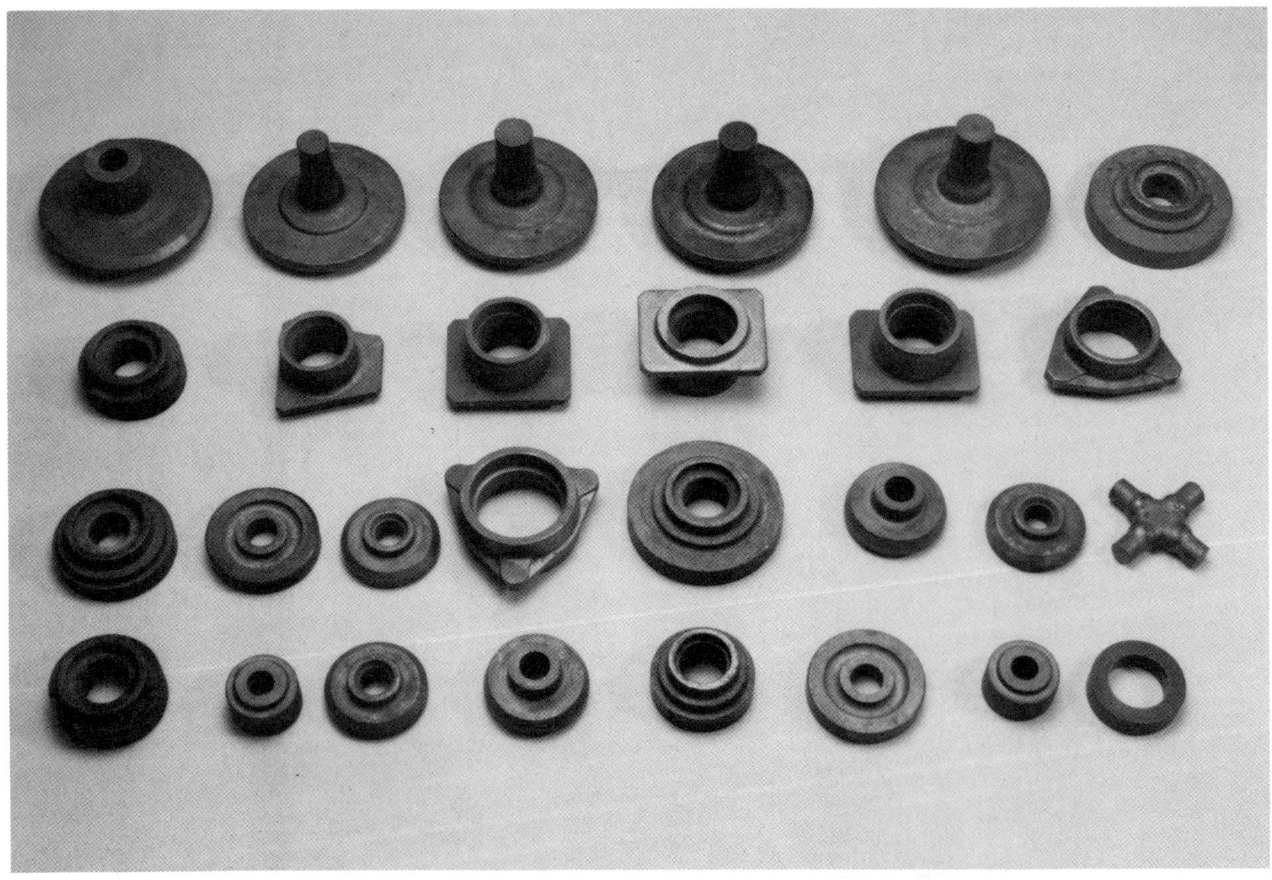

Figure 2 Sample Forged Parts for Inspection

Typical features to be inspected include inner and outer diameters, part height, flange height, no-fill, and concentricity. A maximum of six scenes per part may be analyzed, at a maximum of eleven features per scene. The vision system can accommodate parts which will be presented in either of two stable states.

In addition to the computer-aided visual inspection, an eddy current tester has been installed near the end of the vision station. This equipment was supplied by another vendor to check steel hardness.

The final accept/reject decision for each part is made by the vision system, based on the results of the visual inspection plus the eddy current test results.

EXTERNAL DEVICES

A number of external devices which the vision system must interface to are mounted on the conveyor, shown in Figure 3. Rejected parts are eliminated by a pneumatic kicker. A paint sprayer is activated to mark eddy current test rejects.

A light gate is mounted at the entrance to the vision station. When the light gate is triggered, it signals the vision computer that a new part has entered the station. Parts are tracked continuously through the vision station from the light gate to the kicker. Part position information is provided by an encoder.

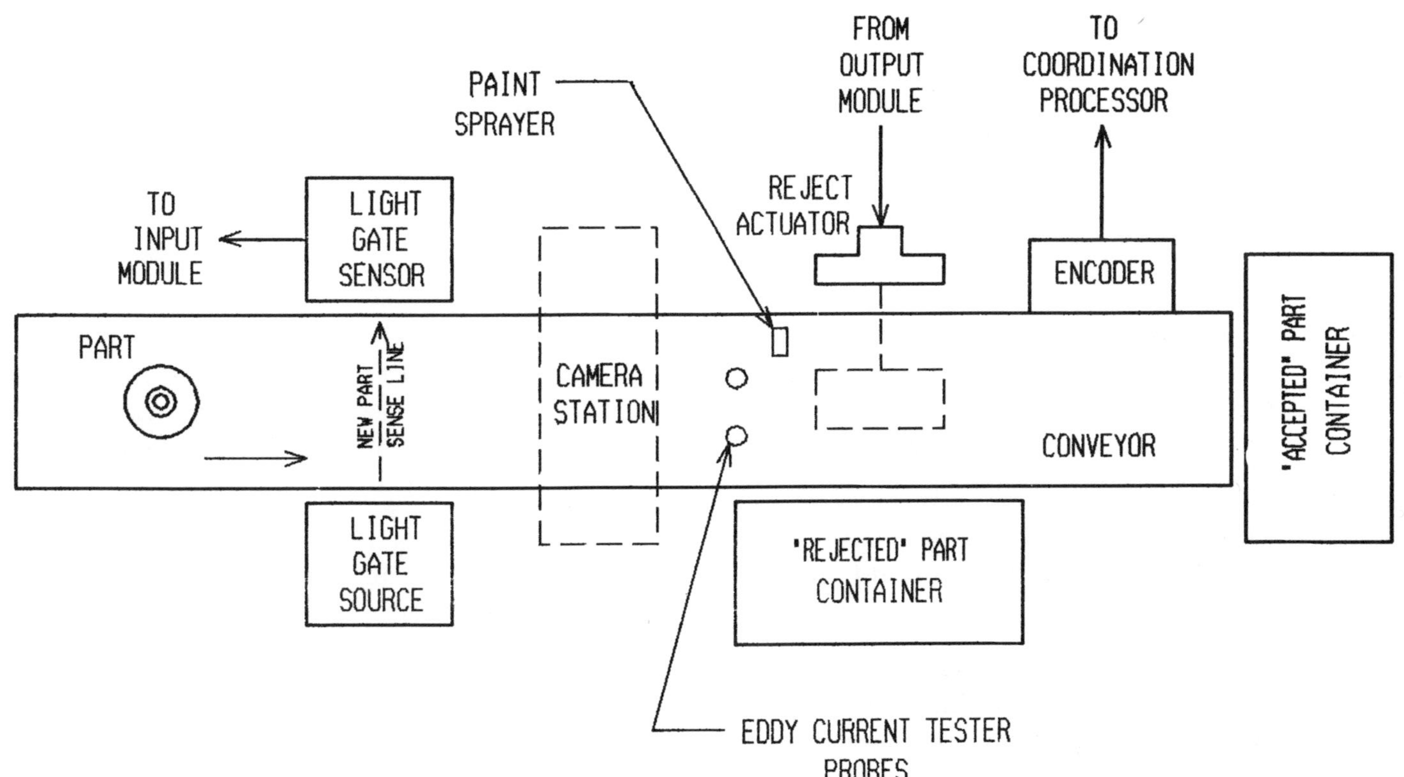

Figure 3 Inspection System Conveyor

The vision system communicates status information to the outside world in two ways. Industrial I/O modules are used to communicate with a Programmable Logic Controller (PLC). A serial communication link connects the vision system to a large computer in the plant's computer room.

The PLC controls loading, unloading, and spacing of parts for three conveyors. It also provides the next model number to the vision system at part changeover time, and monitors the reject rate. It may shut down the line if excessive rejects occur.

The vision systems provide accept/reject summary information via the plant's broadband communication system. This data will be used as the input to Statistical Process Control (SPC) programs.

OPTICS

2048 element line cameras were selected for this application. Each 2048×2048 scene is averaged down to 512×512 'on the fly'. As mentioned earlier, the vision system is capable of analyzing six simultaneous scenes. Cameras are multiplexed four per scene; so the system can accommodate up to 24 cameras to provide the number of (field of view/depth of field/focus) combinations required to handle the variety of parts. Some views are redundant to accommodate two stable states.

Special lighting was required to provide adequate contrast. High intensity quartz halogen lamps were selected for this application. Some scenes are back-lit, some lit from the top, some are side-lit. The wide variety of optical set-ups needed to inspect every part presented one of the greatest challenges of this project.

Interactive color graphics were used to simplify the scene (camera) selection when a new part is taught to the system. A top view of the conveyor (Figure 4) is presented to the 'teacher' and each lighting group or camera is highlighted, one at a time, for his/her selection.

The image data rates provide a clue to the overall complexity of this system. Eight million pixels are processed per second per scene. A part that requires six different scenes makes processing demands of forty-eight million pixels per second!

HARDWARE

Figure 5 is a block diagram of the vision system. It is packaged in an air-conditioned NEMA 12 enclosure. An isolation transformer and EMI filter are included. Two sixteen-bit industrial I/O strips and a modem were installed for data communication. A color monitor and membrane keyboard comprise the user interface. Two forty megabyte Winchester disks were installed to provide non-volatile storage of models and parameters.

All computer hardware except the A/D boards is housed in a 24 slot card cage. The A/D boards, which digitize camera inputs to 256 levels of grey, are located in a remote enclosure near the cameras.

Only three boards were purchased off-the-shelf: a 68010-based CPU card, the disk controller, and a 512×512 color graphics controller. The vision processing hardware developed for this application will be integrated into future systems.

A group of three custom high-speed image processors is required for each of the six scenes. These boards are based on 16-bit bipolar microprocessor technology, running at 24-28 MHz clock rates. The microcode for each board was developed on a proprietary Microcode Development System.

The first board in the set controls the cameras and performs data acquisition and edge detection. The second board acts as an image buffer, and activates the edge detection process by scanning the image radially or linearly. The last board segments the information, saving the data for each feature

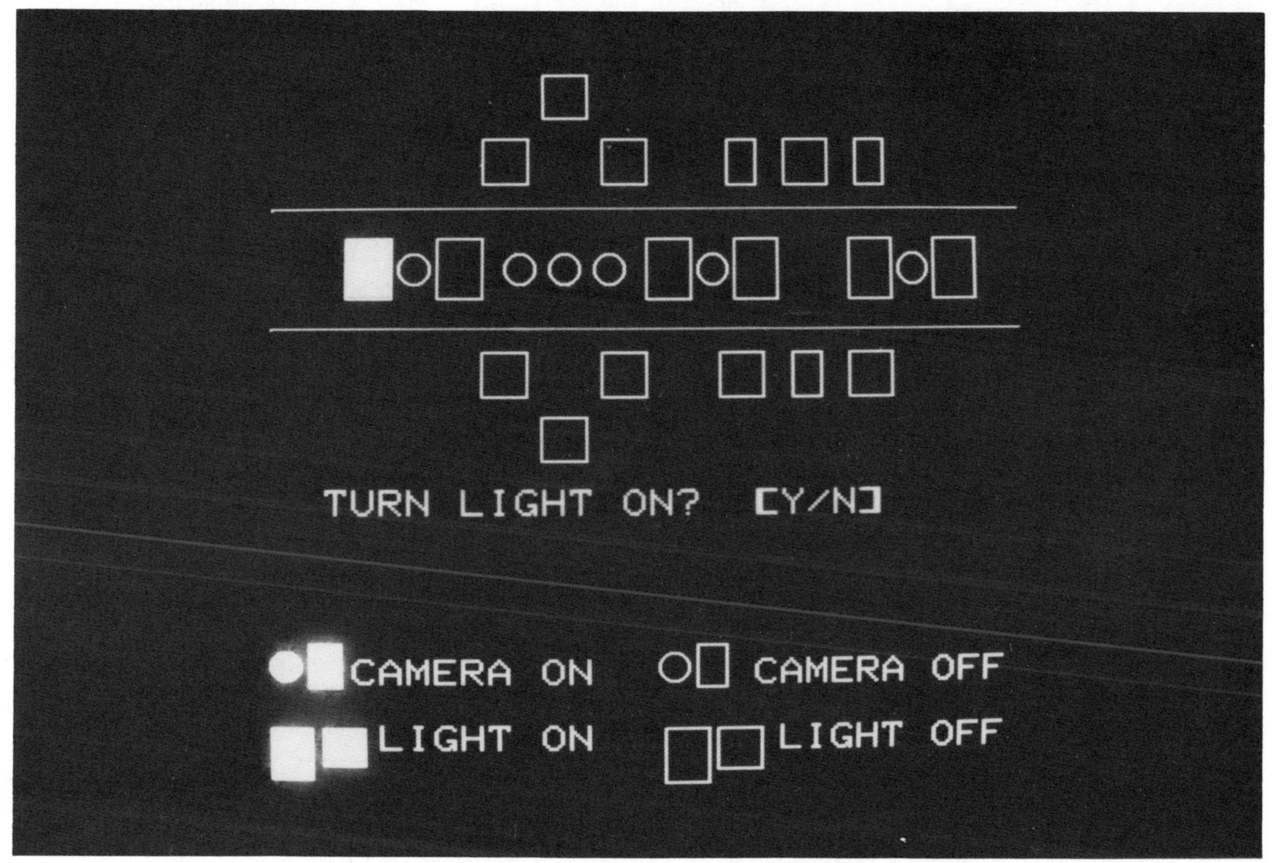

Figure 4 Camera and Lighting Selection

in a table of 1024 X and Y coordinates. Features must be translated or rotated to match the model before a point-by-point inspection may be performed.

SOFTWARE

All system software is menu-driven, using color coded menu screens and color graphics. The software was written in Pascal and 68000 Assembly Language. The total package is approximately 128K bytes.

There are two operating modes: RUN (production mode) and TEACH (off-line mode). The TEACH mode provides the following functions:

1. Board Level Diagnostics - memory, monitor, keyboard, disk
2. Display/Edit System Parameters
3. Disk Utilities - initialize, select, directory, delete
4. Run Time Diagnostic - simulates RUN mode
5. Show Parts - used to set up the system
6. Forget Parts - used to delete obsolete models
7. Teach Parts - used to create/modify models

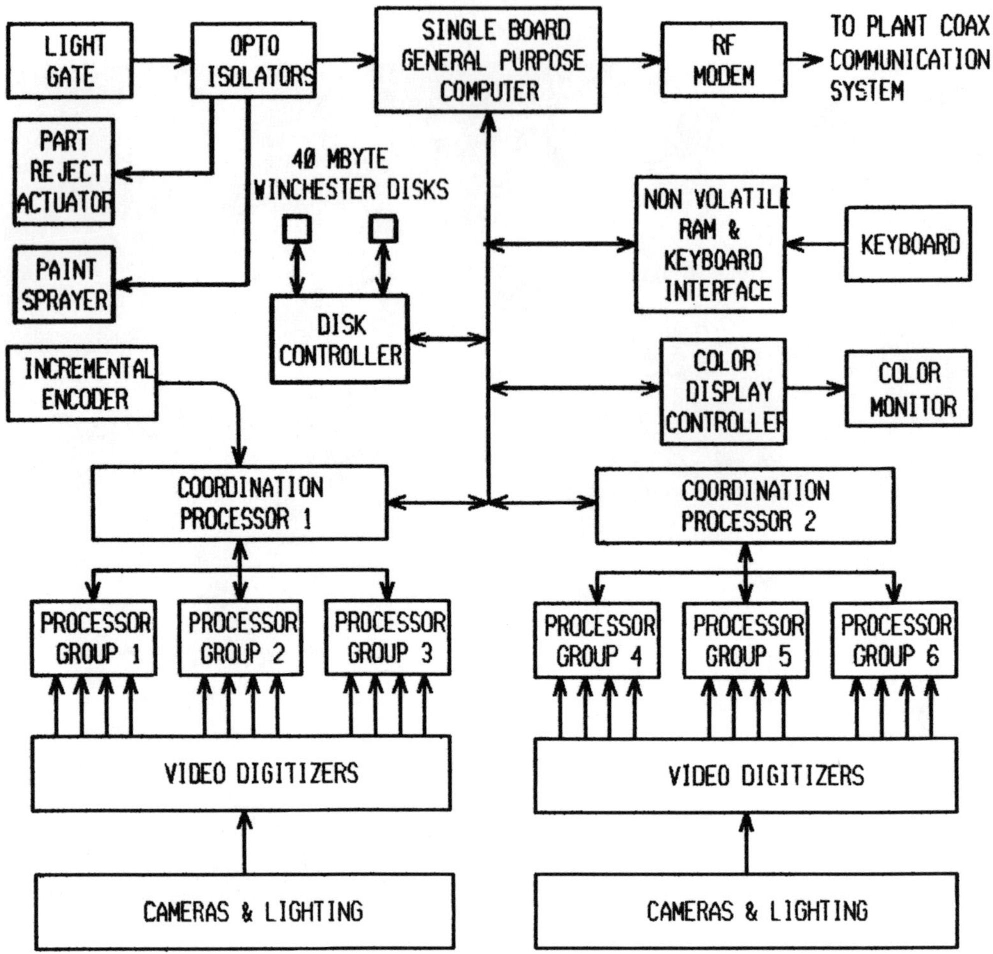

Figure 5 Block Diagram of Vision System

TRAINING

This is a model-based vision system which is taught by showing it a number of representative samples of each part to be inspected. It can learn up to fifty different models, and will easily accommodate product line changes as long as the new parts are similar in size, shape, and reflectivity. The teacher is required to provide a number of inputs such as:

1. Model number
2. Eddy current test on/off
3. PLC code number
4. Family code
5. Number of samples for training

When the first sample has passed in front of the cameras, all recognizable features and their centroids are displayed, scene by scene. Each feature is displayed in a different color, and is represented in memory as 1024 X and 1024 Y points, plus its centroid (X,Y). The features are highlighted in red, one at a time, and the operator may select one or all of them to be inspected in RUN mode. Using interactive graphics, the teacher may select all 1024 points or any subset of points to be inspected, as shown in Figure 6.

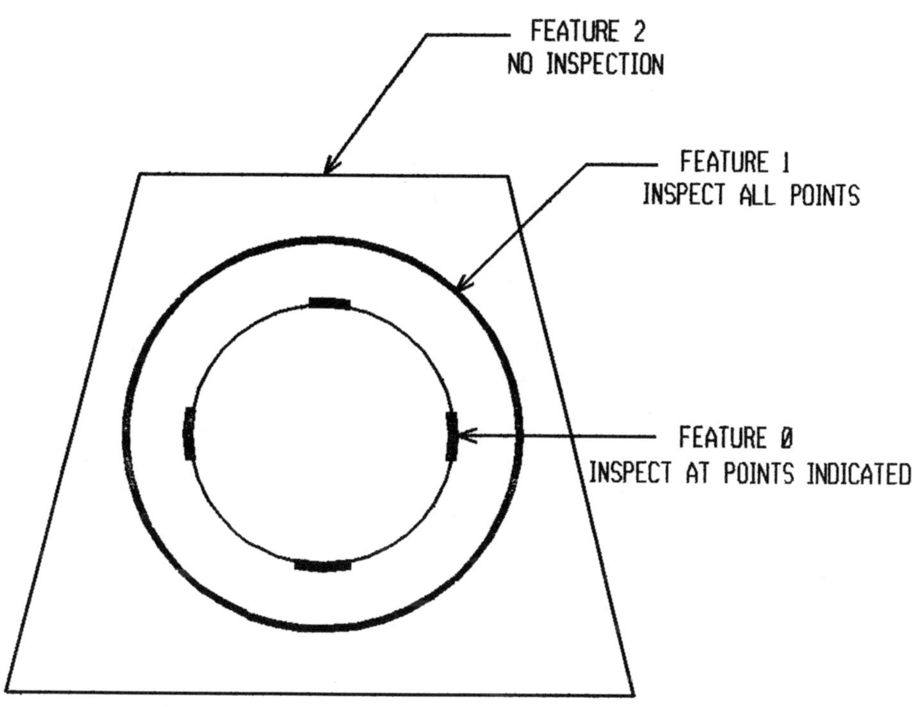

Figure 6 Feature Selection

As subsequent samples are passed in front of the cameras, each is subjected to a simulated inspection. If the sample is 'accepted', it will be averaged into the final model (Figure 7). Each sample may also be rejected by the teacher. One limit table is created per feature, representing the maximum X and Y deviations from the average feature, based on all 'accepted' samples. Fixed limits may also be inserted.

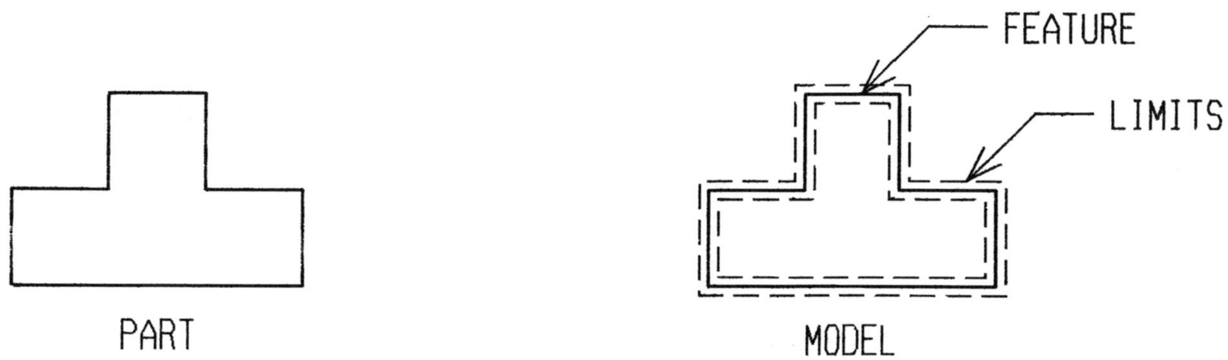

Figure 7 Side View of Spindle

For parts which may be presented in one of two stable states, the training process is then repeated with the part in the opposite state. Redundant scenes are required, one for each state. Interactive graphics are used to select one state for each scene in the final model.

The model for each scene is stored in a file on disk. The filename is the concatenation of four parameters:

1. Model Number
2. Sample Number (final model is always sample 0)
3. Processing Group (scene) Number
4. X or Y (coordinates) or P (parameters)

DATA DISPLAY MODES

A number of data display modes are provided to provide insight to the inspection process. It is easy to display the 'raw' data, as captured by the cameras in 'pseudo-color'; mapping the darkest colors to black and dark greys and the lightest colors to light greys and white. Radial scan lines may be displayed and zoomed for detailed analysis. The output of the edge detector may be shown, as well as the features of a given part, and any model with its limits. A 'calibration' display mode provides moveable cursors which may be used to 'measure' any feature in pixel units.

PART CHANGEOVER

Part changeover for the vision system is totally automatic. It is initiated when the PLC provides a new model number to the vision system, which it obtained from a larger computer upstream. After processing all the parts in its queue, the vision system will automatically activate the new pattern of lights and cameras and load the new model and parameters into the vision processing boards. Guides and/or rails on the conveyor may require adjustment at this time.

PRODUCTION MODE

While running production, the vision system may reject parts based on the total number of mismatches allowed per feature, the number of consecutive mismatches allowed per feature, or concentricity test failure. A display on the monitor shows the current model number and part family. The communication links are active at this time.

RUN TIME DIAGNOSTIC

The Run Time Diagnostic may be used to simulate RUN mode, although not at line rates. It will analyze one part at a time and provide graphics and messages to the operator. The model is displayed in green. Current part features that passed are shown in yellow, failed features in red. Messages include:

1. Part Passed/Failed
2. Concentricity Test Passed/Failed/Not Used
3. Feature Number 'N' Failed
4. Passed All Features
5. Eddy Current Test Override/Passed/Failed/Not Used
6. Failed First/Second Stable State

DEVELOPMENT CHALLENGES

Some of the difficulties to overcome, such as the high data rates and the large variety of parts to be inspected with a single optical set-up, have already been mentioned. Several more were encountered.

One problem which became obvious during testing was the conveyor belt itself. The belt supplied with the conveyor was steel. Abrasion from the rough forgings soon scratched the belt's surface to a shiny finish, creating unwanted reflections. A black plastic belt was installed to relieve this problem.

The hot, noisy forge environment was another obstacle to overcome. Dirt quickly accumulated on lights and lenses. An enclosure was eventually installed over the vision station to alleviate this problem.

The major development challenge was to inspect these parts to very high accuracy at a high production rate. The parts are somewhat imprecise, exhibiting flash and other characteristics, due to the nature of the process. A combination of advanced technologies was required to make this material handling and inspection system a success.

CONCLUSION

The obvious outcome of this project is a new tool for the manufacturing sector to inspect a variety of parts at a high rate of speed. Most importantly, it provides high quality parts for NDH Tonawanda's customers, with financial benefits for both manufacturer and customer. The plant will also receive spin-off benefits, such as the availability of a vast amount of statistical data which will provide dimensional trends for tooling control. Two identical systems are presently installed, with plans for additional units as volume dictates.

ACKNOWLEDGEMENTS

The vision system described in this paper would not have been possible without the extraordinary contributions of many engineers and technicians. The authors are particularly indebted to the Project Engineer, Rene Iadipaolo, for his creativity and foresight.

Presented at the MVA/SME Vision '86 Conference, June 1986

Noncontact Inspection of Sheet Metal Fabrication

by Alan W. Lewis
General Electric Company

1. INTRODUCTION

The oven cavity inspection system is an automatic non-contact gaging system currently in use in General Electric's range manufacturing facility in Louisville, Kentucky. The system combines three automation technologies: laser, vision, and robotics. For comparison this machine can be thought of as an automatic coordinate measuring machine, especially since a cartesian robot is used. The oven cavity inspection system performs nineteen dimensional checks on four different oven cavity models. Three of these models are made on one assembly line known as the "custom" line and the other is made on the "standard" line. The "custom" cavities are 17 inch, 19 inch, and 21 inch models. The standard cavity is the 23 inch model.

An oven cavity begins as a piece of steel sheet metal that is pressed, formed, and spot welded into a four sided box with rounded corners. The oven rack support structures have been pressed into the sides of this box. Another piece of steel sheet metal is pressed and forms the back of the oven cavity. The back is also spot welded into place. The final part of the fabrication process is to spot weld the mounting brackets projecting from the back of the cavity.

This system is designed to aid in the statistical process control of the above described fabrication system. Several reasons make this system a necessity in this manufacturing area. First, of which, is product quality. Prior to installation of the gaging system problems in the fabrication of these oven cavities were not realized until the cavities were well on down in the production process. And sometimes, this caused quality problems in the field after shipping. The gaging system is also needed in order to reduce in house scrap. However, the most important reason the oven cavity gaging system is needed is to gain some control over the fabrication system, since it is primarily automatic manufacturing. An automated gaging system will also provide more repeatable inspection measurements, compared to the manual gaging data previously utilized.

2. OVEN CAVITY SYSTEM

The oven cavity gaging system consists of eight primary pieces of hardware. These are:

- two rangefinding lasers and controllers
- a General Electric Optomation II vision system with two cameras
- two ringlights
- a G.E. A12 Allegro cartesian two arm robot
- two G.E. Series One Programmable Controllers (P.C.)
- a G.E. Operator Interface Terminal

- a push button control panel
- a marking station

The inspection system is installed between two parallel production lines. This enables the gaging system to maintain a queue of cavities from both fabrication lines. The flow of cavities begins in this cavity queue. A cavity is then transfered by conveyor into the inspection station and the cavity is inspected. Inspection takes place in two different positions. The first is a low lift position, where the cavity is lifted off the conveyor and measurements around the top of the oven cavity are made. After those are completed then the cavity is lifted to a high position and the remainder of the measurements are taken. After inspection the cavity proceeds to a marking station, is marked, either good or bad, and then moved onto an exit conveyor. An artist's rendering of the system is shown in Figure 1.

The vision system controls the inspection process and maintains communication with the P.C.s, the robot and the lasers. Communication to the P.C.s and the robot consist of 24 VDC lines while control of the laser controllers is accomplished over RS232 lines. The interconnections between the various pieces of hardware are shown in Figure 2.

The end effectors on each of the robot arms consist of an Optomation camera, a ringlight, and a laser. There is also a guarding system for the end effector to prevent damage in the event that one of the robot arms contacts an oven cavity.

3. MEASUREMENTS

The oven cavity gaging system performs nineteen dimensional checks on three different oven cavity models. Twelve of the dimensions that are taken by the gaging system involve the width, height, and depth of the cavity. These measurements are taken at each of the four corners of the cavity. Other points are gaged to solve particular production or quality problems. Production problems include ear and seam clip hole locations. The ears on the cavity are later used in production, while the seam clip hole identifies how well the cavity has been formed and welded. Primary quality concerns are width and parallelism of the oven rack ribs. The squareness at both the front and the back of the cavity is a problem for both production and product quality.

4. OPERATIONAL CONTROL

The operation of the system can be divided into three distinct parts. These are system calibration, model identification, and testing.

System calibration is necessary for accurate operation of the system and is performed by showing the system what a cavity looks like in terms of laser and camera

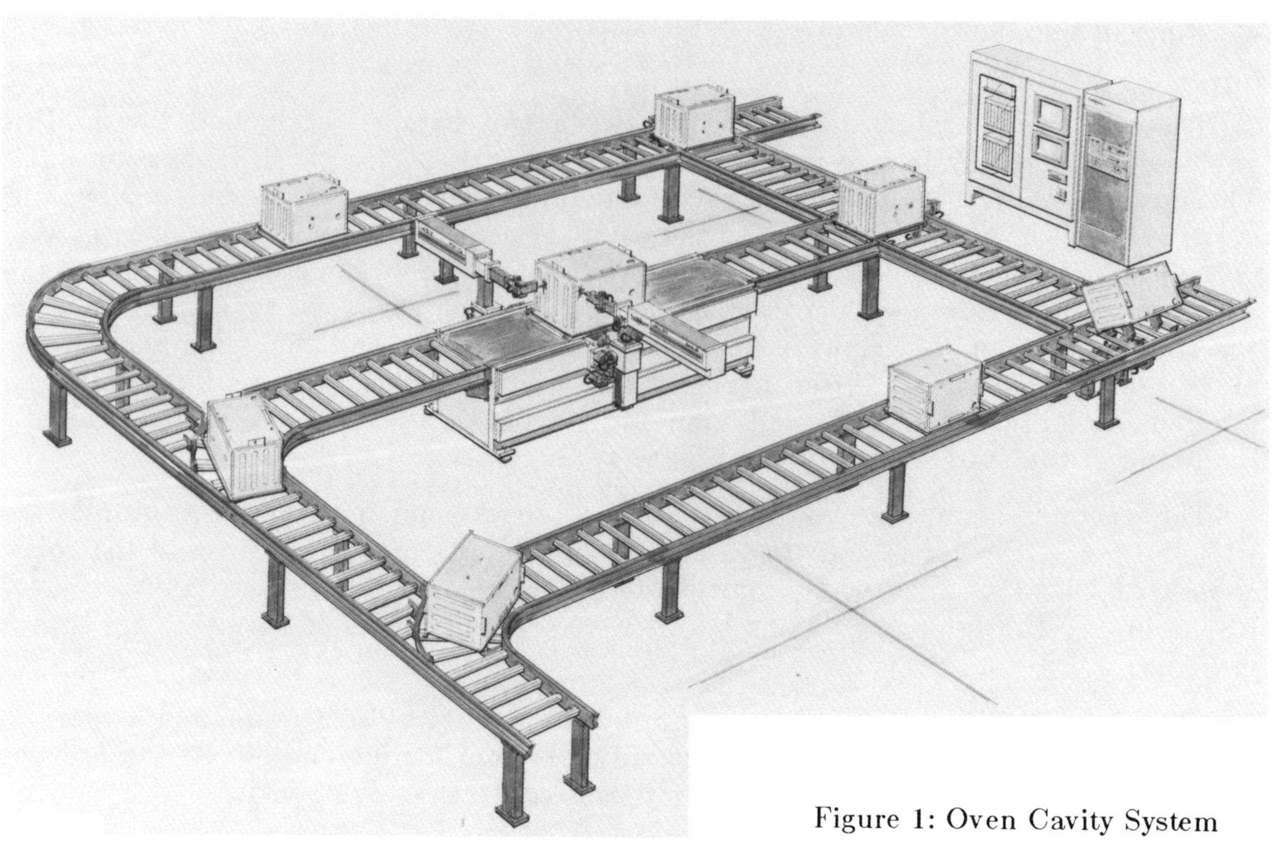

Figure 1: Oven Cavity System

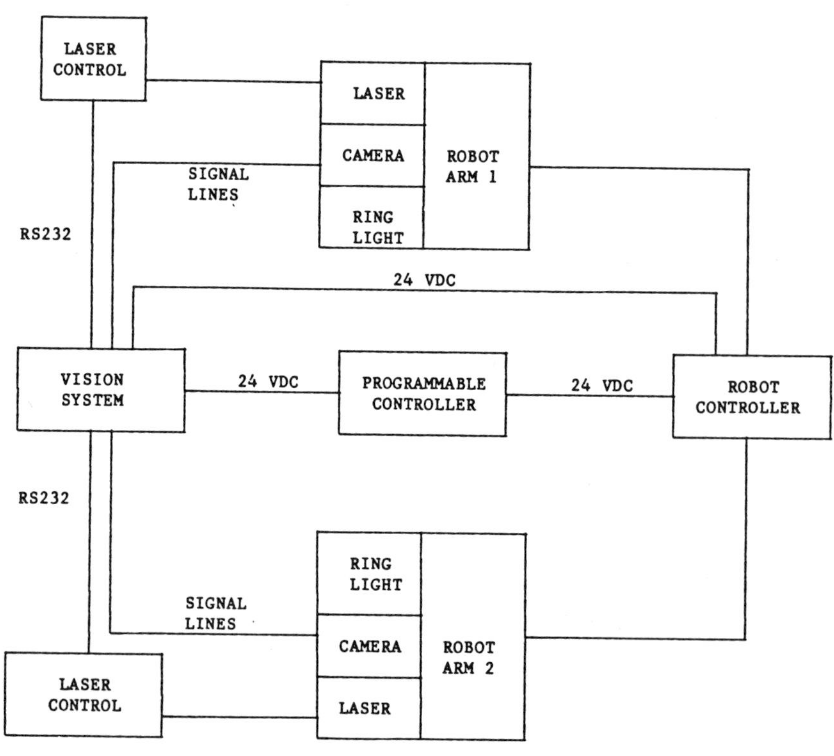

Figure 2: Oven Cavity Block Diagram

readings. Calibration of the oven cavity system is performed by factory personnel and is accomplished by inserting a cavity, that was previously measured on a coordinate measuring machine, into the cavity queue awaiting inspection. This cavity is known as the calibration cavity. When the calibration cavity enters the inspection station the operator stops the testing operation and sets up a new database for that cavity if it is not currently being tested. The only requirements of this cavity is that its actual dimensions be known. Prior to calibration, a few dimensions are entered into the gaging system and calibration is begun. After the calibration run the deviations of the calibration cavity from the "perfect part" are input to the system. These dimensions are entered as deviations of the known part relative to the drawing of the part. Thus, during operation, the results given on the Operator Interface Terminal are given in deviations from the nominal rather than absolute values of the dimensions for the cavities.

The operation of the gaging system is completely automatic and begins with the queuing of a cavity from each fabrication line. One of the cavities is then brought into the inspection station by the conveyor system. First the integrity of the system is checked. This is done by moving each arm of the robot to a calibration plate where laser and camera functional tests are made. Provided these tests are completed the results are compared to setup values to check for possible errors in the movement of the robot. Next, the size of the cavity is read by a series of proximity probes and is lifted to a low lift position by a pneumatic table. After a moment to let the system settle out, the cavity is checked for size by the robot and lasers, and whether or not the cavity has been damaged in the fabrication process. This is known as the find model sequence.

When the proximity probes and the find model sequence agree then the cavity is ready for testing. The testing sequence is a series of preprogrammed robot movements where the vision system tells the robot to move and upon arrival, the robot tells the vision system that the system is ready for a measurement. In the laser case, a range reading is taken perpendicular to the surface of the oven cavity and in the camera case a thresholded image is taken of holes in the surface of the cavity. In these thresholded images the centroid of the hole is found and along with laser measurements combine to give dimensional data about the cavity. Due to the fabrication process, some of the centroids are found based on the expected area of the hole, as a result of the placement of some of the spot welds.

After all the laser and camera measurements are taken, they are compared to the results of the calibration cavity and the inspection data for the cavity is logged into the database. This data is then displayed, by a latest reading, x bar and r chart..

5. USER INTERFACE

The user interface consists of the push button control panel and the Operator Interface Terminal. The push button control panel controls the programmable controllers and gives the operator control over which cavities are tested, either standard or custom or both. Through menu based selections, the terminal gives the operator control over the gaging system itself and access to the database that contains all the test results. The terminal is used for a variety of tasks including, report generation and printing, parameter editing, calibration, and password access.

The report generation feature allows quality personnel to extract the database in several forms. These are:

- a reject chart
- a latest reading, x bar, and r chart
- a trending (or control) chart

The reject chart displays the number and type of reject for the last fifty cavities tested and all the cavities tested since the database was last reset. The second chart displays the most recently tested cavity, the average reading and the range of the last fifty cavities. Dimensions that are out of control limits for the last tested cavity are displayed as flashing numbers. This report is always displayed for the last cavity tested. Since the vision system is a single user system, this minimizes the amount of time that the gager is taken offline to access the database for the information that the quality personnel consider the most vital. The control chart displays the readings for the last fifty cavities tested for a particular dimension. The control chart also displays where the shutdown and control limits are in relation to the measured readings. Any of the above reports can be printed out at the operators discretion.

Any of the control and shutdown limits can be changed, for either of the lines, by the quality engineers through the editor. These parameters can be set to accommodate the production tendencies of each of the fabrication lines. The database can also be reset for either line of inspected cavities.

System calibration is also performed through the terminal. This consists mostly of entering some measured parameters about the calibration cavity and executing a calibration test. At the end of the calibration sequence the operator is reminded to change the deviations of the master cavity if necessary.

The last feature of the terminal is the password access mode. This prevents unauthorized personnel from taking the gager offline or otherwise corrupting data the machine generates. The password system is implemented as a level based system enabling the quality personnel to allocate access at their discretion.

6. DIAGNOSTICS

The diagnostic capabilities of the oven cavity inspection system include laser and camera functional tests prior to each gaging test of a cavity. Provided that the lasers and cameras are operating, this also provides data on the performance of the robot. Anytime there is an operational error a system status light indicates this to draw attention to the fact that the gager is down. These operational errors are displayed on the terminal. These operational errors include laser read errors, camera read errors, and timeout features on the communication lines as well as many others.

Other diagnostic features include the fact that it is easy for factory personnel to verify system repeatability and performance. This also helped bring the system up quickly and enabled the quality personnel to trust system results in a more timely manner.

7. SUMMARY

The oven cavity gaging system has been designed and built in a manner that makes it easy for the factory personnel to maintain. Prior to the installation of this machine, all of the dimensional data used was gathered by use of a manual Coordinate Measuring Machine. This data took a great deal of time to obtain and as a result many parts could be made that were out of production tolerances. This resulted in assembly fit problems further down the line, after additional money had been put into the defective part. This machine will now give the fabrication personnel the data that they need to make timely adjustments to the fabrication process before major problems occur. The statistical process control will minimize manufacturing costs, by holding part dimensions to print mean and minimizing process variations.

Presented at the RI/SME Applied Machine Vision Conference
February 1984

Flexible Dimensional Gaging System

by Michael M. Bahn
and
John Harned
Autoflex, Inc.

I. INTRODUCTION

Gauging systems can be broadly classified into "fixed inspection" and "flexible inspection" systems. Fixed inspecton utilizes multiple contacting or non-contacting sensors mounted in a test fixture that holds the part to be inspected. This approach lends itself to the inspection of parts at very high through-put rates. However, to switch from one part to another requires changing the test and inspection fixture. Flexible inspection utilizes sensors that are moved about the part being inspected along a programmed path trajectory. This approach lends itself to processing at moderate through-put rates. Changing from one part to another can be accomplished quickly, by downloading to the machine controller, a new path trajectory program.

Flexible inspection was first introduced into the flexible manufacturing work cell by mating coordinate measuring machines with CNC machine tools. In recent years, the role of flexible inspection has been expanded by using robots with laser probes to inspect large assemblies such as car and truck bodies (1,2)*.

With the recent introduction of sophisticated machine vision systems into the workplace (3,4), it is now possible to greatly expand the role of robots in flexible inspection. Recent advances in CAD/CAM technology now make possible the integration of CAD/CAM into robotic systems (5). This marriage of CAD/CAM with robotics significantly improves the productivity and economies of robot inspection systems. It allows the accumulation of large data bases containing the measurement vectors obtained from the flexible non-contact gauging systems. Error vectors obtained from these data bases can be used to construct the x-bar (arithmetic mean) and R (range) charts needed to implement statistical process control (SPC).

The purpose of this paper is to present an overview of robotic gauging and inspection systems in the context of the flexible manufacturing environment. Particular emphasis will be given to the application and use of the newer technologies.

* Numbers in parenthesis identify references listed in the bibliography.

II. WHY ROBOT-BASED INSPECTION

Robot-based inspection systems, at the present time, provide the maximum in flexibility. A variety of sensors can be mounted on the robot wrist. Robots with articulated arms can reach inside of parts with cavities, such as car bodies, appliances, etc. Robots can be programmed to measure reference datum points and planes on a part. Algorithms, using these reference point measurements, can calibrate the part coordinate system, which allows the CAD/CAM database to be integrated into the robotic measuring system. This integration eliminates the need for precision part fixturing, allows real-time adaptive robot path trajectory control, and provides off-line programming of robot path trajectories. Control of robotic inspection systems by powerful minicomputers, allows quick downloading of path trajectory programs to recalibrate a current part coordinate system, or to reprogram robots to inspect a new part.

Robots equipped with optical and vision sensors provide the maximum flexibility to carry out a wide range of inspection tasks. Optical sensors provide accurate, non-contact measurement of distances. They can accurately measure coordinates on plane and curved surfaces that would require structured lighting for vision measurements. Vision systems allow the inspection of part features having a variety of geometrical shapes. They can measure hole and stud center coordinates that optical sensors have trouble with.

Inspection robots have some limitations. As previously pointed out, they can inspect parts at moderate through-put rates; about 100 parts per hour would be a nominal rate. They are, therefore, not well suited for very high speed production lines where 1000 parts per hour are the rule. Gauging accuracy is another concern. At the present time, robot positioning accuracy is the main source of errors in robot inspection systems. Positioning repeatabilities of commercially available robots, surtable for inspection applications, range from +/-.025 mm (.001 inch) to +/-0.2 mm (.008 inch).

Therefore, the application areas for robotic inspection systems can be generalized as follows:

1. Moderate through-put rates;
2. Frequent part or model changes;
3. Off-line part inspection, requiring a large number of measurements;
4. Large parts with complex geometry, such as cavities;
5. Low through-put rates, with parts that require a large number of measurements.

III. ROBOTIC-BASED INSPECTION SYSTEM

In selecting a robot-based inspection system for a particular application, the major technical factors that need to be considered are:

1. Robot size;
2. Sensor selection;
3. Accuracy requirements;
4. Process inspection times.

Robot size is determined by the inspection volume that the application requires. It is important to consider design trends in the parts to be inspected, so as to allow for future increases in part size and/or geometry. Sensor selection is largely dependant on the geometrical characteristics of the part features to be inspected.

Inspection accuracies that the system must provide are determined by the CAD/CAM database specifications. In selecting equipment, it is important to consider the effects that aging and wear will have on inspection system accuracies. Process inspection times are principally governed by the number of inspection locations, length of robot path trajectories and robot settling times. Other delays, such as vision image processing times, can be handled in parallel with other system functions, using distributed data processing.

The Robot: Robots are available in a variety of arm coordinate geometries (6). These designs fulfill a wide range of needs in material handling and assembly applications. For inspection applications, the revolute coordinate geometry design appears best for articulated arm robots. The other type of geometry suitable for inspection work is the gantry robot. Due to its geometric design, the gantry robot provides higher repeatable positioning accuracy than the articulated arm robot. Experience has shown that a 5-axis robot is adequate for most inspection work. This is because coordinate transformation algorithms can be used to correct sensor measurements for orientation misalignments.

Robot positioning accuracy is determined by its ability to match its actual position in 3-dimensional space to the command position called for by the program's position instruction. The error between actual and command positions is caused by a number of factors, including servo component design, structural natural frequencies, bearing friction, gear backlash and load torques. This error is a summation of static and dynamic error components. Bearing friction and steady-state load torques produce static position errors.

Structural natural frequencies, inertia load torques, gear backlash and closed-loop servo gain contribute to dynamic errors. As the robot approaches its command position, these dynamic errors produce an effect called "settling time".

Robot settling time is a very important performance specification, because it causes a direct trade-off between measurement accuracy and process inspection time. This trade-off is clearly shown by the data in Figure 1, comparing robot positioning error distributions for two different settling times. These data were obtained with the robot executing an inspection path program representative of a typical process inspection cycle for a production application. Increasing the settling time by 117%, reduced the variability in robot position by 36%; but process inspection cycle time was increased by 19%.

<u>Sensors</u>: Vision systems are composed of a video camera, frame grabber and image processor, ranging from the simple to the highly sophisticated. Both vidicon and solid state matrix type cameras are used. Vidicon cameras have a wider illumination intensity working range than matrix cameras, but they lack the ruggedness, light weight, and simplicity of solid state cameras. As a consequence, solid state cameras are widely used in the industrial work place. These cameras can be obtained in line scan and matrix models. The matrix type is best for general robot inspection applications. Matrix arrays from 128 x 128 pixels up to 512 x 512 pixels are presently available in commercial cameras.

Image processing time can create undesirable delays. The frame grabber requires from 15 to 40 milliseconds to scan an image and store it. Image processing can take from 20 milliseconds to 2.0 seconds to obtain the desired measurements (depending upon the level of software sophistication). Binary images allow the use of simple algorithms for image processing and, thus, introduce the least delays into the process inspection cycle. With good silouette lighting and binary image processing, reliable measurements can be obtained to a one pixel accuracy. This is illustrated by the data shown in Figure 2. Pixel calibration was chosen so as to obtain maximum vision measurement accuracy. This accuracy is limited by the robot positioning accuracy. A good rule-of-thumb is to set pixel calibration equal to 2.3 times the standard deviation of the robot positioning error distribution. This was the pixel calibration used to obtain the Figure 2 data. These data were obtained with the robot carrying out a typical process inspection cycle. The variations shown in Figure 2 are due wholly to robot position variability.

The newer vision systems allow gray scale image processing, at up to 256 gray levels.

Optical sensors employ a light beam to illuminate the measurement location and use reflected light and triangulation techniques to measure distance. Light frequencies in the near infrared range are used to allow measurements on surfaces with poor reflectivities. Detectors used to sense light beam deflection, caused by the trigonometric relationships, are of either the line scan matrix type or a semiconductor film employing analog circuitry. For the latter type detector, close-loop control of light beam intensity is used to maintain constant reflected light intensity.

Measurement accuracy of optical probes is high, but it can be radically effected by probe design. This phenomenon is illustrated by the data presented in Figure 3. One manufacturer of optical sensors provides two different models; one is large and heavy while the other is about one-third of this size and substantially lighter. This compactness and light weight is achieved at a reduction in meaurement accuracy as illustrated by the comparison in Figure 3. The heavier sensor is 300% more accurate at the same measurement frequency of 80 measurements per second.

One advantage of optical sensors, is their high speed measurement capabilities. Maximum measurement rates can vary between 10,000 and 16,000 readings per second depending upon sensor design. The dispersion characteristics of reflected light cause considerable variability in optical sensor readings. This variability can be reduced, to achieve higher accuracy, by averaging of multiple sensor readings. Trade-off between variability and sample size is illustrated in Figure 4. By increasing the optical sensor reading sample size from 200 to 1000 individual sensor readings, measurement accuracy is improved by 82%, but measurement rate is reduced from 80 measurements per second to 16.

Computer/Control Systems: A robot inspection system is composed of three or four small computers that are required to perform different distributed processing functions. These computers are interconnected by a data communication and control network. This configuration allows a host computer to coordinate and control the robot and sensor(s). For example, a robot vision inspection system would be composed of a host computer interconnected to a robot controller and a vision image processor.

The communication and control links can be either parallel or parallel and serial. Parallel links provide handshaking signals for synchronizing control activities and to handle high speed transfer of measurement data. Serial data links provide two-way communications to allow the host computer to carry on a dialogue with other processors in the system. This capability provides many advantages and it makes possible many of the advanced control features that will be discussed below, under "System Software".

System Software: The host computer required by the robot inspection system can range from a small micro to a powerful minicomputer, depending upon the degree of flexibility desired and level of data processing and storage required. Consequently, the system software can range from a 64,000 byte package, for a small micro, to several million bytes for a large minicomputer. Even the smallest software package should provide, as a minimum, the following functions:

1. Initiate and coordinate the process inspection cycle;
2. Read in and store sensor measurement data during inspection cycle execution;
3. Perform minimal data processing;
4. Output process data results in some acceptable report format;
5. Provide a limited operator interface to handle calibration and diagnostic requirements;
6. Output warning and diagnostic messages to indicate system malfunction and type of error(s).

An example of the flexibility and increased system functions that can be provided by a large software package, running on a powerful minicomputer, is illustrated by the software functional diagram in Figure 5. This configuration is used to operate a robot vision inspection system. In addition to the functions described above, the software operating system provides the following functions:

1. Uploading/downloading of robot path programs;
2. Robot off-line programming;
3. Real-time robot program debugging through the computer terminal;
4. Integration of the CAD/CAM database into the measuring system;
5. Adaptive vision image processing;
6. Adaptive real-time robot path trajectory control;
7. Measurement of 3-dimensional part features using a 2-dimensional vision sensor.

The application-specific software can, depending upon application requirements, be quite extensive. This software coordinates and controls the robot and sensor(s) to run reference point programs (required for setting up the CAD/CAM coordinate transformation system), and process inspection programs. Also, the required data processing algorithms are implemented in this software. Another function that it usually provides, is the database manager for storing processed inspection data. This manager can then retrieve selected portions of this data for outputting, in the proper report format, to a CRT terminal or hard copy printer.

IV. APPLICATION EXAMPLES

The use of robot inspection systems in several manufacturing quality control applications, will now be described, to illustrate the flexibility that these systems can provide.

Tube Assembly Inspection: This application requires the non-contact measurement of tube end locations in X, Y, Z coordinates, relative to the tube assembly's coordinate system, as specified in a CAD/CAM database file. Tube assemblies with many different tube configurations have to be inspected. Change over from producing one tube assembly design to another is rapid. Inspection is done off-line at a process cycle rate of 60/hour. Accuracy requirements are +/- 0.25 mm to +/- 1.00 mm depending upon the length of the tube assemblies.

A major problem in this application is the inability to use a so called "master part" to calibrate the inspection system. A reliable master part can not be produced due to two factors:

1. The tubes are very flexible and bend easily (tube length/diameter ratioes range from 70 to 120);
2. Inaccuracies introduced by the NC tube bending machine.

This problem can be solved by integrating the CAD/CAM database into the measuring system. This allows the CAD/CAM design dimensions for a particular tube assembly part number to serve as the "logical master part".

The selection of a robot vision inspection system with a CAD/CAM-based software control system is a good choice for this inspection application. Optical sensors are unsuited for measuring tube end X, Y, Z coordinates. On the other hand, vision provides an accurate means for making these measurements.

Figure 6, shows a camera, positioned by a robot, in position to obtain an image of one tube end. The resulting processed binary image of this tube end is shown in Figure 7. This is a good example of the use of silouette lighting to obtain a sharp, well defined image. Measurements made from processing the tube end image in the camera's field of view are used to construct a vector locating the tube end coordinates in the camera coordinate system. This measurement vector is then mapped, through the use of coordinate transformations, into CAD/CAM coordinates to produce a measurement vector that can be directly compared to the CAD/CAM dimension vector that locates the tube end in the tube assembly coordinate system.

The process inspection cycle time is determined by the number of test points, dwell time at each test point, and robot path trajectory time. For an average tube assembly, there are five tube ends to be inspected. Each requires two camera views, making the total number of test points equal to ten. Dwell time at each test point is 2.7 seconds. Average robot transition time for one cycle time for one tube assembly is 43.5 seconds.

Data processing can take a variety of forms. For example, an error vector, defining how far and in what direction the actual tube end is from its CAD/CAM location can be obtained by subtracting the measurement vector from the CAD/CAM dimension vector. Error vectors for successive tube assemblies can be stored in a database. This data can then be processed by statistical algorithms, commonly known as Statistical Process Control (SPC).

Statistical Process Control (SPC): Statistical process control (SPC) is the application of statistical methods (analysis of central tendency and dispersion) to aid in the control of various processes over time. Moreover, successful application of SPC is dedicated to the detection and elimination of the assignable causes of variation. Statistical process control contributes significantly to the quality of the product. As Deming says (Ref. 9), "Quality is achieved by improvement of the process, improvement of the process increases uniformity of output of the product, reduces mistakes and reduces waste of manpower, machine time and materials. Reduction of waste transfers man hours and machine hours from the manufacture of defectives to the manufacture of good quality parts."

The database of error vectors obtained from the flexible non-contact gauging systems, provides us with complete documentation of the X-bar (arithmetic mean) and R (range) charts (Ref. 7).

X-bar = (1/n) Sum X

and R (range) = Xmax - Xmin

X-bar and R charts are very effective in controlling the variation of manufactured parts.

These charts allow a decision rule to be applied based on the fact that parts falling within the limits, which are calculated from the data itself, would reasonably be expected to be due to natural variation but parts outside the limits would probably be due to an assignable cause, thus telling us when assignable cause (trouble) is present. For further clarification of how the error vector data base and statistical process control can help quality, the reader is advised to consult the references (7,9).

Car Body Inspection: Another application for robots is in the inspection of car bodies, Figure 8. A body production line handles a mix of different designs corresponding to different car models and styles. In this application, an inspection system is required to carry out detailed inspection of studs and stud holes on-line to achieve 100 percent quality control. Inspection cycle time is 30 seconds so that a through-put rate of 75 bodies per hour can be maintained. Accuracy requirements range from +/- 0.05mm (.002 inches) to +/- 0.2mm (.008 inches). All inspection data has to be stated in CAD/CAM coordinates. Location accuracy of a car body relative to the robot coordinate system at the inspection station is +/- 3.0 mm (0.12 inches) along each coordinate axis.

Accurate measurement of body dimensions is dependent upon the integration of the body's CAD/CAM database into the measuring system. A fixed inspection system has to use sensors which can accommodate the +/- 3.0mm shift in the body coordinate system axes while maintaining a +/- 0.05mm measuring accuracy. A flexible inspection system,using robots, can take a completely different approach to compensate for shifts in the body coordinate axes. The Autoflex "3-D Vision" system can accurately measure reference points on the body's datum surfaces to +/- 0.070mm while accommodating shifts in the datum surfaces relative to the robot of up to +/- 1.5mm (0.60 inches). The host computer's "application program" then alters the robot path program, for the process inspection cycle, to compensate each robot test point's world coordinates for the shift in the body coordinate axes.

This compensation of the robot path program for shifts in body coordinate axes is of particular value when vision sensors are being employed. Image processing algorithms are very sensitive to large shifts in the part features being viewed. While these algorithms can be expanded to accommodate these large shifts, image processing time also increases, which lenghtens process inspection cycle times.

Robot inspection systems, using both optical and vision sensors, are a good choice for body inspection. Figure 8 shows a close-up view of the Autoflex "3-D - Vision" system inspecting a front door pillar. Vision is used to measure two coordinates of stud and hole center point while the optical sensor is used to obtain the third location in 3-D space. The large gauge hole shown in the lower left of Figure 8 is inspected by using vision to measure three points on the hole's circumference. The binary image used to measure one of these points is shown in Figure 9. Use of vision to measure a stud hole center point is illustrated in Figure 10. The indentations around the hole circumference are serations caused by insertion and removal of a stud. Figure 10 illustrates the sharpness that can be obtained with binary images.

Measurement of stud center point coordinates are carried out using front lighting provided by a fiber optic light ring mounted over the camera lens. A typical histogram of a stud gray scale image is shown in Figure 11. The stud itself is represented by the black pixel distribution between grey scale values of 10 and 120. Stud circumference is well defined at a grey scale level of 45. First appearance of small noise dots in the image occurs at a grey scale value of 100. This large separation between stud definition and first appearance of noise allows a direct conversion to be made, in the vision processor, from the gray scale image to a binary image. The vision processor software then processes the resultant binary image to obtain the stud coordinates.

V. OTHER APPLICATIONS

New and more demanding areas of manufacturing quality control are a logical extension, for future application of robot inspection systems. These new applications will come about as the accuracy and capabilities of robot and sensors improve, and their costs come down. The primary limitation of robot-based inspection systems is through-put rates. The new generation of robot controllers allow sensor measurements to be synchronized with the recording of instantaneous robot position while the robot is moving along a path trajectory. Therefore, the robot does not have to stop at each test point before a measurement can be taken.

An estimate, made by the authors for a robot body inspection system, showed that through-put rates could be increased by a factor of three. Also, this new capability of robot controllers, makes it possible to use robot inspection systems to obtain detailed measurements of surface profiles, at rapid scanning rates.

An application presently under study at Autoflex is in manufacturing assembly operations. Here, an inspection robot would be teamed up with one or more assembly robots. This robot team would be coordinated by a host computer (the brain), to carry out complex assembly operations not currently possible. The inspection robot would provide the "eyes" for coordinating the mating of parts. This technique offers two advantages in robotic assembly work. First, it can move the "eyes" around to accommodate for blockage of view caused by movements of the assembly robot arms, and to view the parts being assembled from different perspectives. Secondly, it can provide precise measurements during the assembly process to provide important feedback signals to the host computer.

VI. CONCLUSIONS

It is clear that combining robots with either vision sensors or multiple sensors opens up new vistas for improving quality control, in flexible manufacturing operations. The flexibility of these robot-based inspection systems is further enhanced by integrating CAD/CAM databases into the measuring system. With newer robots there is potential for inspection accuracies to +/- 0.025 mm (0.001 inches). Through-put rates will increase substantially when the latest generation of robot controllers come into wider use, and allow inspection measurements to be made "on-the-fly".

BIBLIOGRAPHY

1. DiPietro, F.A.; "Automated Body Systems - From the Ground Up"; SME Conf. Proc. Vol. 1, 13th Int'l Symp. on Ind. Robots & Robots 7.

2. Hawkins, S.; "Laser Detects Body Fit-Up Flaws"; Amer. Machinist, Apr., 1982.

3. Villers, P.; "Recent Proliferaton of Industrial Artificial Vision Applications"; SME Conf. Proc. Vol. 1, 13th Int'l Symp. on Ind. Robots & Robots 7.

4. Pryor, T.; Pastorius, W.; "Applications of Machine Vision to Parts Inspection and Machine Control in the Piece Part Manufacturing Industries"; SME Conf. Proc. Vol. 1, 13th Int'l Symp. on Ind. Robots & Robots 7.

5. Simon, R.L.; "The Marriage Between CAD/CAM Systems and Robotics"; Paper published by Computer-Vision Corp., Bedford, Mass.

6. Engelberger, J.F.; "Robotics In Practice"; AMACOM, Div. of Amer. Mngt. Associations, 1980.

7. Bickerstaff, D.J.; "The Benefits In Application of Statistical Process Controls on Quality in the Powdered Metal Industry"; Paper published by Ferro Manufacturing Corp., Southfield, Michigan.

8. Harned, J.; Holcman, S.B.; "Gauging And Inspection Using Robots And Vision Systems"; Paper No. MS83-889; SME MACH-TEC Conference, Dec. 7, 1983, Chicago, Ill.

9. Dr. W. Edward Deming; "On Some Statistical Aids Towards Economic Production"; Interfaces, August, 1975; "Management of Statistical Techniques for Quality and Productivity", July, 1981.

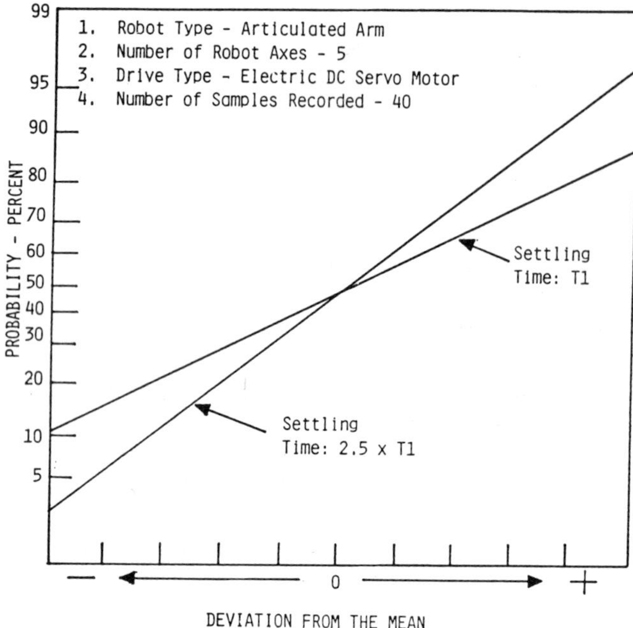

FIGURE 1 - Comparison of robot position repeatability for two different settling times.

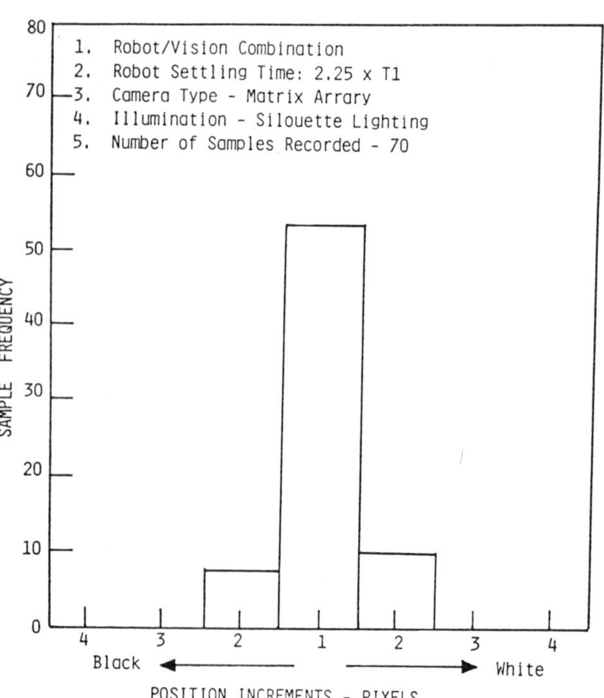

FIGURE 2 - Dispersion of vision measurements caused by robot position variability.

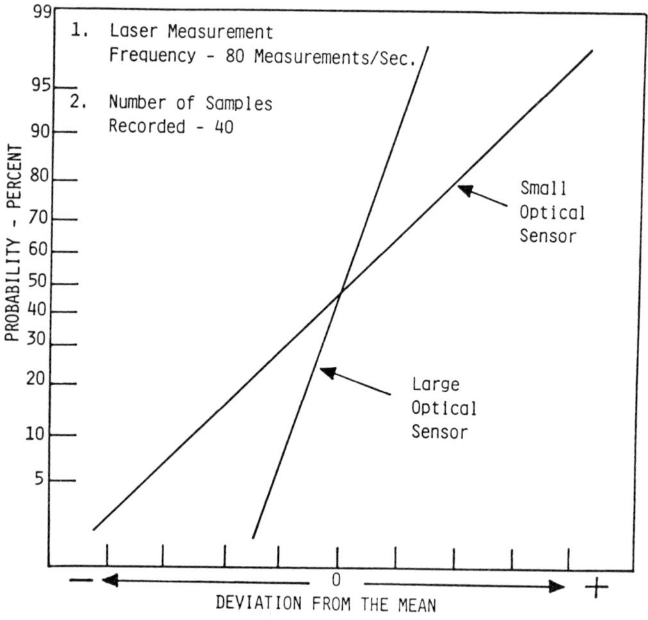

FIGURE 3 - Comparison of measurement dispersion for two optical sensors from the same manufacturer.

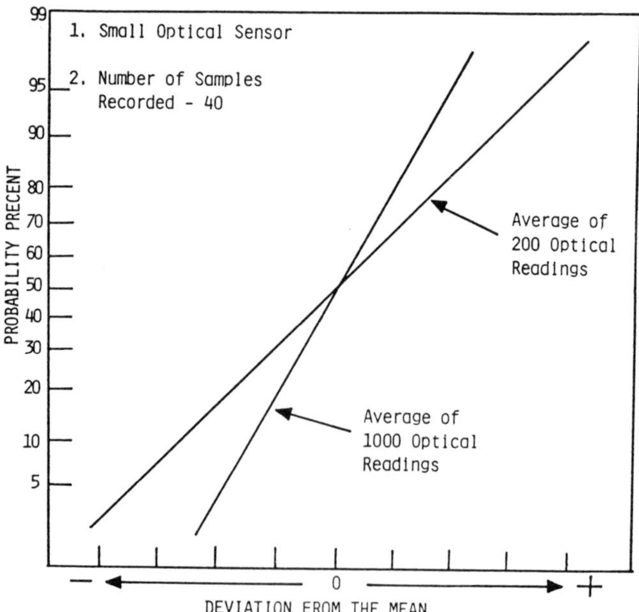

FIGURE 4 - Comparison of measurement dispersion as a function of optical reading sample size.

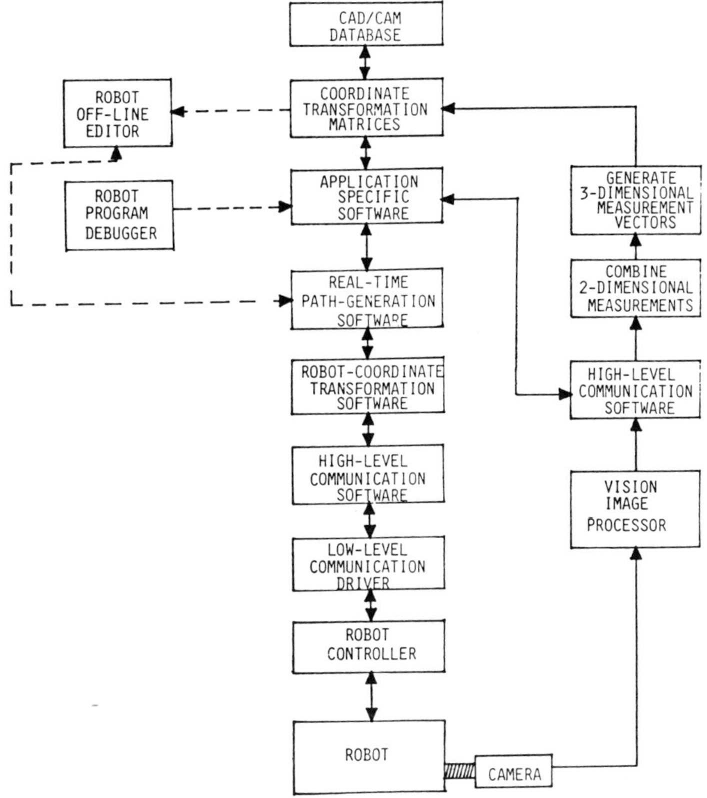

FIGURE 5 - Functional diagram of the software package needed to operate a robot inspection system using vision sensors.

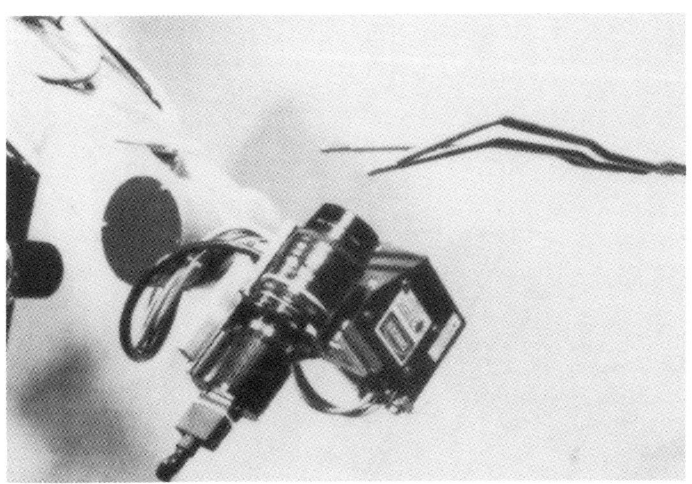

FIGURE 6 - COMPACT ARRANGEMENT OF OPTICAL/VISION SENSORS SHOWN IN POSITION TO INSPECT TUBE ASSEMBLY.

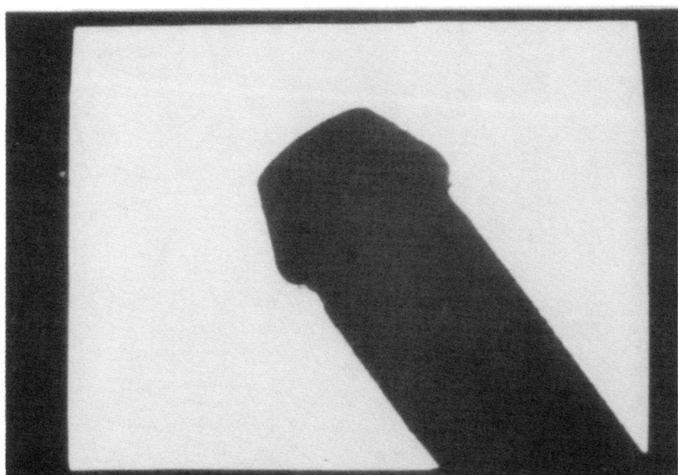

FIGURE 7 - BINARY IMAGE OF A TUBE END OBTAINED WITH SILOUETTE LIGHTING.

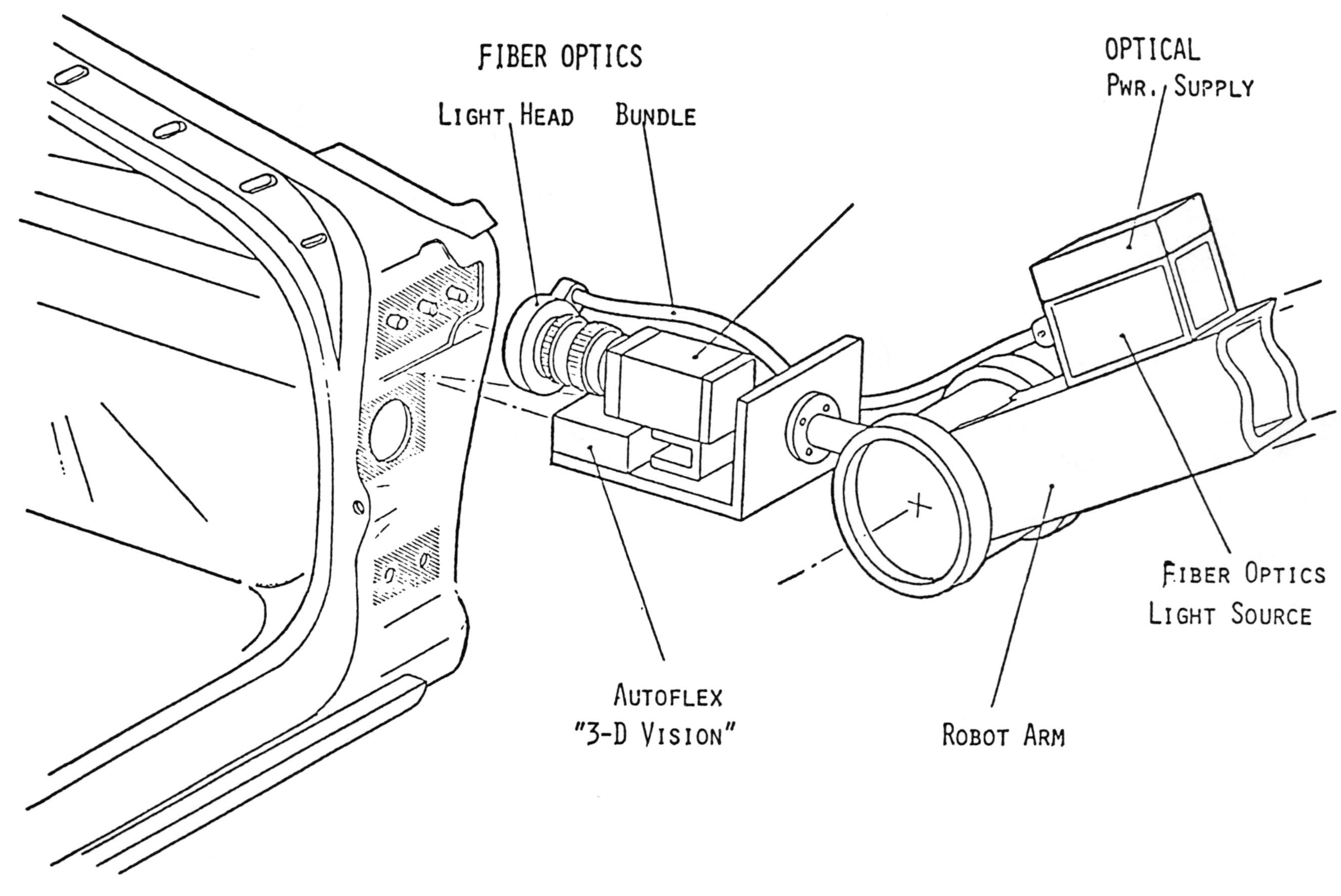

FIGURE 8 - USE OF OPTICAL AND VISION SENSORS MOUNTED ON ROBOT WRIST TO INSPECT CAR BODIES

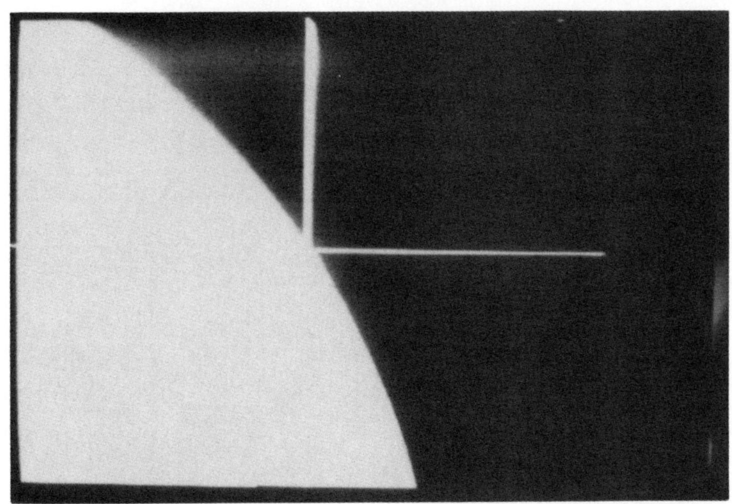

Figure 9 - Binary image of a circumference segment of a large chassis hole.

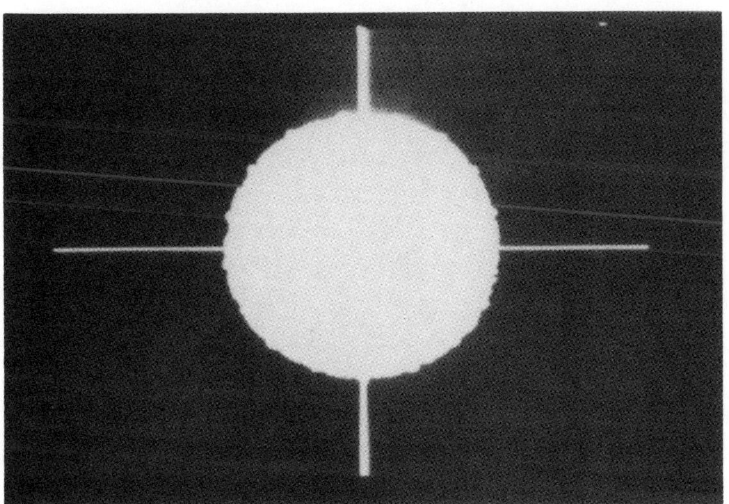

Figure 10 - Binary image of a chassis stud hole showing serrations after removing stud.

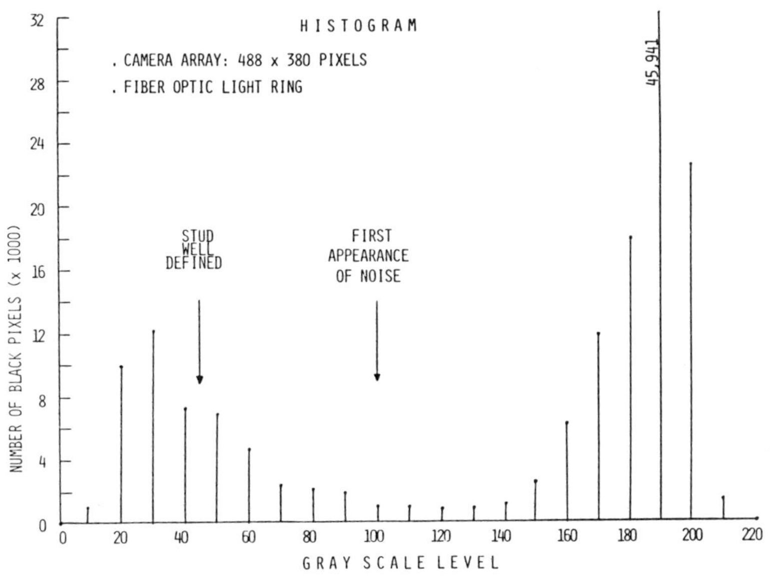

Figure 11 - Histogram of a representative stud image using front lighting.

INDEX

A

B

C

D

E

F

G

H

I

J

K

L

M

N

O

P

Q

R

S

T

U

V

W

X

Y

Z